D1190148

Nonlinear Systems Analysis

PRENTICE-HALL NETWORKS SERIES

ROBERT W. NEWCOMB, *editor*

ANDERSON AND MOORE, *Linear Optimal Control*
ANDERSON AND VONGPANITLERD, *Network Analysis and Synthesis: A Modern Systems Theory Approach*
ANNER, *Elementary Nonlinear Electronic Circuits*
CARLIN AND GIORDANO, *Network Theory: An Introduction to Reciprocal and Nonreciprocal Circuits*
CHIRLIAN, *Integrated and Active Synthesis and Analysis*
DEUTSCH, *Systems Analysis Techniques*
HERRERO AND WILLONER, *Synthesis of Filters*
HOLTZMAN, *Nonlinear System Theory: A Functional Analysis Approach*
HUMPHERYS, *The Analysis, Design, and Synthesis of Electrical Filters*
KIRK, *Optimal Control Theory: An Introduction*
KOHONEN, *Digital Circuits and Devices*
MANASSE, EKISS AND GRAY, *Modern Transistor Electronics Analysis and Design*
NEWCOMB, *Active Integrated Circuit Synthesis*
SAGE, *Optimum Systems Control*
SU, *Time-Domain Synthesis of Linear Networks*
VAN DER ZIEL, *Introductory Electronics*
VAN VALKENBURG, *Network Analysis,* 2nd ed.
VIDYASAGAR, *Nonlinear Systems Analysis*

असतोमा सद्गमय
तमसोमा ज्योतिर्गमय
मृत्योर्मा अमृतंगमय

Lead me from evil to virtue.
Lead me from darkness to light.
Lead me from death to eternal life.

(Vedic Prayer)

Nonlinear Systems Analysis

M. VIDYASAGAR

Professor of Electrical Engineering
Concordia University
Montreal, Canada

Prentice-Hall, Inc. *Englewood Cliffs, New Jersey 07632*

Library of Congress Cataloging in Publication data

VIDYASAGAR, M 1947–
Nonlinear systems analysis.

(Prentice-Hall electrical engineering series)
Bibliography: p.
Includes index.
1. System analysis. 2. Differential equations, nonlinear. I. Title.
QA402. V53 003 77-24379
ISBN 0–13–623280–9

Printed in the United States of America

10 9 8 7 6 5 4 3 2 1

PRENTICE-HALL INTERNATIONAL, INC., *London*
PRENTICE-HALL OF AUSTRALIA PTY. LIMITED, *Sydney*
PRENTICE-HALL OF CANADA, LTD., *Toronto*
PRENTICE-HALL OF INDIA PRIVATE LIMITED, *New Delhi*
PRENTICE-HALL OF JAPAN, INC., *Tokyo*
PRENTICE-HALL OF SOUTHEAST ASIA PTE. LTD., *Singapore*
WHITEHALL BOOKS LIMITED, *Wellington, New Zealand*

Contents

Preface

This book is intended as a text for a one-term or one-quarter course in nonlinear systems, at either the first-year graduate or senior-graduate level; it is almost self-contained and hence suitable for self-study. The only prerequisite for using the book is a course in ordinary differential equations. It is generally not necessary for the reader to have had a course in linear systems, though it is perhaps helpful to have an understanding of the concept of the *state* of a system. The contents of the book should be of interest to engineers from all branches who are interested in the *systems* approach, as well as to applied mathematicians, mathematical economists, biologists, *et cetra*. The results developed in the book are of a sufficiently general nature as to be applicable to all of these disciplines. Most of the important techniques for the analysis of nonlinear systems are covered in the book, though the coverage is by no means encyclopaedic. One of the novel features of the book is a chapter on input-output stability, presented at an elementary level for the first time.

The first version of this book was written in 1973, while I was visiting UCLA. Subsequent drafts were classroom-tested at both Concordia University and UCLA. In addition, portions of the book were also used at Berkeley for one quarter. Generally, the classes consisted of graduate students in both engineering and mathematics. This experience revealed that the entire book can be covered in about fifty classroom hours, while most of it can be covered in forty hours.

The book contains five chapters besides the introduction. Chapter 2 contains a discussion of various phase-plane techniques for the analysis of second-order systems. In chapter 3, the reader is introduced to some basic mathematical tools such as normed spaces, contraction mapping theorem, etc; this is followed by statements and proofs of the basic existence and uniqueness theorems for nonlinear differential equations, and some useful solution estimates. Chapter 4 consists of an introduction to several commonly used *approximate* analysis techniques. Chapter 5 contains a thorough treatment of Liapunov stability, including the Lur'e problem. Finally, chapter 6 comprises an elementary discussion of input-output stability, including the Nyquist, circle, and Popov criteria for feedback systems. There are numerous examples and exercises throughout. Two appendixes close the book.

It is now my pleasure to acknowledge all those who helped me in the writing of this book. I would like, first of all, to thank my wife Shakunthala for her encouragement and complete moral support throughout this project. Thanks are also due to Professor Charles A. Desoer for his thorough review of the manuscript and numerous constructive comments, as well as to Professor E. I. Jury for class-testing the manuscript and for several useful suggestions. I thank Professor M. N. S. Swamy and Dean J. C. Callaghan, both of Concordia University, for providing excellent logistic support as well as for their moral support. Professor A. V. Balakrishnan is to be thanked for making possible my visit to UCLA, during which this project was started. Finally, thanks to Veronica Markowitz and June Anderson for their excellent typing.

Montreal M. VIDYASAGAR

Notes to the Reader

1. All items within each section of each chapter (equations, theorems, examples, etc.) are numbered consecutively. A reference such as "Theorem (17)" refers to the 17th item *within the same section.* If a reference is made to an item in another section, the full number is given, e.g. example [5.1(13)] means example (13) in Sec. 5.1.
2. In some places, we write, e.g.

   $$\phi = \tan^{-1} \frac{x_2}{x_1}$$

   This means that ϕ is the unique number in $[0, 2\pi)$ such that

   $$\sin \phi = \frac{x_2}{(x_1^2 + x_2^2)^{1/2}}, \qquad \cos \phi = \frac{x_1}{(x_1^2 + x_2^2)^{1/2}}$$

   Thus $\tan^{-1}$ is a function of *both* variables x_1 and x_2, and not just of the ratio x_2/x_1. Note that $\tan^{-1}$ is well-defined everywhere in R^2 except at $(0, 0)$.

Nonlinear Systems Analysis

1 Introduction

1.1 GENERAL CONSIDERATIONS

Nonlinear physical systems, that is, systems that are not ~~necessarily~~ linear, differ from linear systems in two important respects:

1. Generally speaking, one can usually obtain *closed-form* expressions for solutions of linear systems, whereas this is not always possible in the case of nonlinear systems. More often, one is forced to be content with obtaining sequences of approximating functions that converge to the true solution or with generating estimates for the true solution. As a result, one may not have a good "feel" for what makes a nonlinear system "tick," compared with a linear system.
2. The analysis of nonlinear systems generally involves mathematics that is more advanced in concept and more messy in detail than is the case with linear systems.

A mathematical model that describes a wide variety of physical nonlinear systems is an nth-order ordinary differential equation of the type

$$\frac{d^n y(t)}{dt^n} = h\left[t, y(t), \dot{y}(t), \ldots, \frac{d^{n-1}y(t)}{dt^{n-1}}, u(t)\right], \qquad t \geq 0 \tag{1}$$

where t is the time parameter, $u(\cdot)$ is the input function (the terms *control function* and *forcing function* are also used), and $y(\cdot)$ is the output function (or *response function*). If we define the auxiliary functions

$$x_1(t) = y(t) \tag{2}$$

$$x_2(t) = \dot{y}(t) \tag{3}$$

$$\vdots$$

$$x_n(t) = \frac{d^{n-1}y(t)}{dt^{n-1}} \tag{4}$$

then the single nth-order equation (1) can be equivalently expressed as a system of n first-order equations:

$$\dot{x}_1(t) = x_2(t) \tag{5}$$

$$\dot{x}_2(t) = x_3(t) \tag{6}$$

$$\vdots$$

$$\dot{x}_{n-1}(t) = x_n(t) \tag{7}$$

$$\dot{x}_n(t) = h[t, x_1(t), x_2(t), \ldots, x_n(t), u(t)] \tag{8}$$

Finally, if we define n-vector-valued functions $\mathbf{x}(\cdot): R_+ \longrightarrow R^n$ and $\mathbf{f}: R_+ \times R^n \times R \longrightarrow R^n$ by

$$\mathbf{x}(t) = [x_1(t), x_2(t), \ldots, x_n(t)]' \tag{9}$$

$$\mathbf{f}(t, \mathbf{x}, u) = [x_2, x_3, \ldots, x_n, h(t, x_1, \ldots, x_n, u)]' \tag{10}$$

then the n first-order equations (5)–(8) can be combined into a first-order vector differential equation, namely,

$$\dot{\mathbf{x}}(t) = \mathbf{f}[t, \mathbf{x}(t), u(t)], \qquad t \geq 0 \tag{11}$$

For the system described by (1), the n quantities x_1 through x_n constitute a set of *state variables*, and the vector $\mathbf{x}$ constitutes a *state vector*.

Similarly, suppose a system with p inputs and k outputs is described by a set of k ordinary differential equations of the form

$$\frac{d^{m_i}y_i(t)}{dt^{m_i}} = h_i[t, y_1(t), \dot{y}_1(t), \ldots, y_1^{(m_1-1)}(t), y_2(t), \ldots, y_2^{(m_2-1)}(t), \ldots, y_k(t), \ldots, y_k^{(m_k-1)}(t), u_1(t), \ldots, u_p(t)], \qquad i = 1, \ldots, k \tag{12}$$

where $u_1(\cdot), \ldots, u_p(\cdot)$ are the input functions and $y_1(\cdot), \ldots, y_k(\cdot)$ are the output functions. As before, define

$$x_{m_i+j}(t) = \frac{d^{j-1}y_{i+1}(t)}{dt^{j-1}}, \qquad j = 1, \ldots, m_{i+1}, i = 0, \ldots, k-1 \tag{13}$$

$x_{\sum_{k=0}^{i} m_k + j}(t)$

14 $$\mathbf{x}(t) = [x_1(t) \dots x_n(t)]'$$

15 $$\mathbf{u}(t) = [u_1(t) \dots u_p(t)]'$$

where we take $m_0 = 0$, and define

16 $$n = m_1 + \dots + m_k$$

Then $\mathbf{x}(t)$ is a state vector for the system described by (12), and the system of equations (12) can once again be equivalently represented by a single first-order vector differential equation of the form

17 $$\dot{\mathbf{x}}(t) = \mathbf{f}[t, \mathbf{x}(t), \mathbf{u}(t)], \qquad t \geq 0$$

With this background in mind, we shall devote much of this book to the study of systems described by an equation of the form (17).[1] For (17) to truly represent a physical system, we would expect that, corresponding to each input $\mathbf{u}(\cdot)$,

1. (17) has at least one solution (existence).
2. (17) has exactly one solution (uniqueness).
3. (17) has exactly one solution that is defined over the entire half-line $[0, \infty)$.
4. (17) has exactly one solution over $[0, \infty)$, and this solution depends continuously on the initial condition $\mathbf{x}(0)$.

Statements 1–4 are progressively stronger. Unfortunately, without some restrictions on the nature of the function $\mathbf{f}$, none of these statements may be true, as illustrated by the following examples.

18 **Example.** Consider the scalar differential equation

19 $$\dot{x}(t) = -\text{sign}\, x(t), \quad t \geq 0; \qquad x(0) = 0$$

where the "sign" function is defined by

20 $$\text{sign}\; x(t) = \begin{cases} 1 & \text{if } x \geq 0 \\ -1 & \text{if } x < 0 \end{cases}$$

It is easy to verify that no continuously differentiable function $x(\cdot)$ exists such that (19) is satisfied. Thus statement 1 does not hold for this system.

21 **Example.** Consider the scalar differential equation

22 $$\dot{x}(t) = \frac{1}{2x(t)}, \quad t \geq 0; \qquad x(0) = 0$$

This equation admits two solutions, namely,

---

[1]The exception is Chap. 6, where we shall study distributed systems, e.g., systems containing time delays.

23 $$x_1(t) = t^{1/2}$$

24 $$x_2(t) = -t^{1/2}$$

Thus statement 1 is true, but 2 is false.

25 **Example.** Consider the scalar differential equation

26 $$\dot{x}(t) = 1 + x^2(t), \quad t \geq 0; \qquad x(0) = 0$$

Then over the interval $[0, \frac{\pi}{2})$, this equation has the unique solution

27 $$x(t) = \tan t$$

but there is no continuously differentiable function $x(\cdot)$ *defined over all of* $[0, \infty)$ such that (26) holds. Thus for this system statements 1 and 2 are true, but 3 fails.

It is therefore clear that the questions of existence and uniqueness of solutions to (17), and their continuous dependence on the initial condition, are very important. These questions are studied in Chap. 3.

In the last two examples, it was possible to derive the closed-form solutions to the equations under study, because they were of an extremely simple nature. However, in most cases, one cannot obtain an exact solution to the differential equation describing the system behavior. In such cases, one must be content either with generating "approximate" solutions or with solution *bounds*, which tell us that the solution at any time lies in a certain region of the state space. Both of these are studied in Chaps. 3 and 4.

An important question is that of the *well-behavedness*, in some suitable sense, of the solutions to (17). This is usually called the question of stability. Ideally, we would like to know whether or not the solutions to (17) are well behaved *without actually solving the system equations* (17). The stability question is studied in depth in Chaps. 5 and 6.

Finally, as a prelude to these more advanced subjects, we shall study second-order systems in Chap. 2. As we shall see there, a "geometric" approach to second-order systems yields much intuition and insight.

Problem 1.1. Determine whether or not each of the following differential equations has a unique solution over $[0, \infty)$, and if so, whether this solution depends continuously on the initial condition.

(a) $\dot{x}(t) = [x(t)]^{1/3}; \qquad x(0) = 0$

(b) $\dot{x}(t) = -x^2(t), \qquad x(0) = -1$

(c) $\dot{x}(t) = \begin{cases} -x(t) & \text{if } x(t) < 0 \\ x^2(t) & \text{if } x(t) \geq 0 \end{cases}; \qquad x(0) = 0$

1.2 AUTONOMY, EQUILIBRIUM POINTS

In this section, we shall introduce two definitions that are frequently used in the sequel. Before proceeding to these definitions, we shall clear up one small point. Many of the definitions, theorems, etc., that follow are stated for differential equations of the type

$$\dot{\mathbf{x}}(t) = \mathbf{f}[t, \mathbf{x}(t)] \tag{1}$$

Comparing (1) with (17) of Sec. 1.1,[2] we see that in [1.1(17)] the dependence of the right-hand side on an input $\mathbf{u}(\cdot)$ is explicitly identified, whereas this dependence, if any, is suppressed in (1). This might mislead one into thinking that [1.1(17)] describes a "forced" system, whereas (1) describes an "unforced" system. However, this is not necessarily the case. In problems of system *analysis*, as opposed to optimal control problems, one is generally concerned with the behavior of a system of the form [1.1(17)] under a fixed known input. Thus suppose that, in [1.1(17)], $\mathbf{u}(\cdot)$ is a *known fixed* function, and define $\mathbf{f}_{\mathbf{u}}: R_+ \times R^n \longrightarrow R^n$ by

$$\mathbf{f}_{\mathbf{u}}(t, \mathbf{x}) = \mathbf{f}[t, \mathbf{x}, \mathbf{u}(t)] \tag{2}$$

Then [1.1(17)] can be rewritten as

$$\dot{\mathbf{x}}(t) = \mathbf{f}_{\mathbf{u}}[t, \mathbf{x}(t)] \tag{3}$$

which is of the form (1). Therefore (3) can represent either an "unforced" system or a system with a fixed input.

We shall now introduce two concepts.

4 *[definition]* The system described by (1) is said to be *autonomous* if $\mathbf{f}(t, \mathbf{x})$ is independent of t and is said to be *nonautonomous* otherwise.

5 *[definition]* A vector $\mathbf{x}_0 \in R^n$ is said to be an *equilibrium point* at time $t_0 \in R_+$ of (1) if

$$\mathbf{f}(t, \mathbf{x}_0) = \mathbf{0}, \qquad \forall\, t \geq t_0 \tag{6}$$

If $\mathbf{x}_0$ is an equilibrium point of (1) at time t_0, then it is clear that $\mathbf{x}_0$ is also an equilibrium point of (1) at all times $t_1 \geq t_0$. Furthermore, if (1) is autonomous, then $\mathbf{x}_0 \in R^n$ is an equilibrium point of (1) at *some* time if and only if it is an equilibrium point of (1) at *all* times. Therefore we may speak of an equilibrium point of an autonomous system without specifying the time.

[2]Hereafter referred to as [1.1(17)].

The physical significance of an equilibrium point is as follows: Suppose $\mathbf{x}_0 \in R^n$ is an equilibrium point of (1) at time t_0. Then, whenever $t_1 \geq t_0$, the equation

7 $$\dot{\mathbf{x}}(t) = \mathbf{f}[t, \mathbf{x}(t)], \quad t \geq t_1; \qquad \mathbf{x}(t_1) = \mathbf{x}_0$$

has the unique solution

8 $$\mathbf{x}(t) = \mathbf{x}_0, \qquad \forall\, t \geq t_1$$

Conversely, if an element $\mathbf{x}_0 \in R^n$ has the property that the unique solution of (7) is given by (8) whenever $t_1 \geq t_0$, then it follows by simple differentiation that $\mathbf{x}_0$ satisfies (6), i.e., that $\mathbf{x}_0$ is an equilibrium point of (1) at time t_0. Thus, in other words, $\mathbf{x}_0$ is an equilibrium point of (1) at time t_0 if, should any solution $\mathbf{x}(\cdot)$ of (1) assume the value $\mathbf{x}_0$ at some time $t_1 \geq t_0$, it then remains at that value $\mathbf{x}_0$ for all $t \geq t_1$. The terms *stationary point* and *singular point* are also used in place of *equilibrium point*.

9 **Example.** Consider the motion of a frictionless simple pendulum, and let θ denote the angle of the pendulum from the vertical. Then the motion of the pendulum is described by

10 $$\ddot{\theta}(t) + \frac{g}{l} \sin \theta(t) = 0$$

where g is the acceleration due to gravity and l is the length of the pendulum. If we define

11 $$\mathbf{x}(t) = \begin{bmatrix} x_1(t) \\ x_2(t) \end{bmatrix} \triangleq \begin{bmatrix} \theta(t) \\ \dot{\theta}(t) \end{bmatrix}$$

then the dynamics of the system are described by the state variable equations

12 $$\begin{bmatrix} \dot{x}_1(t) \\ \dot{x}_2(t) \end{bmatrix} = \begin{bmatrix} x_2(t) \\ -(g/l) \sin x_1(t) \end{bmatrix}$$

Notice first of all that the system is autonomous. Next, we have that $x_0 = [x_{10} \quad x_{20}]'$ is an equilibrium point of (12)[3] if and only if

13 $$x_{20} = 0$$

14 $$\sin x_{10} = 0$$

i.e., the set of equilibrium points of (12) is the set of points in R^2 of the form

15 $$(n\pi, 0), \qquad \text{where } n = 0, \pm 1, \pm 2, \ldots$$

---

[3]Notice that we need not specify the time because the system is autonomous.

Because we commonly identify two values of θ that differ by a multiple of 2π, this system has basically two equilibrium points, namely (0, 0) and $(0, \pi)$. Of these, the first equilibrium point corresponds to the pendulum hanging straight down, while the second equilibrium point corresponds to the pendulum being at rest pointing straight up. Of course, we do not ever expect to find a real pendulum at rest pointing straight up, because the slightest perturbation (such as wind drafts present in the room) would knock the pendulum out of this equilibrium position. This is intimately connected with the question of the stability of an equilibrium point, which is studied in Chap. 5.

16 **Example.** Consider the one-dimensional motion of a particle in a potential field. Let r denote the position of the particle, m the mass of the particle, and $p(r)$ the potential energy at r. We assume that $p(r)$ is a continuously differentiable function of r. The motion of the particle is described by

17 $$\ddot{r}(t) = \frac{1}{m} \frac{dp(\xi)}{d\xi}\bigg|_{\xi=r(t)} = \frac{f[r(t)]}{m}$$

where $f(r) = dp(r)/dr$ denotes the force at r. To obtain a state variable description, define

18 $$\mathbf{x}(t) = \begin{bmatrix} x_1(t) \\ x_2(t) \end{bmatrix} \triangleq \begin{bmatrix} r(t) \\ \dot{r}(t) \end{bmatrix}$$

Then the state equations are

19 $$\begin{bmatrix} \dot{x}_1(t) \\ \dot{x}_2(t) \end{bmatrix} = \begin{bmatrix} x_2(t) \\ \frac{1}{m} f[x_1(t)] \end{bmatrix}$$

From (19), we see that the set of equilibrium points of this (autonomous) system consists of all points of the form $(r_0, 0)$, where $f(r_0) = 0$. Therefore this system is in an equilibrium state if the particle has zero velocity and is at a position where the force is zero, i.e., if the potential energy is stationary.

20 *[definition]* An equilibrium point $\mathbf{x}_0$ at time t_0 of (1) is said to be *isolated* if there exists a neighborhood N of $\mathbf{x}_0$ in R^n such that N contains no equilibrium points at time t_0 of (1) other than $\mathbf{x}_0$.

21 **Example.** Both the equilibrium points of the system in Example (9) are isolated. In the system of Example (16), an equilibrium point $(r_0, 0)$ is isolated if and only if r_0 is an isolated zero of the function $f(\cdot)$, i.e., if there exists a $\delta > 0$ such that $f(r) \neq 0$ whenever $0 < |r - r_0| < \delta$.

22 **Example.** Consider the linear vector differential equation

23 $$\dot{\mathbf{x}}(t) = \mathbf{A}(t)\mathbf{x}(t), \qquad t \geq 0$$

Clearly $\mathbf{0}$ is an equilibrium point of (23) at all times $t_0 \geq 0$. Suppose now that $\mathbf{A}(t_0)$ is nonsingular for some t_0. This means that $\mathbf{A}(t_0)\mathbf{x} = \mathbf{0}$ implies $\mathbf{x} = \mathbf{0}$. In this case, $\mathbf{0}$ is the only equilibrium point at time t_0 of (23) and is hence isolated.

24 **fact** Consider the system (1), and suppose $\mathbf{x}_0$ is an equilibrium point at time t_0 of (1); i.e., suppose (6) holds. Suppose further that $\mathbf{f}(t_0, \cdot)$ is continuously differentiable, and define

25 $$\mathbf{A}(t_0) = \left.\frac{\partial \mathbf{f}(t_0, \mathbf{x})}{\partial \mathbf{x}}\right|_{\mathbf{x}=\mathbf{x}_0}$$

If $\mathbf{A}(t_0)$ is nonsingular, then $\mathbf{x}_0$ is an isolated equilibrium point at time t_0 of (1).

proof For each $\mathbf{x} = [x_1 \ldots x_n]'$ in R^n, define

26 $$\|\mathbf{x}\|_2 = \left(\sum_{i=1}^{n} x_i^2\right)^{1/2}$$

The real number $\|\mathbf{x}\|_2$ is known as the Euclidean norm of the vector $\mathbf{x}$.[4] If $\mathbf{A}(t_0)$ is nonsingular, then there exists a *positive* constant c such that

27 $$\|\mathbf{A}(t_0)\mathbf{x}\|_2 \geq c\|\mathbf{x}\|_2, \qquad \forall\, \mathbf{x} \in R^n$$

Because $\mathbf{f}(t_0, \cdot)$ is continuously differentiable, we can expand $\mathbf{f}(t_0, \mathbf{x})$ in the form

28 $$\mathbf{f}(t_0, \mathbf{x}) = \mathbf{f}(t_0, \mathbf{x}_0) + \mathbf{A}(t_0)(\mathbf{x} - \mathbf{x}_0) + \mathbf{r}(t_0, \mathbf{x})$$

where the "remainder" term $\mathbf{r}(t_0, \cdot)$ satisfies the condition

29 $$\lim_{\|\mathbf{x}-\mathbf{x}_0\|_2 \to 0} \frac{\|\mathbf{r}(t_0, \mathbf{x})\|_2}{\|\mathbf{x} - \mathbf{x}_0\|_2} = 0$$

However, because $\mathbf{x}_0$ is an equilibrium point at time t_0 of (1), we have $\mathbf{f}(t_0, \mathbf{x}_0) = \mathbf{0}$; therefore,

30 $$\mathbf{f}(t_0, \mathbf{x}) = \mathbf{A}(t_0)(\mathbf{x} - \mathbf{x}_0) + \mathbf{r}(t_0, \mathbf{x})$$

Now, pick a number $d > 0$ such that

31 $$\frac{\|\mathbf{r}(t_0, \mathbf{x})\|_2}{\|\mathbf{x} - \mathbf{x}_0\|_2} \leq \frac{c}{2} \text{ whenever } \|\mathbf{x} - \mathbf{x}_0\|_2 < d$$

Such a choice for d is always possible in view of the limit condition (29). Let N be the neighborhood of $\mathbf{x}_0$ defined by

32 $$N = \{\mathbf{x} \in R^n : \|\mathbf{x} - \mathbf{x}_0\|_2 < d\}$$

---

[4] A detailed discussion of norms, including the explanation for the subscript 2, is found in Chap. 3. For the present, it is enough to note that $\|\mathbf{x}\|_2 > 0$ whenever $\mathbf{x} \neq \mathbf{0}$.

We shall show that N contains no equilibrium points at time t_0 of (1) other than $\mathbf{x}_0$. By definition (20), this is enough to show that $\mathbf{x}_0$ is isolated.

Accordingly, suppose $\mathbf{x} \in N$ and $\mathbf{x} \neq \mathbf{x}_0$; we shall show that $\mathbf{f}(t_0, \mathbf{x}) \neq \mathbf{0}$. We have, whenever $\|\mathbf{x} - \mathbf{x}_0\|_2 < d$, that

$$\begin{aligned}\|\mathbf{f}(t_0, \mathbf{x})\|_2 &= \|\mathbf{A}(t_0)(\mathbf{x} - \mathbf{x}_0) + \mathbf{r}(t_0, \mathbf{x})\|_2 \\ &\geq \|\mathbf{A}(t_0)(\mathbf{x} - \mathbf{x}_0)\|_2 - \|\mathbf{r}(t_0, \mathbf{x})\|_2 \\ &\geq c\|\mathbf{x} - \mathbf{x}_0\|_2 - \frac{c}{2}\|\mathbf{x} - \mathbf{x}\|_2 \\ &= \frac{c}{2}\|\mathbf{x} - \mathbf{x}_0\|_2 \\ &> 0 \text{ whenever } \mathbf{x} \neq \mathbf{x}_0\end{aligned} \tag{33}$$

Hence, whenever $\mathbf{x} \in N$ and $\mathbf{x} \neq \mathbf{x}_0$, we have $\|\mathbf{f}(t_0, \mathbf{x})\|_2 > 0$, i.e., $\mathbf{f}(t_0, \mathbf{x}) \neq \mathbf{0}$. Thus N contains no equilibrium points at time t_0 of (1) other than $\mathbf{x}_0$, and therefore $\mathbf{x}_0$ is isolated. ■

Problem 1.2. For the Volterra predator-prey equations

$$\dot{x}_1 = a_1 x_1 + b_1 x_1 x_2$$

$$\dot{x}_2 = a_2 x_2 + b_2 x_1 x_2$$

(a) Show that (0, 0) is an equilibrium point.

(b) Show that (0, 0) is an isolated equilibrium point if and only if both a_1 and a_2 are nonzero.

Problem 1.3. Consider the tunnel-diode circuit of Figure 1.1, where

$$i_d = v_d - 2v_d^2 + v_d^3 \triangleq f(v_d)$$

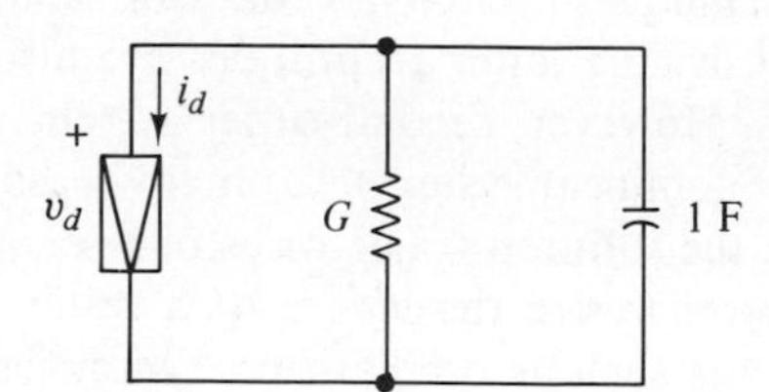

FIG. 1.1

(a) Show that the voltage v_d is governed by the equation

$$\frac{dv_d}{dt} = -Gv_d - f(v_d)$$

(b) Find all the equilibrium points of this system, when (i) $G = 0$, (ii) $G = 0.1$, (iii) $G = 1$.

2 Second-order systems

2.1 PRELIMINARIES

In this chapter, we shall study various specialized techniques that are available for the analysis of second-order systems. In subsequent chapters, we shall remove this restriction on the order of the system and study some techniques of analysis that can be applied to systems of any order. Obviously, the latter techniques are also applicable to second-order systems. However, second-order systems occupy a special place in the study of nonlinear systems, for many reasons. The most important reason is that the solution trajectories of a second-order system can be represented by curves in the *plane*. As a result, many of the nonlinear systems concepts such as oscillations, vector fields, etc., have a simple geometric interpretation, in the case of second-order systems. (All the technical terms used above will be defined shortly.) For these and other reasons, second-order systems, by themselves, have been the subject of much research, and in this chapter we shall present some of the simpler results that are available.

In general, a second-order system under study is represented by two scalar differential equations:

$$\dot{x}_1(t) = f_1[t, x_1(t), x_2(t)] \tag{1}$$

$$\dot{x}_2(t) = f_2[t, x_1(t), x_2(t)] \tag{2}$$

A basic concept in the analysis of second-order systems is the so-called state-plane plot. The *state plane* is the usual two-dimensional plane with the horizontal axis labeled x_1 and the vertical axis labeled x_2. Suppose $[x_1(t), x_2(t)]$, $t \in R_+$, denotes a solution of (1)–(2). Then a plot of $x_1(t)$ versus $x_2(t)$, as t varies over R_+, is called a *state-plane plot* or *state-plane trajectory* of the system (1)–(2). In such a plot, the time t is a parameter that can be either explicitly displayed or omitted. In the special case where (1) is of the form

$$\dot{x}_1(t) = x_2(t) \tag{3}$$

it is customary to refer to the state plane as the *phase plane*. Correspondingly, in this case one also refers to *phase-plane plots* or *phase-plane trajectories*. This special case arises quite commonly in practice. In particular, if the system under study is governed by a scalar differential equation of second order of the form

$$\ddot{y}(t) = g[t, y(t), \dot{y}(t)] \tag{4}$$

then a natural choice for the state variables is to select

$$x_1(t) = y(t) \tag{5}$$

$$x_2(t) = \dot{y}(t) \tag{6}$$

In this case, the system equation (4) is equivalent to the following two first-order equations:

$$\dot{x}_1(t) = x_2(t) \tag{7}$$

$$\dot{x}_2(t) = g[t, x_1(t), x_2(t)] \tag{8}$$

Phase-plane plots also have another useful feature, namely, that it is easy to reconstruct the implicit parameter t from a phase-plane plot. Suppose we are given a phase-plane plot, which we denote by $\mathcal{C}$. Suppose we know that a particular point (x_{10}, x_{20}) on $\mathcal{C}$ corresponds to time t_0 (see Fig. 2.1). Typically, t_0 might be the initial time and (x_{10}, x_{20})

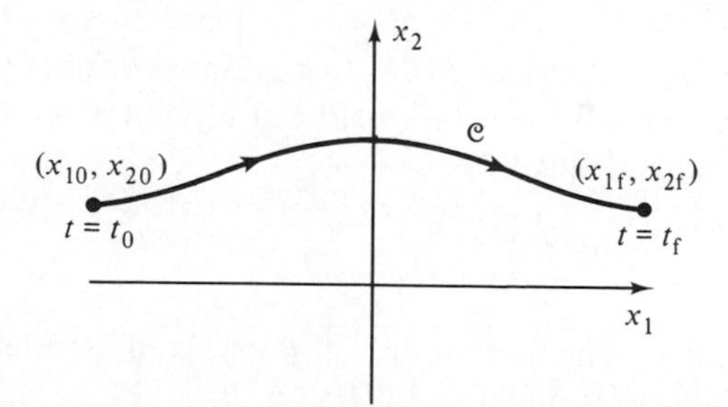

FIG. 2.1

the initial state of the system. Now, if (x_{1f}, x_{2f}) is another point on $\mathcal{C}$, and if it is desired to determine the value of t (say t_f) corresponding to (x_{1f}, x_{2f}), we proceed as follows: If x_2 does not change sign along $\mathcal{C}$

between (x_{10}, x_{20}) and (x_{1f}, x_{2f}), then

9 $$t_f = t_0 + \int_{\mathcal{C}} \frac{dx_1}{x_2}$$

where the integral in (9) is taken along $\mathcal{C}$. If x_2 does change sign along $\mathcal{C}$, then the integral in (9) has to be evaluated as the sum of several integrals, one corresponding to each segment of $\mathcal{C}$ along which x_2 does not change sign (see Fig. 2.2). Note that, as $x_{2f} \longrightarrow 0$, the integral (9) becomes an improper integral. The proof of the relationship (9) is easily obtained starting from (7) and is left as an exercise for the reader (see Problem 2.1).

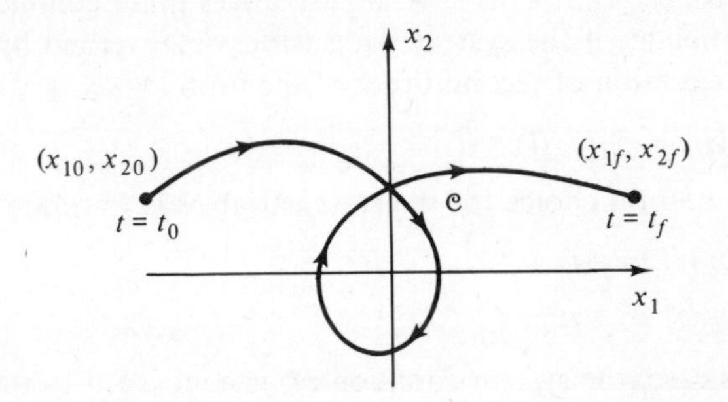

FIG. 2.2

Another very important concept for *autonomous* second-order systems is that of a vector field. Consider the autonomous system

10 $$\dot{x}_1(t) = f_1[x_1(t), x_2(t)]$$

11 $$\dot{x}_2(t) = f_2[x_1(t), x_2(t)]$$

For this system, it is possible to associate, with each vector (x_1, x_2), a corresponding vector $[f_1(x_1, x_2), f_2(x_1, x_2)]$. The latter vector $[f_1(x_1, x_2), f_2(x_1, x_2)]$ is known as a vector field. This is made precise in the following definition.

12 *[definition]* A *vector field* is a continuous function $\mathbf{f}: R^2 \longrightarrow R^2$. The *direction* of the vector field $\mathbf{f}$ at a point $\mathbf{x} \in R^2$ is denoted by $\theta_{\mathbf{f}}(\mathbf{x})$ and is defined by

$$\theta_{\mathbf{f}}(\mathbf{x}) = \tan^{-1} \frac{f_2(x_1, x_2)}{f_1(x_1, x_2)}$$

13 REMARKS: The definition of $\theta_{\mathbf{f}}(\mathbf{x})$ is illustrated in Fig. 2.3. Clearly $\theta_{\mathbf{f}}(\mathbf{x})$ is undefined if $\mathbf{f}(\mathbf{x}) = \mathbf{0}$.

The utility of the vector field concept is immediately apparent from (10)–(11). Suppose $\mathbf{x} = (x_1, x_2)$ is a point in R^2; then it is easy to see from (10)–(11) that if $\mathcal{C}$ is a solution trajectory of (10)–(11) passing

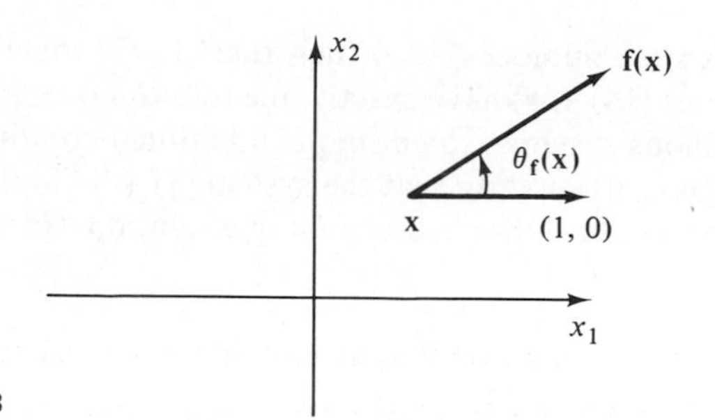

FIG. 2.3

through **x**, then the vector **f**(**x**) is tangent to $\mathcal{C}$ at **x**. Hence, in principle at least, it is possible to construct graphically the trajectories of (10)–(11) by plotting the vector field **f**(**x**). Actually, the concept is very deep and has many applications, only a few of which are touched upon in this book. Furthermore, the concept of a vector field is applicable to (autonomous) systems of any order; however, the geometrical visualization is particularly simple for second-order systems. A reader interested in a deeper study of vector fields in differential equations may refer to [1] Arnold.[1]

Note that it is quite common to refer to **f**(**x**) as the *velocity vector field* associated with the system of equations (10)–(11).

Our objective in this chapter is to present some ways of either finding the state-plane trajectory of a given system to a reasonably high degree of accuracy or determining some qualitative features of the state-plane trajectory without too much work. Throughout this chapter, our attention will be confined to autonomous systems, because, even though the general concept of a state-plane trajectory is valid even for nonautonomous systems, most of the significant results are applicable only to autonomous systems. For example, in an autonomous system, we have an oscillatory or periodic solution [i.e., the solution $\mathbf{x}(t)$ is periodic in t] if and only if the corresponding solution trajectory is a closed curve in R^2. A similar statement for nonautonomous systems is false in general.

Finally, in this chapter, we shall bypass the questions of existence and uniqueness of solutions to (1)–(2) by assuming that the functions f_1 and f_2 in (1)–(2) are continuously differentiable. This ensures (as shown in Chap. 3) that (1)–(2) have a unique solution at least locally in t. That is, given the equations (1)–(2) together with some arbitrary initial conditions

$$x_1(0) = x_{10} \tag{14}$$

$$x_2(0) = x_{20} \tag{15}$$

---

[1]References section located at end of book, before Index. Reference number is in brackets, followed by last name of author.

there exists a number $T > 0$ such that (1)–(2) together with the initial conditions (14)–(15) have exactly one solution over $[0, T]$. In the case of autonomous systems, some simple additional conditions on the vector field $\mathbf{f}$ [basically stating that the system (1)–(2) "looks linear" for large x_1, x_2] will ensure that in fact a unique solution exists over $[0, \infty)$.

Problem 2.1. Prove the relationship (9). *Hint:* Use (7) to write

$$x_1(t + \Delta t) = x_1(t) + \Delta t\, x_2(t) + o(\Delta t) \tag{16}$$

Problem 2.2. Show that if $\mathcal{C}$ is a solution trajectory of (10)–(11) passing through $\mathbf{x}$, then the vector field $\mathbf{f}(\mathbf{x})$ is tangent to $\mathcal{C}$ at $\mathbf{x}$. *Hint:* Express (10) and (11) in difference approximation form as

$$x_1(t + \Delta t) = x_1(t) + \Delta t\, f_1[x_1(t), x_2(t)] + o(\Delta t) \tag{17}$$

$$x_2(t + \Delta t) = x_2(t) + \Delta t\, f_2[x_1(t), x_2(t)] + o(\Delta t) \tag{18}$$

Eliminate Δt as $\Delta t \longrightarrow 0$.

Problem 2.3. Does the function $\mathbf{f}\colon R^2 \longrightarrow R^2$ defined by

$$f_1(x_1, x_2) = x_2 + [1 - (x_1^2 + x_2^2)^{1/2}]$$

$$f_2(x_1, x_2) = -x_1 + [1 - (x_1^2 + x_2^2)^{1/2}]$$

constitute a vector field? Justify your answer. (*Hint:* Consider the behavior of $\mathbf{f}$ near the origin.)

2.2 LINEAR SYSTEMS

We shall begin by studying linear systems, which are simpler to analyze than nonlinear systems and yet provide much insight into the behavior of nonlinear systems. The general form for a second-order autonomous linear system is

$$\dot{x}_1(t) = a_{11}x_1(t) + a_{12}x_2(t) \tag{1}$$

$$\dot{x}_2(t) = a_{21}x_1(t) + a_{22}x_2(t) \tag{2}$$

together with the initial conditions

$$x_1(0) = x_{10} \tag{3}$$

$$x_2(0) = x_{20} \tag{4}$$

or, in matrix notation,

$$\dot{\mathbf{x}}(t) = \mathbf{A}\mathbf{x}(t) \tag{5}$$

$$\mathbf{x}(0) = \mathbf{x}_0 \tag{6}$$

To better understand the behavior of solutions to (5)–(6), it is helpful to make a transformation of variables. Accordingly, let

$$\mathbf{z}(t) = \mathbf{M}^{-1}\mathbf{x}(t) \tag{7}$$

where $\mathbf{M}$ is a constant nonsingular 2×2 matrix with *real* coefficients. In terms of the transformed variables $\mathbf{z}$, (5)–(6) become

$$\dot{\mathbf{z}}(t) = \mathbf{M}^{-1}\mathbf{A}\mathbf{M}\mathbf{z}(t) \tag{8}$$

$$\mathbf{z}(0) = \mathbf{M}^{-1}\mathbf{x}_0 \tag{9}$$

It is known (see, for example, [4] Bellman) that by appropriately choosing the matrix $\mathbf{M}$, the matrix $\mathbf{M}^{-1}\mathbf{A}\mathbf{M}$ can be made to have one of the following forms:

1. *Diagonal form.* In this case,

$$\mathbf{M}^{-1}\mathbf{A}\mathbf{M} = \begin{bmatrix} \lambda_1 & 0 \\ 0 & \lambda_2 \end{bmatrix} \tag{10}$$

where λ_1 and λ_2 are the *real* (and not necessarily distinct) eigenvalues of $\mathbf{A}$.

2. *Jordan form.* In this case,

$$\mathbf{M}^{-1}\mathbf{A}\mathbf{M} = \begin{bmatrix} \lambda & 1 \\ 0 & \lambda \end{bmatrix} \tag{11}$$

where λ is the *repeated real* eigenvalue of $\mathbf{A}$.

3. *Complex conjugate form.* In this case,

$$\mathbf{M}^{-1}\mathbf{A}\mathbf{M} = \begin{bmatrix} \alpha & \beta \\ -\beta & \alpha \end{bmatrix} \tag{12}$$

where $\alpha + j\beta$, $\alpha - j\beta$ are the *complex conjugate* eigenvalues of $\mathbf{A}$ (and we choose $\beta < 0$ to be definite).

We shall study each of these cases in detail.

Case 1 *Diagonal form:* In this case (8)–(9) assume the form

$$\dot{z}_1(t) = \lambda_1 z_1(t) \tag{13}$$

$$\dot{z}_2(t) = \lambda_2 z_2(t) \tag{14}$$

$$z_1(0) = z_{10} \tag{15}$$

$$z_2(0) = z_{20} \tag{16}$$

which has the solution

$$z_1(t) = z_{10}e^{\lambda_1 t} \tag{17}$$

$$z_2(t) = z_{20}e^{\lambda_2 t} \tag{18}$$

We may assume at this point that either λ_1 or λ_2 is nonzero, because if λ_1 and λ_2 are both zero, then clearly $z_1(t)$ and $z_2(t)$ are constants, and the state-plane plot (in the z_1–z_2 plane) consists of the single point (z_{10}, z_{20}). Thus suppose that $\lambda_1 \neq 0$. Then we can eliminate the parameter t from (17)–(18) to get

19 $$z_2 = z_{20} \cdot \left(\frac{z_1}{z_{10}}\right)^{\lambda_2/\lambda_1}$$

Equation (19) describes the state-plane trajectories of (13)–(16) in the z_1–z_2 plane. If λ_1 and λ_2 are of the same sign, the trajectories have the characteristic shape shown in Fig. 2.4, but if λ_1 and λ_2 are of opposite signs, the trajectories are as in Fig. 2.5. The arrowheads in Fig. 2.4 correspond to the case where $\lambda_2 < \lambda_1 < 0$; if λ_1 and λ_2 are both posi-

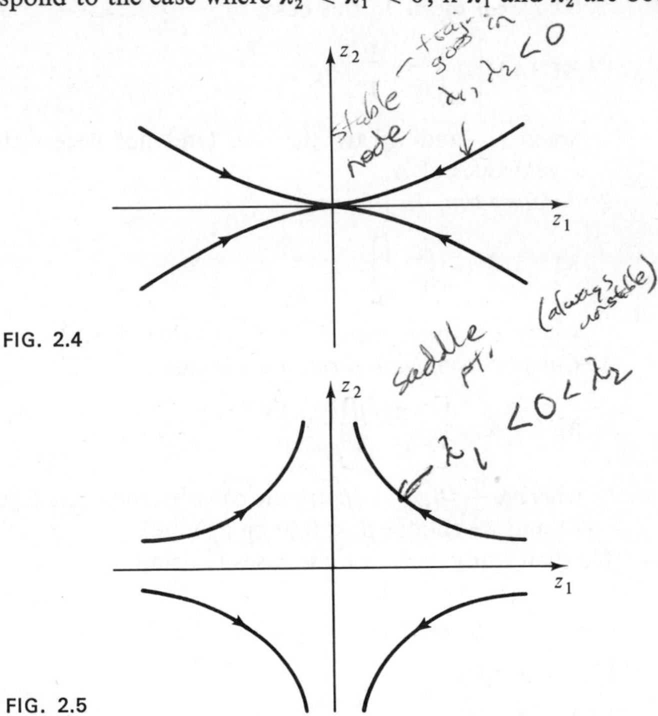

FIG. 2.4

FIG. 2.5

tive, the direction of the arrowheads is reversed, and the trajectories go *away* from the origin as t increases. Similarly, the arrowheads in Fig. 2.5 correspond to the case $\lambda_1 < 0 < \lambda_2$. It should be emphasized that the trajectories depicted in Figs. 2.4 and 2.5 are in the z_1–z_2 coordinate system; the corresponding trajectories in the x_1–x_2 coordinate system, although they will have the same general appearance as those in the z_1–z_2 coordinate system, will be a little distorted. This can be seen in Figs. 2.6 and 2.7, where the trajectories in the x_1–x_2 coordinate system

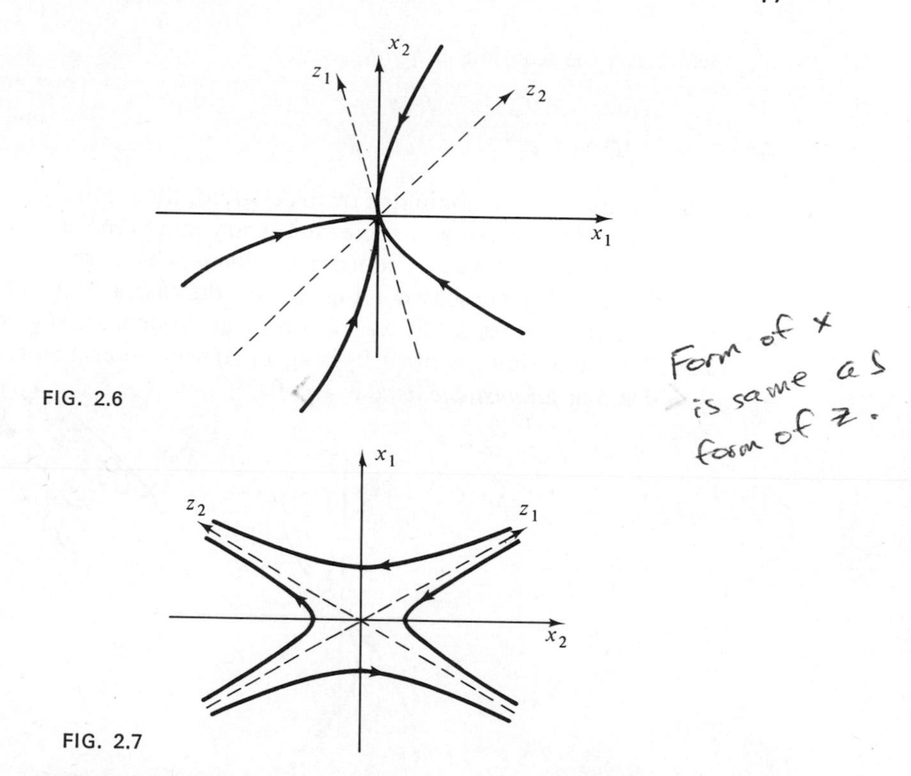

FIG. 2.6

FIG. 2.7

are illustrated for, respectively, (1) λ_1 and λ_2 of the same sign and (2) λ_1 and λ_2 of the opposite sign. In the case where λ_1 and λ_2 are both of the same sign, the equilibrium point (0, 0) is referred to as a *stable node* if λ_1 and λ_2 are both negative and as an *unstable node* if λ_1 and λ_2 are both positive. The rationale is that if λ_1 and λ_2 are both negative, the trajectories of the system move toward the origin as t increases, whereas if λ_1 and λ_2 are both positive, the trajectories of the system move away from the origin as t increases. In the case where λ_1 and λ_2 are of opposite signs, the equilibrium point (0, 0) is referred to as a *saddle point*, because, in such a case, if one were to make a three-dimensional plot with x_1, x_2, and t (or z_1, z_2, and t) as the three axes, the resulting solution surface resembles a saddle.

Case 2 *Jordan form:* In this case (8)–(9) assume the form

$$\dot{z}_1(t) = \lambda z_1(t) + z_2(t) \tag{20}$$

$$\dot{z}_2(t) = \lambda z_2(t) \tag{21}$$

$$z_1(0) = z_{10} \tag{22}$$

$$z_2(0) = z_{20} \tag{23}$$

which has the solution

$$z_1(t) = z_{10}e^{\lambda t} + z_{20}te^{\lambda t} \tag{24}$$

$$z_2(t) = z_{20}e^{\lambda t} \tag{25}$$

Once again, t can be eliminated from (24)–(25); the resulting expression describing the trajectory is somewhat messy and is left as a problem. The trajectories in the z_1–z_2 coordinate system, which are easily obtained from (24)–(25), are shown in Fig. 2.8 (for the case $\lambda < 0$). The corresponding trajectories in the x_1–x_2 coordinate system are shown in Fig. 2.9. The equilibrium point (0, 0) is again referred to as a *stable node* if $\lambda < 0$ and as an *unstable node* if $\lambda > 0$.

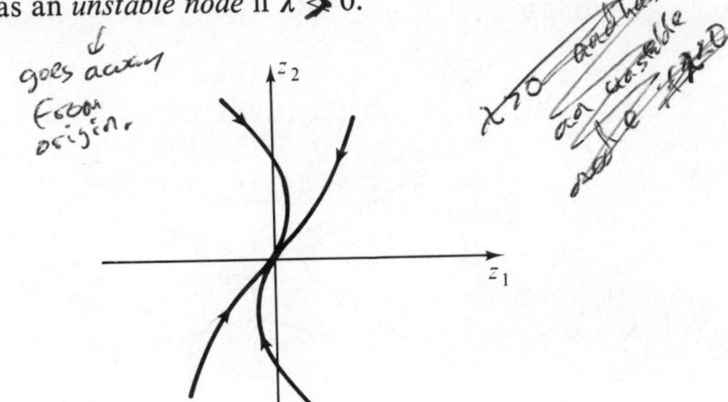

FIG. 2.8

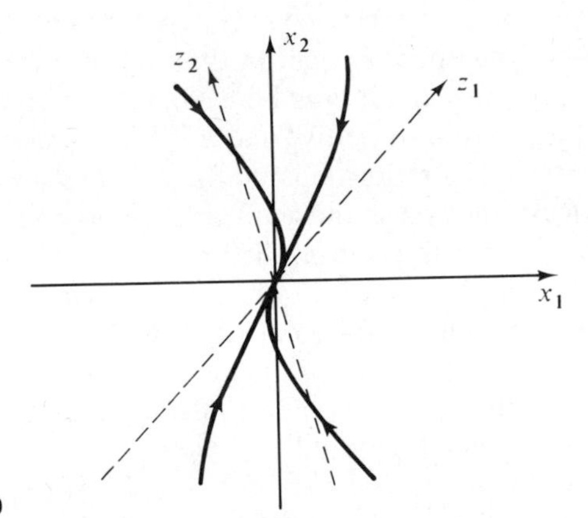

FIG. 2.9

Case 3 *Complex conjugate form:* In this case, (8)–(9) become

$$\dot{z}_1(t) = \alpha z_1(t) + \beta z_2(t) \tag{26}$$

$$\dot{z}_2(t) = -\beta z_1(t) + \alpha z_2(t) \tag{27}$$

$$z_1(0) = z_{10} \tag{28}$$

$$z_2(0) = z_{20} \tag{29}$$

To further simplify matters, let us introduce the polar coordinates

$$r = (z_1^2 + z_2^2)^{1/2} \tag{30}$$

$$\phi = \tan^{-1} \frac{z_2}{z_1} \tag{31}$$

where, as usual, ϕ is chosen in the interval $[0, \pi]$ if z_2 is positive and in $[\pi, 2\pi]$ if z_2 is negative. Then (26)–(27) are transformed into

$$\dot{r}(t) = \alpha r(t) \tag{32}$$

$$\dot{\phi}(t) = -\beta \tag{33}$$

which has the solution

$$r(t) = r_0 e^{\alpha t} \tag{34}$$

$$\phi(t) = \phi_0 - \beta t \tag{35}$$

In the z_1–z_2 coordinate system, (34)–(35) represent an exponential spiral. If $\alpha > 0$, the spiral expands as t increases, while if $\alpha < 0$, the spiral shrinks as t increases; if $\alpha = 0$, the trajectory is a circle. The equilibrium point (0, 0) is referred to as an *unstable focus* if $\alpha > 0$, as a *stable focus* if $\alpha < 0$, and as a *center* if $\alpha = 0$.

The trajectories in the z_1–z_2 coordinate system, corresponding to each of these three cases, are depicted in Figs. 2.10, 2.11, and 2.12.

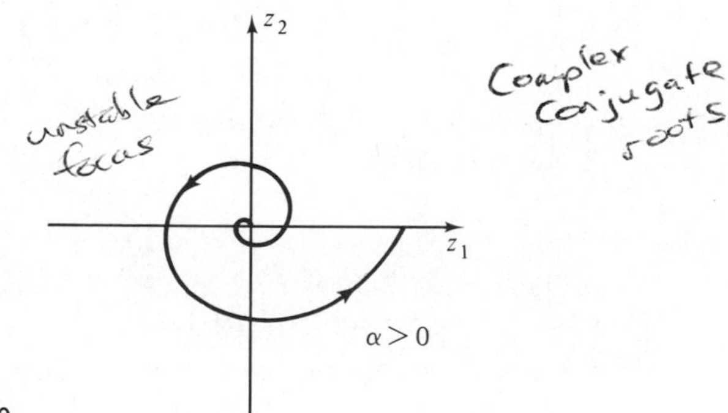

FIG. 2.10

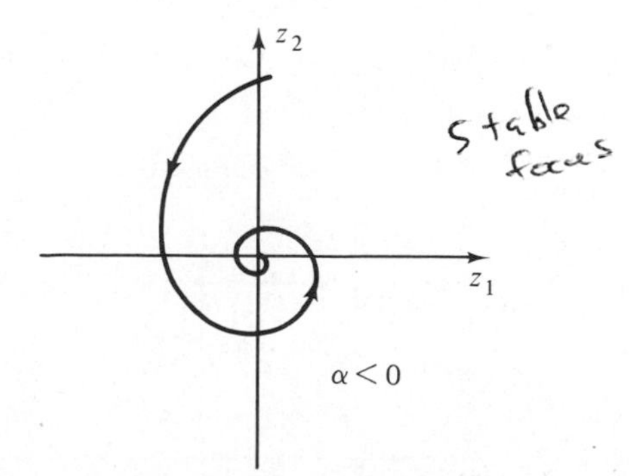

FIG. 2.11

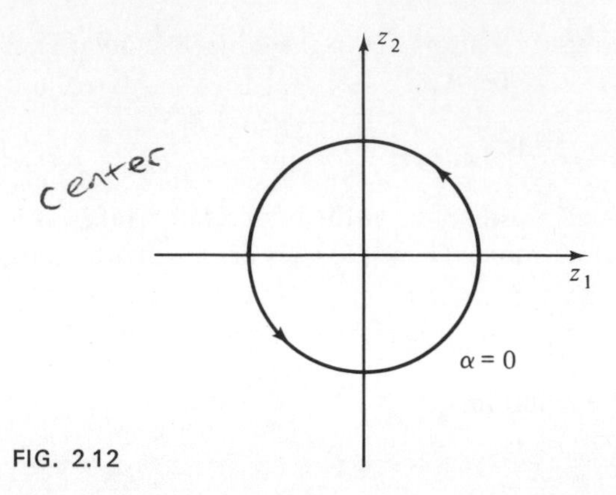

FIG. 2.12

Table 2.1 summarizes the various kinds of equilibrium points for second-order linear systems.

TABLE 2.1

Eigenvalues of **A**	*Type of Equilibrium Point*
λ_1, λ_2 real, $\lambda_1 < 0$, $\lambda_2 < 0$	Stable node
λ_1, λ_2 real, $\lambda_1 > 0$, $\lambda_2 > 0$	Unstable node
λ_1, λ_2 real, $\lambda_1\lambda_2 < 0$	Saddle point
λ_1, λ_2 complex conjugates, Re $\lambda_1 > 0$	Unstable focus
λ_1, λ_2 complex conjugates, Re $\lambda_1 < 0$	Stable focus
λ_1, λ_2 imaginary	Center

Problem 2.4. Eliminate t from (23)–(25), and obtain an expression for the state-plane trajectory involving only z_1, z_2, z_{10}, and z_{20}.

Problem 2.5. Consider the electrical circuit shown in Fig. 2.13.

(a) Select the capacitor voltage x_1 and the inductor current x_2 as the state variables, and show that the network is described by the equations

$$\dot{x}_1(t) = -2x_1(t) - x_2(t) + 2v(t)$$

$$\dot{x}_2(t) = x_1(t) - x_2(t)$$

(b) Suppose $v(t) \equiv 0$. Determine the nature of the equilibrium point

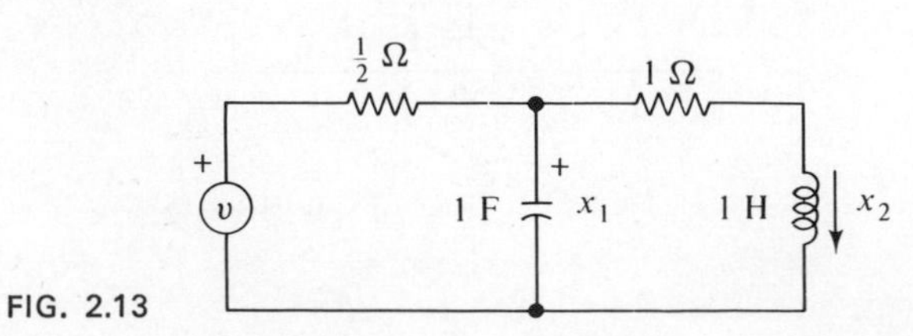

FIG. 2.13

(0, 0), and find the matrix **M** that transforms the equations into the appropriate canonical form.

Problem 2.6. Suppose the $\frac{1}{2}\,\Omega$ resistor in Fig. 2.12 is replaced by a general resistance R.

(a) Write the state equations for the network, with $v(t) \equiv 0$.

(b) For what values of R is the equilibrium point (0, 0) of this system (i) a node, (ii) a focus, (iii) a saddle?

Problem 2.7. For each of the **A** matrices below:

(a) Determine the matrix **M** that transforms **A** into the appropriate canonical form;

(b) Sketch the state-plane trajectories in both the z_1–z_2 and the x_1–x_2 coordinate system; and

(c) Classify the equilibrium point (0, 0) as to its type.

$$\text{(i)}\ \mathbf{A} = \begin{bmatrix} 0 & 1 \\ -2 & -3 \end{bmatrix} \qquad \text{(ii)}\ \mathbf{A} = \begin{bmatrix} 0 & -1 \\ 1 & 2 \end{bmatrix} \qquad \text{(iii)}\ \mathbf{A} = \begin{bmatrix} 1 & 1 \\ 0 & -1 \end{bmatrix}$$

$$\text{(iv)}\ \mathbf{A} = \begin{bmatrix} 1 & 5 \\ -1 & -1 \end{bmatrix} \qquad \text{(v)}\ \mathbf{A} = \begin{bmatrix} 2 & -1 \\ 2 & 0 \end{bmatrix} \qquad \text{(vi)}\ \mathbf{A} = \begin{bmatrix} 0 & -1 \\ 2 & -2 \end{bmatrix}$$

2.3 NONLINEAR SYSTEMS

In this section, we shall study four methods for obtaining the state-plane trajectories of a second-order autonomous system of the type

$$\dot{x}_1(t) = f_1[x_1(t), x_2(t)] \tag{1}$$

$$\dot{x}_2(t) = f_2[x_1(t), x_2(t)] \tag{2}$$

2.3.1 Linearization Method

The linearization method, as the name implies, consists of linearizing the given system in the neighborhood of an equilibrium point and determining the behavior of the *nonlinear* system's trajectories by studying the resulting *linear* system. The power of the method lies in the fact that, except for special cases to be specified later, the method yields definitive results that are valid in some neighborhood of an equilibrium point.

The method can be summarized as follows: Suppose that (0, 0) is an equilibrium point of (1)–(2)[2] and that both f_1 and f_2 are continuously differentiable in some neighborhood of (0, 0). Define

[2]There is no loss of generality in assuming that (0, 0) is an equilibrium point of (1)–(2), because if some other point (x_{10}, x_{20}) is an equilibrium point of (1)–(2), then we can always *translate* the coordinates x_1 and x_2 in such a way that (x_{10}, x_{20}) becomes the point (0, 0) in the new coordinate system: More precisely, let $\tilde{x}_1 = x_1 - x_{10}$, $\tilde{x}_2 = x_2 - x_{20}$.

3 $$a_{ij} = \left.\frac{\partial f_i}{\partial x_j}\right|_{x_1=0, x_2=0}, \quad i, j = 1, 2$$

4 $$\mathbf{A} = \begin{bmatrix} a_{11} & a_{12} \\ a_{21} & a_{22} \end{bmatrix}$$

Then, by Taylor's theorem, we can expand f_1 and f_2 in the form

5 $$\begin{aligned} f_1(x_1, x_2) &= f_1(0, 0) + a_{11}x_1 + a_{12}x_2 + r_1(x_1, x_2) \\ &= a_{11}x_1 + a_{12}x_2 + r_1(x_1, x_2) \end{aligned}$$

6 $$f_2(x_1, x_2) = a_{21}x_1 + a_{22}x_2 + r_2(x_1, x_2)$$

where r_1 and r_2 are the remainder terms. [Note that we have used the fact that $f_1(0, 0) = f_2(0, 0) = 0$.] Now, associated with the nonlinear system (1)–(2), define the *linear* system

7 $$\dot{\xi}_1(t) = a_{11}\xi_1(t) + a_{12}\xi_2(t)$$

8 $$\dot{\xi}_2(t) = a_{21}\xi_1(t) + a_{22}\xi_2(t)$$

Clearly $\xi_1 = 0$, $\xi_2 = 0$ is an equilibrium point of the system (7)–(8), which is commonly known as the *linearization* of the system (1)–(2) around the equilibrium point (0, 0). The linearization method is based on the fact (proved in Chap. 5) that in most cases the trajectories of the nonlinear system (1)–(2) have, in some suitably small neighborhood of the origin, the same qualitative features as the trajectories of the linear system (7)–(8). Table 2.2 summarizes the situation.

TABLE 2.2

Equilibrium Point $\xi_1 = 0, \xi_2 = 0$ *of the Linearized System (7)–(8)*	*Equilibrium Point* $x_1 = 0, x_2 = 0$ *of the Nonlinearized System (1)–(2)*
Stable node	Stable node
Unstable node	Unstable node
Saddle point	Saddle point
Stable focus	Stable focus
Unstable focus	Unstable focus
Center	?

In other words, if the matrix $\mathbf{A}$ does not have any eigenvalues with zero real parts, then the trajectories of the *nonlinear* system (1)–(2) in the vicinity of the equilibrium point $x_1 = 0$, $x_2 = 0$ have the same characteristic shape as the trajectories of the *linearized* system (7)–(8) in the vicinity of the equilibrium point $\xi_1 = 0$, $\xi_2 = 0$. The meaning of the last entry in the table can be explained as follows: If the equilibrium point $\xi_1 = 0, \xi_2 = 0$ of the linearized system (7)–(8) is a center, then the linearized system exhibits perfect oscillations that neither grow nor decay with time. Under these conditions, it is logical that, in the original nonlinear system, the remainder terms $r_1(x_1, x_2)$ and $r_2(x_1, x_2)$ (which are neglected in the linearization process) are the ones that actually determine the trajectory behavior—depending on the nature of these remainder terms, the oscillations in the nonlinear system can either grow or decay with time. This is why in this case studying the linearized system provides no definitive answers about the nonlinear system.

9 **Example.** Consider the following second-order equation, generally known as Van der Pol's equation:

$$\ddot{y}(t) - \mu[1 - y^2(t)]\dot{y}(t) + y(t) = 0 \tag{10}$$

where $\mu > 0$ is a constant. By defining the state variables

$$x_1(t) = y(t) \tag{11}$$

$$x_2(t) = \dot{y}(t) \tag{12}$$

(10) is transformed into the pair of first-order equations

$$\dot{x}_1(t) = x_2(t) \tag{13}$$

$$\dot{x}_2(t) = -x_1(t) + \mu[1 - x_1^2(t)]x_2(t) \tag{14}$$

The associated linear system is

$$\dot{\xi}_1(t) = \xi_2(t) \tag{15}$$

$$\dot{\xi}_2(t) = -\xi_1(t) + \mu\xi_2(t) \tag{16}$$

The eigenvalues of the associated **A** matrix satisfy the characteristic equation

$$\lambda^2 - \mu\lambda + 1 = 0 \tag{17}$$

Hence, for all positive values of μ, the roots of (17) are complex with positive real parts, so that the equilibrium point $\xi_1 = 0, \xi_2 = 0$ of (15)–(16) is an unstable focus. Referring to Table 2.1, we see that the equilibrium point $x_1 = 0, x_2 = 0$ of (13)–(14) is also an unstable focus. The Van der Pol equation is studied further in a subsequent example.

2.3.2 Graphical Euler Method

The simple method presented below is quite well suited for constructing a single trajectory of (1)–(2), starting from a given initial point, and is especially suitable for implementation on a digital computer. This method actually amounts to a graphical interpretation of the forward Euler numerical integration procedure and is an example of a so-called first-order method. More refined numerical techniques, which are applicable to nonautonomous systems of any order, are discussed in Chap. 4.

For the system (1)–(2), we can write

$$x_1(t + \Delta t) = x_1(t) + \Delta t \cdot f_1[x_1(t), x_2(t)] + o(\Delta t) \tag{18}$$

$$x_2(t + \Delta t) = x_2(t) + \Delta t \cdot f_2[x_1(t), x_2(t)] + o(\Delta t) \tag{19}$$

Because we want ultimately to eliminate t, let us denote $x_1(t)$ by x_1 and $x_1(t + \Delta t)$ by $x_1 + \Delta x_1$ and similarly define x_2 and $x_2 + \Delta x_2$. Then, *because the system (1)–(2) is autonomous*, we can eliminate t from (18)–(19) to get

$$\begin{aligned} x_2 + \Delta x_2 &= x_2 + \frac{f_2(x_1, x_2)}{f_1(x_1, x_2)} \Delta x_1 \\ &= x_2 + s(x_1, x_2)\, \Delta x_1 \end{aligned} \tag{20}$$

where

$$s(x_1, x_2) \triangleq \frac{f_2(x_1, x_2)}{f_1(x_1, x_2)} \tag{21}$$

The relationship (20) can be interpreted to mean that whenever a trajectory of (1)–(2) passes through the point (x_1, x_2), its slope at that point is given by $s(x_1, x_2)$. If $f_1(x_1, x_2) \neq 0$, the quantity $s(x_1, x_2)$ is well defined. If $f_1(x_1, x_2) \neq 0$ and $f_2(x_1, x_2) = 0$, then the tangent at (x_1, x_2) to the trajectory of (1)–(2) passing through (x_1, x_2) is horizontal, while if $f_1(x_1, x_2) = 0$ and $f_2(x_1, x_2) \neq 0$, then the tangent is vertical. Finally, if $f_1(x_1, x_2) = f_2(x_1, x_2) = 0$, the quantity $s(x_1, x_2)$ is undefined, but in this case (x_1, x_2) is an equilibrium point of (1)–(2), so that the trajectory consists of the single point (x_1, x_2).

The numerical method consists therefore of starting with the given initial point (x_{10}, x_{20}), calculating the slope $s(x_{10}, x_{20})$, drawing a "short" line segment through (x_{10}, x_{20}) with the slope $s(x_{10}, x_{20})$, picking the end of the line segment as the new starting point, and repeating the procedure. There is only one small point to be cleared up. The quantity $s(x_1, x_2)$ specifies the slope of the trajectory but does not specify its direction. But this is easily determined by examining the signs of the quantities $f_1(x_1, x_2)$ and $f_2(x_1, x_2)$. We can assume without loss of generality that not both $f_1(x_1, x_2)$ and $f_2(x_1, x_2)$ are zero, because if $f_1(x_1, x_2) = f_2(x_1, x_2) = 0$, then (x_1, x_2) is an equilibrium point of (1)–(2), and no further analysis is necessary. Thus suppose $f_i(x_1, x_2) \neq 0$, where i is either 1 or 2. If $f_i(x_1, x_2)$ is positive, then $\dot{x}_i$ is positive when the trajectory passes through the point (x_1, x_2). Because we are interested in the evolution of the solution to (1)–(2) as *t increases* from 0, it follows that the direction of the trajectory at (x_1, x_2) should be so chosen that x_i increases. Similarly, if $f_i(x_1, x_2)$ is negative, the direction of the trajectory at (x_1, x_2) should be so chosen that x_i decreases. If both $f_1(x_1, x_2)$ and $f_2(x_1, x_2)$ are nonzero, one will reach the same conclusions regarding the direction of the trajectory at (x_1, x_2) whether he examines the sign of $f_1(x_1, x_2)$ or that of $f_2(x_1, x_2)$.

The situation is illustrated in Figs. 2.14 and 2.15. Suppose that at some point (x_1, x_2) both $f_1(x_1, x_2)$ and $f_2(x_1, x_2)$ are positive. Then the slope $s(x_1, x_2)$ is also positive. To determine the direction of the trajectory as it passes through (x_1, x_2), we observe that because $f_1(x_1, x_2)$ is positive, the trajectory direction at (x_1, x_2) corresponds to increasing x_1, as in Fig. 2.14. Alternatively, we could have reasoned that because $f_2(x_1, x_2)$ is positive, the trajectory direction at (x_1, x_2) corresponds to increasing x_2. The conclusion is the same as before.

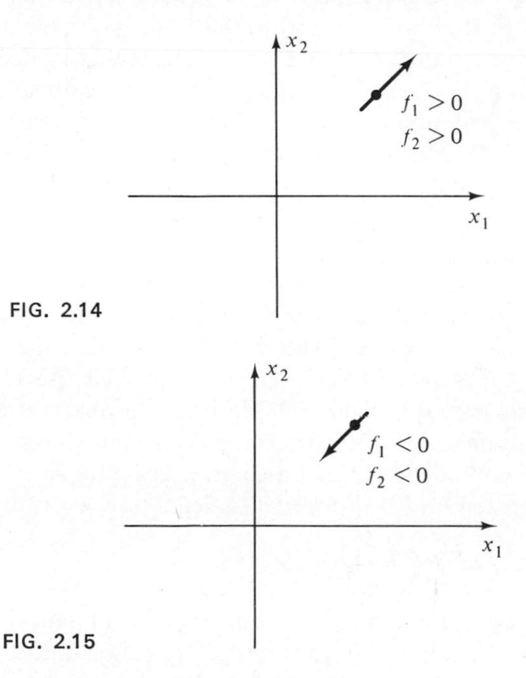

FIG. 2.14

FIG. 2.15

Now suppose that at (x_1, x_2) both $f_1(x_1, x_2)$ and $f_2(x_1, x_2)$ are negative. Then $s(x_1, x_2)$ is still positive, so that the trajectory has a positive slope as it passes through the point (x_1, x_2). However, the *direction* of the trajectory is as shown in Fig. 2.15.

In the special case where $x_2 = \dot{x}_1$, i.e., when the state plane is actually a phase plane, the considerations of the direction of the trajectory can be somewhat simplified. In this case, (1)–(2) simplify to

$$\dot{x}_1 = x_2 \tag{22}$$

$$\dot{x}_2 = f_2(x_1, x_2) \tag{23}$$

and the slope function $s(x_1, x_2)$ assumes the form

$$s(x_1, x_2) = \frac{f_2(x_1, x_2)}{x_2} \tag{24}$$

Thus the first thing to be noticed is that whenever $x_2 = 0$ we have $s(x_1, x_2) = \infty$ unless $f_2(x_1, x_2) = 0$. Geometrically, this means that whenever a trajectory of (22)–(23) meets the x_1-axis [at say $(x_1, 0)$] either $(x_1, 0)$ is an equilibrium point of the system (22)–(23) or else the tangent to the trajectory of (22)–(23) at $(x_1, 0)$ is vertical. In addition, the direction of the trajectories of (22)–(23) can be determined by inspection: Whenever the point (x_1, x_2) lies in the first or second quadrants, we have $x_2 > 0$, so that the direction of the trajectory of (22)–(23) should be so chosen as to correspond to increasing x_1. Similarly, whenever (x_1, x_2) lies in the third or fourth quadrants, we have $x_2 < 0$, so the direction of the trajectory at (x_1, x_2) corresponds to decreasing x_1. If (x_1, x_2) lies on the x_1-axis, i.e., if $x_2 = 0$, then the direction of the trajectory is vertical upward if $f_2(x_1, 0) > 0$ and vertical downward if $f_2(x_1, 0) < 0$. Of course, if $f_2(x_1, 0) = 0$, then $(x_1, 0)$ is an equilibrium point of (22)–(23).

2.3.3 Isocline Method

The isocline method is a procedure for sketching trajectories that is not particularly efficient if one wishes to sketch a single trajectory of (1)–(2) but that is quite useful if one wants to sketch several trajectories starting from several initial points, in order to obtain a understanding of the overall behavior of the trajectories. In some cases, one is quickly able to spot potential periodic trajectories using this method.

The procedure is as follows: The equation

$$s(x_1, x_2) = \frac{f_2(x_1, x_2)}{f_1(x_1, x_2)} = \text{constant} = c \tag{25}$$

determines, for each value of the constant c, a curve in the x_1–x_2 plane along which the solution trajectories of (1)–(2) have the slope c. Thus, whenever a solution trajectory of (1)–(2) crosses the curve defined by $s(x_1, x_2) = c$, it must do so with a slope of c. The procedure is to plot the curve $s(x_1, x_2) = c$ in the x_1–x_2 plane and along this curve draw "short" line segments having the slope c. Such a curve is known as an isocline. The directions of these line segments are determined as in Sec. 2.3.2. The procedure is repeated for sufficiently many values of the constant c, so that the x_1–x_2 plane is filled with isoclines, and one is then able to rapidly sketch the trajectories of (1)–(2) starting from any initial point.

26 **Example.** Consider again the equations of the Van der Pol oscillator, namely,

$$\dot{x}_1 = x_2 \tag{27}$$

$$\dot{x}_2 = -x_1 + \mu(1 - x_1^2)x_2 \tag{28}$$

The "slope function" $s(x_1, x_2)$ associated with this system of equations is

$$s(x_1, x_2) = \frac{-x_1 + \mu(1 - x_1^2)x_2}{x_2} \tag{29}$$

The curves $s(x_1, x_2) = c$ corresponding to $\mu = 1$ and for various values of c are shown in Fig. 2.16. With the aid of this construction, one can easily sketch the trajectory of the Van der Pol oscillator starting from, say, $x_1(0) = -2$, $x_2(0) = 3$. Also one can see that there may possibly be a periodic solution as indicated by the bold line.

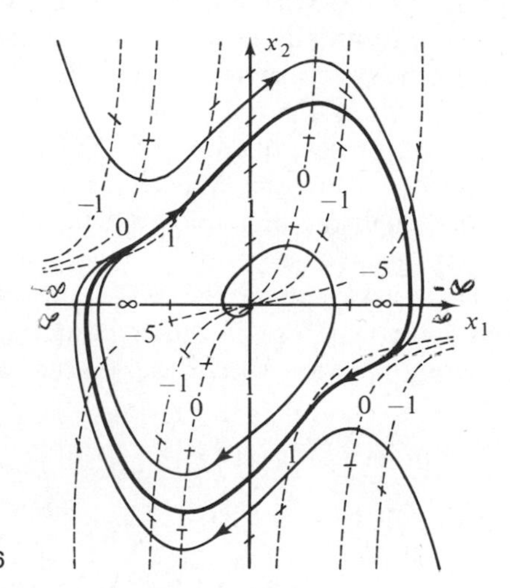

FIG. 2.16

2.3.4 Vector Field Method

The vector field method is quite similar to the isocline method, except that one plots the vector field itself rather than isoclines. Given the system (1)–(2), one plots, at each point (x_1, x_2), the corresponding vector $[f_1(x_1, x_2), f_2(x_1, x_2)]$ (of course with suitable scaling). In the end, one obtains a *vector field diagram* that clearly indicates the nature of the trajectories of the system. The procedure is illustrated in the following example.

30 **Example.** Consider the system of equations

$$\dot{x}_1 = -ax_1 + bx_1x_2 \tag{31}$$

$$\dot{x}_2 = -bx_1x_2 \tag{32}$$

The system (31)–(32) is a very elementary model for the spreading of disease within a population. Here x_1 denotes the number of infected people, while x_2 denotes the number of noninfected or "susceptible" people. Equation (32) states that noninfected people become infected at a rate proportional to x_1x_2, which is a measure of the interaction between the two groups. The expression (31) for $\dot{x}_1$ consists of two terms: (1) $-ax_1$, which is the rate at which people die from disease or survive and become forever immune, and (2) bx_1x_2, which is the rate at which previously noninfected people become infected. Note that the system (31)–(32) is a special case of the predator–prey equations of Example [2.4 (30)].

With regard to (31)–(32), any point of the form $(0, x_2)$ is an equilibrium point; i.e., if initially there is no infection, there will be none thereafter. Hence the system exhibits a continuum of equilibrium points on the x_2-axis. To study the behavior of the solution trajectories, we plot the velocity vector field of (31)–(32), with $a = 2$, $b = 1$; the result is shown in Fig. 2.17. Note that the velocity vector field is plotted for all x_1, x_2, even though the first quadrant (i.e., $x_1 \geq 0$, $x_2 \geq 0$) is the only practically relevant case.

From Fig. 2.17, we can see the *qualitative* behavior of the solution trajectories. Suppose (x_{10}, x_{20}) in the first quadrant is the initial condition. By examining the vector field, we see that as $t \longrightarrow \infty$, $x_1(t) \longrightarrow 0$

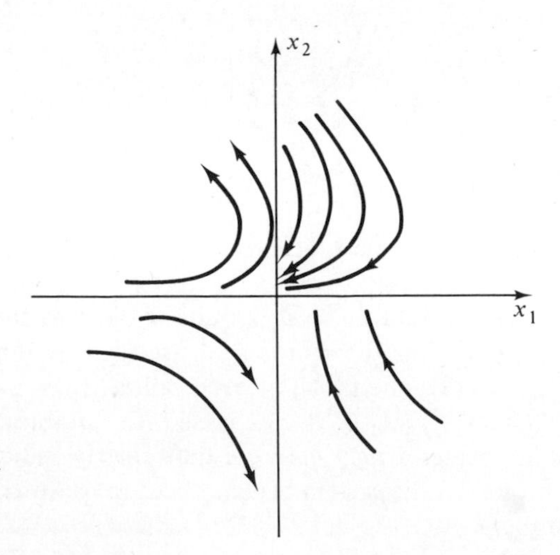

FIG. 2.17

while $x_2(t)$ approaches a nonzero limit. In other words, eventually the infection does die down. However, the larger the initial ratio x_{10}/x_{20}, the smaller the final ratio $x_2(\infty)/x_{20}$.

Problem 2.8. Find all the equilibrium points of the Volterra predator–prey equations

$$\dot{x}_1 = -x_1 + x_1x_2$$

$$\dot{x}_2 = x_2 - x_1x_2$$

Linearize the system around each of the equilibrium points, and determine, if possible, the nature of each equilibrium point. (*Answer:* one center, one saddle, two indeterminates.)

Problem 2.9. Plot the isoclines for Rayleigh's equation

$$\dot{x}_1 = x_2$$

$$\dot{x}_2 = -x_1 + \varepsilon\left(x_2 - \frac{x_2^3}{3}\right)$$

for (a) $\varepsilon = 1$, (b) $\varepsilon = 0.1$

Problem 2.10. Plot the velocity vector field for the pendulum equation

$$\dot{x}_1 = x_2$$

$$\dot{x}_2 = -\sin x_1$$

Problem 2.11. Consider the nonlinear circuit in Fig. 2.18. Suppose the voltage–current relationship of the nonlinear resistor is given by

$$v_r = i_r^3 - 3i_r^2 + 3i_r \triangleq f(i_r)$$

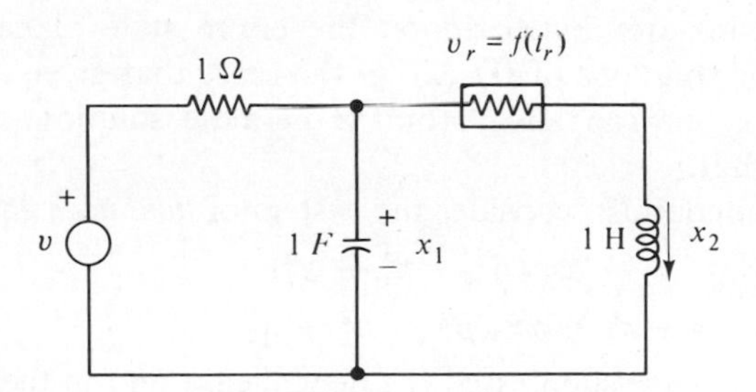

FIG. 2.18

(a) Select the capacitor voltage x_1 and the inductor current x_2 as the state variables, and show that the state equations are

$$\dot{x}_1 = v - x_1 - x_2$$

$$\dot{x}_2 = x_1 - f(x_2)$$

(b) With $v \equiv 0$, calculate the equilibrium points of the system.

(c) Linearize the system around each of the equilibrium points, and determine the nature of the trajectories around each point.

2.4 PERIODIC SOLUTIONS AND LIMIT CYCLES

2.4.1 Introduction

Some autonomous systems exhibit periodic solutions. For example, consider a simple harmonic oscillator, which is described by the linear equations

$$\dot{x}_1(t) = x_2(t) \tag{1}$$

$$\dot{x}_2(t) = -x_1(t) \tag{2}$$

The solution of (1)–(2), subject to the initial conditions

$$x_1(0) = x_{10} \tag{3}$$

$$x_2(0) = x_{20} \tag{4}$$

is given by

$$x_1(t) = r_0 \cos(-t + \phi_0) \tag{5}$$

$$x_2(t) = r_0 \sin(-t + \phi_0) \tag{6}$$

where

$$r_0 = (x_{10}^2 + x_{20}^2)^{1/2} \tag{7}$$

$$\phi_0 = \tan^{-1} \frac{x_{20}}{x_{10}} \tag{8}$$

Thus the solution of (1)–(2) is periodic regardless of what the initial conditions are. Furthermore, the entire state plane is covered with periodic solutions of (1)–(2), in the sense that given an arbitrary point (x_1, x_2), one can always find a periodic solution of (1)–(2) passing through it.

In contrast, consider the system of *nonlinear* equations

$$\dot{x}_1 = x_2 + \alpha x_1(\beta^2 - x_1^2 - x_2^2) \tag{9}$$

$$\dot{x}_2 = -x_1 + \alpha x_2(\beta^2 - x_1^2 - x_2^2) \tag{10}$$

where we have suppressed the dependence on t in the interests of brevity. The reader can easily verify that the velocity vector field of the system (9)–(10) is the sum[3] of two vector fields: (1) the velocity vector field of the system (1)–(2) and (2) a radial vector field that is outgoing for $x_1^2 + x_2^2 < \beta^2$ and incoming for $x_1^2 + x_2^2 > \beta^2$. [Note that a vector field $\mathbf{f}(\mathbf{x})$ is *radial* if the vector $\mathbf{f}(\mathbf{x})$ is always aligned with the vector $\mathbf{x}$, for all $\mathbf{x}$.] Now, if we introduce the polar coordinates

---

[3]The sum of two vector fields $\mathbf{f}(\mathbf{x})$ and $\mathbf{g}(\mathbf{x})$ is defined by

$$(\mathbf{f} + \mathbf{g})(\mathbf{x}) = \mathbf{f}(\mathbf{x}) + \mathbf{g}(\mathbf{x})$$

11 $$r = (x_1^2 + x_2^2)^{1/2}$$

12 $$\phi = \tan^{-1} \frac{x_2}{x_1}$$

then the equations (10)–(11) are transformed into

13 $$\dot{r} = \alpha r(\beta^2 - r^2)$$

14 $$\dot{\phi} = -1$$

It can be easily verified that the solution of (13)–(14) is

15 $$r(t) = \frac{\beta}{(1 + c_0 e^{-2\beta^2 \alpha t})^{1/2}}$$

16 $$\phi(t) = \phi_0 - t$$

where

17 $$c_0 = \frac{\beta^2}{r_0^2} - 1$$

Thus the system (9)–(10) has only one periodic solution, namely $r_0 = \beta$, i.e., $x_1^2 + x_2^2 = \beta^2$. Furthermore, whenever $r_0 \neq 0$, all solutions of (9)–(10) approach this periodic solution as $t \longrightarrow \infty$. This example differs from the example of a simple harmonic oscillator in that the periodic solution in the present case is *isolated;* i.e., there exists a neighborhood of it that does not contain any other periodic solutions.

18 *[definition]* A limit cycle of [2.3(1)]–[2.3(2)] is a periodic solution of [2.3(1)]–[2.3(2)].

By convention, we do not regard an equilibrium point as a periodic solution or as a limit cycle. Also, a simple consequence of definition (18) is that a limit cycle can be either isolated or nonisolated.

In the remainder of this section, we shall present some results pertaining to limit cycles in nonlinear systems.

2.4.2 Bendixson's Theorem

19 ***Theorem*** Suppose D is a simply connected[4] domain in R^2 such that the quantity $\nabla \mathbf{f}(\mathbf{x})$ defined by

20 $$\nabla \mathbf{f}(\mathbf{x}) = \frac{\partial f_1}{\partial x_1}(x_1, x_2) + \frac{\partial f_2}{\partial x_2}(x_1, x_2)$$

[4] A *connected* region can be thought of as a set that is in one piece, i.e., one in which every two points in the set can be connected by a curve lying entirely within the set. A set is *simply connected* if (1) it is connected and (2) its boundary is connected. One can also think of a simply connected set as one that can be obtained by continuously deforming a circle. For example, an annular region is connected but not simply connected.

is not identically zero over any subregion of D and does not change sign in D. Then D contains no closed trajectories of [2.3(1)]–[2.3(2)].

21 REMARKS: Theorem (19) gives conditions under which a region does *not* contain a periodic solution and as such gives a sufficient condition for the nonexistence of a periodic solution.

proof of theorem (19) Suppose J is a closed trajectory of [2.3(1)]–[2.3(2)]. Then at each point $\mathbf{x} = (x_1, x_2) \in J$, the velocity vector field $\mathbf{f}(\mathbf{x}) = [f_1(x_1, x_2), f_2(x_1, x_2)]$ is tangent to J. Let $\mathbf{n}(\mathbf{x})$ denote the outward normal to J at $\mathbf{x}$. Then $\mathbf{f}(\mathbf{x}) \cdot \mathbf{n}(\mathbf{x}) = 0$ for all $\mathbf{x} \in J$. In particular,

22 $$\int_J \mathbf{f}(\mathbf{x}) \cdot \mathbf{n}(\mathbf{x}) \, dl = 0$$

But by the divergence theorem,

23 $$\int_J \mathbf{f}(\mathbf{x}) \cdot \mathbf{n}(\mathbf{x}) \, dl = \iint_S \nabla \mathbf{f}(\mathbf{x}) \, ds \qquad (= 0)$$

where S is the area enclosed by J. Therefore, for (23) to hold, we must have either (i) $\nabla \mathbf{f}(\mathbf{x}) \equiv 0 \; \forall \; \mathbf{x} \in S$ or (ii) $\nabla \mathbf{f}(\mathbf{x})$ changes sign over the region S. But if S is a subset of D, neither can happen. Hence D contains no closed trajectories of [2.3(1)]–[2.3(1)]. Note that if $\nabla f(\mathbf{x})$ is a polynomial in the components of $\mathbf{x}$, then one can systematically check whether or not $\nabla f(\mathbf{x})$ changes sign over a given region; see [15] Jury. ∎

24 **Example.** Consider the application of Theorem (19) to the *linear* system of equations

25 $$\dot{x}_1 = a_{11}x_1 + a_{12}x_2$$

26 $$\dot{x}_2 = a_{21}x_1 + a_{22}x_2$$

or, in matrix form,

27 $$\dot{\mathbf{x}} = \mathbf{A}\mathbf{x}$$

We know from Sec. 2.2 that a necessary and sufficient condition for the system (25)–(26) to have periodic solutions is that the matrix $\mathbf{A}$ has two nonzero imaginary eigenvalues. Because the eigenvalues of $\mathbf{A}$ satisfy the characteristic equation

28 $$\lambda^2 - (a_{11} + a_{22})\lambda + (a_{11}a_{22} - a_{12}a_{21}) = 0$$

it is clear that the system (25)–(26) has periodic solutions if and only if

29 $$a_{11} + a_{22} = 0$$

30 $$a_{11}a_{22} - a_{12}a_{21} > 0$$

Equivalently, a necessary and sufficient condition to have *no* periodic solution is that either (29) or (30) is violated.

Applying Theorem (19) to the present case, we get

$$\nabla \mathbf{f}(\mathbf{x}) \equiv a_{11} + a_{22} \qquad \forall\ \mathbf{x} \in R^2 \tag{31}$$

Hence we conclude that if $a_{11} + a_{22} \neq 0$, then no periodic solutions exist, which is in accordance with the previous discussion.

32 **Example.** Consider the system of nonlinear equations

$$\dot{x}_1 = x_2 + x_1 x_2^2 \tag{33}$$

$$\dot{x}_2 = -x_1 + x_1^2 x_2 \tag{34}$$

The linearization of this system around the equilibrium point (0, 0) is

$$\dot{x}_1 = x_2 \tag{35}$$

$$\dot{x}_2 = -x_1 \tag{36}$$

which exhibits a continuum of periodic solutions. However, for the nonlinear system,

$$\nabla \mathbf{f}(\mathbf{x}) = x_1^2 + x_2^2 > 0 \text{ whenever } (x_1, x_2) \neq (0, 0) \tag{37}$$

Thus over every region in R^2 that does not consist of just a single point, we have that $\nabla \mathbf{f}(\mathbf{x})$ is not identically zero and does not change sign. Therefore this system has no periodic solutions anywhere in R^2 (other than the trivial one $x_1 = x_2 = 0$).

38 **Example.** In applying Theorem (19), the assumption that D is a simply connected region in crucial—it is not enough for D to be just connected. To see this, consider the system (9)–(10), and let D be the annual region

$$D = \left\{(x_1, x_2): \frac{2\beta^2}{3} < x_1^2 + x_2^2 < 2\beta^2\right\} \tag{39}$$

For this example, we have

$$\nabla \mathbf{f}(\mathbf{x}) = 2\alpha\beta^2 - 4\alpha(x_1^2 + x_2^2) \tag{40}$$

which is everywhere nonnegative on D. Yet D contains a periodic solution. The reason, of course, that Theorem (19) is inapplicable here is that D is not simply connected, even though it is connected.

2.4.3 Poincaré–Bendixson Theorem

The Poincaré–Bendixson theorem can be used to prove the existence of a periodic solution, provided a region M satisfying certain conditions can be found. The strength of this theorem is its generality and its simple geometric interpretation. The weakness of the theorem is the necessity of having to find the region M. We shall first introduce a definition.

41 *[definition]* Let $\mathbf{x}(t)$ be a solution trajectory of [2.3(1)]–[2.3(2)]. A point $\mathbf{z} \in R^2$ is said to be a *limit point* of this trajectory if there exists a sequence $(t_n)_1^\infty$ in R_+ such that $t_n \longrightarrow \infty$ as $n \longrightarrow \infty$ and $\mathbf{x}(t_n) \longrightarrow \mathbf{z}$ as $n \longrightarrow \infty$. The set of all limit points of a trajectory $\mathbf{x}(t)$ is called the *limit set* of the trajectory and is denoted by L.

42 REMARKS: Basically, a limit point of the trajectory $\mathbf{x}(t)$ is a point $\mathbf{z}$ which has the property that, as time progresses, the trajectory passes arbitrarily close to $\mathbf{z}$ infinitely many times. We shall encounter limit points and limit sets again in Chap. 5.

43 ***Theorem*** (Poincaré–Bendixson) Let

44 $$S = \{\mathbf{x}(t), t \geq 0\}$$

denote a trajectory in R^2 of the system [2.3(1)]–[2.3(2)], and let L denote its limit set. If L is contained in a closed bounded region M in R^2 and if M contains no equilibrium points of [2.3(1)]–[2.3(2)], then either

(i) S is a periodic solution of [2.3(1)]–[2.3(2)], or
(ii) L is a periodic solution of [2.3(1)]–[2.3(2)].

We shall omit the proof because it is beyond the scope of this book.

45 REMARKS: Roughly speaking, what Theorem (43) states is the following: Suppose we can find a closed bounded region M in R^2 such that M does not contain any equilibrium points of [2.3(1)]–[2.3(2)] and such that all limit points of some trajectory S are contained in M. Then M contains at least one periodic solution of [2.3(1)]–[2.3(2)]. In practice, it is very difficult to verify that M contains all the *limit points* of a trajectory. However, because M is closed, it can be shown that if some trajectory $\mathbf{x}(t)$ is *eventually* contained in M, i.e., there exists a time $t_0 < \infty$ such that $\mathbf{x}(t) \in M \; \forall \; t \geq t_0$, then L is contained in M. Thus the theorem comes down to this: If we can find a closed bounded region M containing no equilibrum points such that some trajectory is eventually confined to M, then M contains at least one periodic solution. Now, a sufficient condition for a trajectory to be eventually confined to M is that, at every point along the boundary of M, the velocity vector field always points *into* M. If this is the case, then any trajectory originating within M must remain in M, and hence M contains at least one periodic solution trajectory. (This is depicted in Fig. 2.19.)

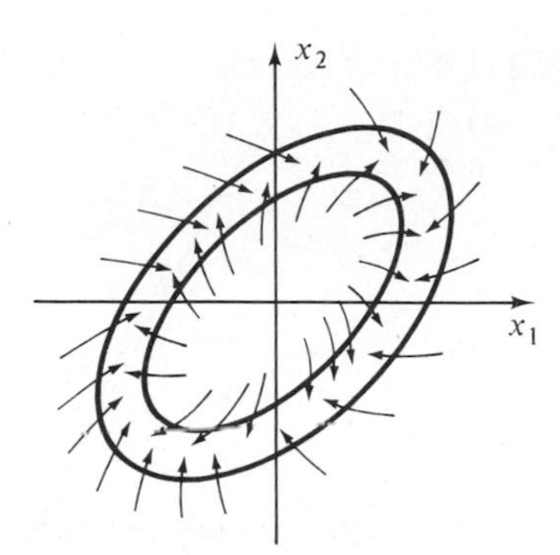

FIG. 2.19

46 **Example.** Consider once again the system (9)–(10), and let M be the ~~annualr~~ annular region defined by

47 $$M = \{(x_1, x_2): 0.9 \leq x_1^2 + x_2^2 \leq 1.1\}$$

Then M contains no equilibrium points of the system (9)–(10). Furthermore, a sketch of the velocity vector field for this system reveals that, all along the boundary of M, the vector field always points *into* M, as depicted in Fig. 2.19. Hence we can apply Theorem (43) and conclude that M contains a periodic solution.

48 **Example.** In applying Theorem (43), the condition that M should not contain any equilibrium points is very important. To see this, consider the system

49 $$\dot{x}_1 = -x_1 + x_2$$

50 $$\dot{x}_2 = -x_1 - x_2$$

The velocity vector field for this system is sketched in Fig. 2.20. If we choose M to be the unit disk centered at the origin, then all along the boundary of M the velocity vector field points into M. Hence all trajec-

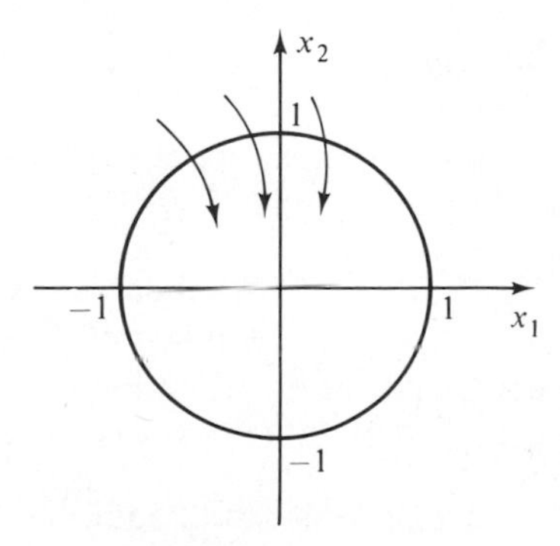

FIG. 2.20

tories originating within M remain within M. The same conclusion can be reached by analytical reasoning, because in polar coordinates, the system (49)–(50) becomes

51 $$\dot{r} = -r$$

52 $$\dot{\phi} = -1$$

which has the solution

53 $$r(t) = r_0 e^{-t}$$

54 $$\phi(t) = \phi_0 - t$$

However, M contains no (nontrivial) periodic solutions.

2.4.4 Index Theorems

The concept of index is a very powerful one, and the results given below only scratch the surface of the many results that are available. Unfortunately, the arguments involved in index theory are well beyond the scope of this book. Hence we shall content ourselves with the few results given below, most of which are given without proof. For further results, see [18] Nemytskii and Stepanov.

The definition below introduces the concept of the index of a vector field.

55 *[definition]* Suppose D is an open, simply connected subset of R^2, and suppose $\underset{\sim}{f}: R^2 \longrightarrow R^2$ is a vector field on R^2. Suppose that D contains only isolated equilibrium points of the system

56 $$\dot{\mathbf{x}}(t) = \underset{\sim}{f}[\mathbf{x}(t)]$$

Let J be a simple, closed, positively oriented Jordan curve in D that does not pass through any equilibrium points of (56). Then the *index of the curve J with respect to the vector field* **f**, denoted by $I_{\mathbf{f}}(J)$, is defined by

57 $$I_{\mathbf{f}}(J) = \frac{1}{2\pi} \int_J d\theta_{\mathbf{f}}(x_1, x_2)$$

where $\theta_{\mathbf{f}}(x_1, x_2)$ is the direction of the vector field **f** at the point (x_1, x_2).

58 REMARKS: A positively oriented curve is one that is traversed in a counterclockwise sense. The index of J with respect to **f** is nothing but the net change in the direction of **f**, as **x** traverses around J, divided by 2π. Clearly $I_{\mathbf{f}}(J)$ is always an integer.

59 *[definition]* Let **p** be an isolated equilibrium point of (56). Then the *index of* **p** is defined as $I_{\mathbf{f}}(J)$, where J is any suitable Jordan

curve such that (i) $\mathbf{p}$ is interior to J and (ii) J contains no equilibrium points in its interior other than $\mathbf{p}$.

We shall now state some facts without proof.

60 **fact** suppose J contains no equilibrium points of (56) in its interior. Then $I_{\mathbf{f}}(J) = 0$.

61 **fact** The index of a center, focus, and node is 1 and that of a saddle is -1. (This fact can be easily verified by a sketch of the velocity vector field near a center, focus, node, and saddle.)

62 **fact** Suppose J contains in its interior a *finite* number of equilibrium points of (56), say $p_1, \ldots, p_n$ (which are of necessity isolated). Then

63 $$I_{\mathbf{f}}(J) = \sum_{i=1}^{n} I_{\mathbf{f}}(p_i)$$

64 **fact** Let $\mathbf{f}$ and $\mathbf{g}$ be two vector fields on R^2, and let $\theta_{\mathbf{f}}$ and $\theta_{\mathbf{g}}$ denote the directions of $\mathbf{f}$ and $\mathbf{g}$, respectively. Let J be a simple, closed, positively oriented Jordan curve such that (i) $|\theta_{\mathbf{f}} - \theta_{\mathbf{g}}| < \pi$ along J (i.e., the vector fields $\mathbf{f}$ and $\mathbf{g}$ are never in opposition on J) and (ii) J does not pass through any equilibrium point of (56) or of the system

65 $$\dot{\mathbf{x}}(t) = \mathbf{g}[\mathbf{x}(t)]$$

Then

66 $$I_{\mathbf{f}}(J) = I_{\mathbf{g}}(J)$$

67 **fact** Let J be a simple, closed, positively oriented trajectory of (56). Then

68 $$I_{\mathbf{f}}(J) = 1$$

(Note that the vector field $\mathbf{f}$ is always tangential to J.)

On the basis of these facts, we can state the following general theorem.

69 ***Theorem*** Every closed trajectory of (56) contains at least one equilibrium point of (56) in its interior. If the system (56) has only isolated equilibrium points, then every closed trajectory of (56) contains in its interior a finite number of equilibrium points such that the sum of their indices is 1.

70 **Example.** We shall study in detail the Volterra predator–prey equations

71 $$\dot{x}_1 = -x_1 + x_1 x_2$$

72 $$\dot{x}_2 = x_2 - x_1 x_2$$

We shall digress briefly to discuss the rationale behind the above model. In (71)–(72), x_1 denotes the number of predators (foxes, say), and x_2 denotes the number of prey (rabbits). If $x_2 = 0$, (71) reduces to $\dot{x}_1 = -x_1$, which implies that in the absence of prey the number of predators will dwindle exponentially to zero. If $x_2 \neq 0$, (71) shows that $\dot{x}_1$ contains an exponential growth term proportional to x_2. The situation is just the opposite in the case of the prey. If $x_1 = 0$, x_2 will grow exponentially, while if $x_1 \neq 0$, $\dot{x}_2$ contains an exponential decay term proportional to x_1.

The velocity vector field for the predator–prey system is shown in Fig. 2.21. Clearly there are four equilibrium points, namely (0, 0), (1, 0), (0, 1), and (1, 1). By linearizing the system (71)–(72) around each of these equilibrium points, we see that (0, 0) is a saddle, while (1, 1) is a center. Therefore the index of (0, 0) is -1, while that of (1, 1) is 1. However, in the linearizations around (1, 0) and (0, 1), the matrix of

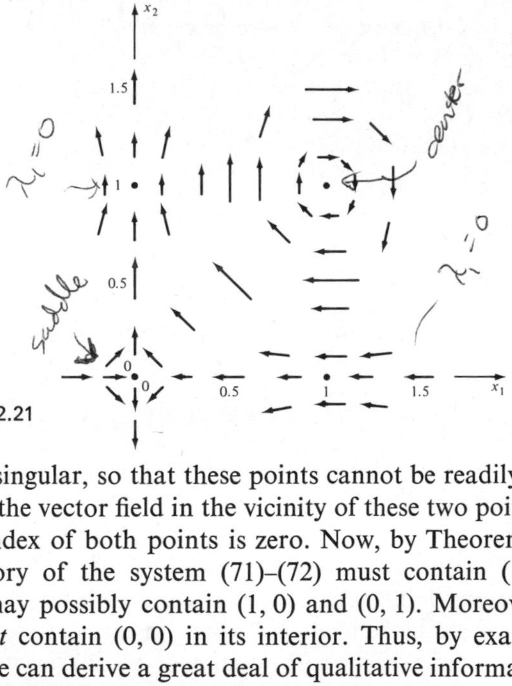

FIG. 2.21

coefficients is singular, so that these points cannot be readily classified. By examining the vector field in the vicinity of these two points, we can see that the index of both points is zero. Now, by Theorem (69), any closed trajectory of the system (71)–(72) must contain (1, 1) in its interior and may possibly contain (1, 0) and (0, 1). Moreover, it definitely does *not* contain (0, 0) in its interior. Thus, by examining the index alone, we can derive a great deal of qualitative information about the (possible) closed trajectories of the system.

2.4.5 An Analytical Method

In this subsection, which could equally well have been entitled "miscellaneous," we shall present a technique for obtaining analytical expressions for the closed trajectories of some nonlinear systems that exhibit a continuum of periodic solutions. Rather than presenting a general theorem, which would have to be rather weak because of all the possible pathological cases, we shall illustrate the method by means of two examples.

The basic idea of the method is as follows: Given the system

$$\dot{x}_1 = f_1(x_1, x_2) \tag{73}$$

$$\dot{x}_2 = f_2(x_1, x_2) \tag{74}$$

suppose we can find a continuously differentiable function $V: D \longrightarrow R$, where D is some open subset of R, such that

$$\dot{V}(x_1, x_2) \triangleq \frac{\partial V}{\partial x_1} f_1(x_1, x_2) + \frac{\partial V}{\partial x_2} f_2(x_1, x_2) \equiv 0 \quad \forall\ (x_1, x_2) \in D \tag{75}$$

The function $\dot{V}(x_1, x_2)$ is known as the *derivative of V* along the trajectories of (73)–(74), because if $[x_1(t), x_2(t)]$ is a trajectory of (73)–(74), then the time derivative of the function $V[x_1(t), x_2(t)]$ is given by $\dot{V}[x_1(t), x_2(t)]$. (The details of this argument can be found in Sec. 5.1.) Now, suppose $(x_{10}, x_{20}) \in D$, and let $\mathcal{C}$ denote the solution trajectory of (73)–(74) originating at (x_{10}, x_{20}). By (75), we see that the time derivative of $V[x_1(t), x_2(t)]$ is zero along $\mathcal{C}$, so that $V[x_1(t), x_2(t)]$ is constant along $\mathcal{C}$. In fact

$$V[x_1(t), x_2(t)] = V(x_{10}, x_{20}) \qquad \forall\ t \geq 0 \tag{76}$$

Let us now consider the set

$$S = \{(x_1, x_2)\colon V(x_1, x_2) = V(x_{10}, x_{20})\} \tag{77}$$

Then $\mathcal{C}$ is a subset of S. In particular, if S is a closed curve, we can conclude (under some relatively mild additional assumptions) that $\mathcal{C}$ itself is a closed trajectory of (73)–(74).

Of course, a lot depends on the choice of the function V. If we choose $V(x_1, x_2) = 1$ for all (x_1, x_2), then (75) is satisfied, but the set S in (77) is the whole space R^2, and as a result no insight has been gained using this particular V function. However, in some cases, by properly choosing V, we can show that the family of sets

$$\{(x_1, x_2)\colon V(x_1, x_2) = c\} \tag{78}$$

defines a continuum of closed trajectories of (73)–(74).

79 Example. We return to the predator–prey equations (71)–(72), and try a function V of the form

$$V(x_1, x_2) = h_1(x_1) + h_2(x_2) \tag{80}$$

where h_1 and h_2 are to be adjusted so that (75) is satisfied. We have

$$\dot{V}(x_1, x_2) = h_1'(x_1)(-x_1 + x_1 x_2) + h_2'(x_2)(x_2 - x_1 x_2) \tag{81}$$

where the prime denotes differentiation with respect to the appropriate argument. For (75) to hold, we must have

82 $$h_1'(x_1)x_1(x_2 - 1) + h_2'(x_2)x_2(1 - x_1) = 0$$

which can be rearranged as

83 $$h_1'(x_1)\frac{x_1}{1 - x_1} = h_2'(x_2)\frac{x_2}{1 - x_2}$$

Because the right-hand side of (83) depends only on x_2 and the left-hand side depends only on x_1, in fact both must equal a constant; in other words,

84 $$h_1'(x_1)\frac{x_1}{1 - x_1} = c$$

85 $$h_2'(x_2)\frac{x_2}{1 - x_2} = c$$

where c is a real constant. It can be easily verified that the solution of (84)–(85) is

86 $$h_1(x_1) = c(\ln x_1 - x_1)$$

87 $$h_2(x_2) = c(\ln x_2 - x_2)$$

Hence (75) is satisfied if we choose

88 $$V(x_1, x_2) = (\ln x_1 - x_1 + \ln x_2 - x_2)$$

where we have dropped the arbitrary constant c without loss of generality. Now, for the above choice of V, any set of the form (78) is actually a closed curve. Hence the family of curves defined by

89 $$\ln x_1 - x_1 + \ln x_2 - x_2 = \text{const.}$$

constitutes closed trajectories of the predator–prey system. Note that V is defined only for $x_1 > 0$, $x_2 > 0$, i.e., in the first quadrant.

90 **Example.** Consider the pendulum equation

91 $$\dot{x}_1 = x_2$$

92 $$\dot{x}_2 = -\sin x_1$$

Let us once again choose $V(x_1, x_2)$ to be of the form (80). Then, for (75) to be satisfied, we must have

93 $$h_1'(x_1)x_2 - h_2'(x) \sin x_1 = 0$$

which implies, as in Example (79), that

94 $$\frac{h_1'(x_1)}{\sin x_1} = \frac{h_2'(x_2)}{x_2} = \text{const.} = c$$

Solving (94) gives

$$h_1(x_1) = -c \cos x_1 \tag{95}$$

$$h_2(x_2) = c\frac{x_2^2}{2} \tag{96}$$

Hence the family of curves

$$\frac{x_2^2}{2} - \cos x_1 = \text{const.} \tag{97}$$

constitute a continuum of closed trajectories for the simple pendulum.

98 REMARKS: Examples (79) and (90) illustrate how the method presented here can sometimes yield good results. However, it should be clear that (1) a function $V(x_1, x_2)$ of the type (80) does not always work and (2) even if it does, there is no guarantee that all closed trajectories of the system are of the form (78). Despite these limitations, however, the method *is* of some value, as indicated by the two examples.

Problem 2.12. Consider the system of equations

$$\dot{x}_1 = x_2$$

$$\dot{x}_2 = -g(x_1) - h(x_2)$$

where $g, h\colon R \longrightarrow R$ are continuously differentiable functions. Using Bendixson's theorem, show that this system has no periodic solutions if $h'(\xi) \neq 0$ $\forall\ \xi \in R$.

Problem 2.13. Sketch a vector field with exactly one node and one saddle point. Show that it is not possible to continuously deform this vector field in such a way that there is a periodic solution enclosing both the node and the saddle point.

Problem 2.14. Using the method of Sec. 2.4.4, derive an expression for the closed trajectories of the system

$$\dot{x}_1 = x_2$$

$$\dot{x}_2 = -g(x_1)$$

2.5 TWO ANALYTICAL APPROXIMATION METHODS

In this section, we shall describe two techniques for obtaining analytical expressions that approximate the periodic solution of second-order nonlinear differential equations. In contrast with the method presented in Sec. 2.4.4 (which gives exact expressions, if it works), the two methods presented here are only approximate. However, they have the advantage of having a wide range of applicability and of enabling one to study the so-called "slowly varying" oscillations. It should be emphasized that, depending on the particular problem to which they are applied, one technique might work better than the other. Moreover, the two methods presented here are only a portion of the many techniques that are available, and the reader is referred to [6] Blaquiere for a more complete account.

2.5.1 Krylov–Boguliubov Method

The Krylov–Boguliubov method is applicable to differential equations of the type

$$\ddot{y}(t) + y(t) = \mu f[y(t), \dot{y}(t)] \tag{1}$$

The class of equations of type (1) includes many of the commonly encountered ones, such as the Van der Pol equation and the pendulum equation. Note that, in (1), the angular velocity of the oscillations corresponding to $\mu = 0$ has been normalized to 1. Clearly this presents no limitation and can always be achieved by scaling the variable t.

If $\mu = 0$, the solution of (1) is of the form

$$y(t) = a \sin (t + \phi) \tag{2}$$

With this in mind, we assume that the solution of (1) (when $\mu \neq 0$) is representable in the form

$$y(t) = a(t) \sin [t + \phi(t)] \tag{3}$$

$$\dot{y}(t) = a(t) \cos [t + \phi(t)] \tag{4}$$

where $a(\cdot)$ and $\phi(\cdot)$ are "slowly varying"; i.e., $\dot{a}(t)$ and $\dot{\phi}(t)$ are "small." Actually, if $y(\cdot)$ is given by (3), we have

$$\dot{y}(t) = \dot{a}(t) \sin [t + \phi(t)] + a(t) \cos [t + \phi(t)][1 + \dot{\phi}(t)] \tag{5}$$

Hence, for (4) to be valid, we must have

$$\dot{a} \sin (t + \phi) + a\dot{\phi} \cos (t + \phi) = 0 \tag{6}$$

where we have suppressed the dependence of a and ϕ on t in the interests of brevity. Substituting for y and $\dot{y}$ from (3) and (4) into (1) gives

$$\dot{a} \cos (t + \phi) - a\dot{\phi} \sin (t + \phi) = \mu f[a \sin (t + \phi), a \cos (t + \phi)] \tag{7}$$

Equations (6) and (7) represent two linear equations in the unknowns $\dot{a}$ and $\dot{\phi}$. Solving for $\dot{a}$ and $\dot{\phi}$ gives

$$\dot{a} = \mu \cos (t + \phi) f[a \sin (t + \phi), a \cos (t + \phi)] \tag{8}$$

$$\dot{\phi} = -\frac{\mu}{a} \sin (t + \phi) \cdot f[a \sin (t + \phi), a \cos (t + \phi)] \tag{9}$$

If we want to find solutions of (1) of the form (3) where $a(\cdot)$ is periodic, we impose an extra condition, namely

10 $$\frac{a(T) - a(0)}{T} = 0$$

or, equivalently,

11 $$\frac{1}{T} \int_0^T \dot{a}(t)\, dt = 0$$

where T is the period of $a(\cdot)$. Unfortunately, (11) cannot be used directly because the period T is in general dependent on μ and is hence unknown. To get around this difficulty, we observe that $a(\cdot)$ goes through one complete period as the phase $\theta = t + \phi(t)$ goes from 0 to 2π. Thus we change the variable of integration in (11) from t to θ; then the limits of integration become 0 and 2π, and the integrand $\dot{a}(t)$ becomes [using (8)]

12 $$\dot{a}(t) \longrightarrow \mu \cos \theta \cdot f(a \sin \theta, a \cos \theta)$$

Finally, we make the approximation

13 $$\frac{d\theta}{2\pi} = \frac{dt}{T}$$

Equation (13) expresses the fact that as t varies by T, θ varies by 2π. Thus (11) becomes

14 $$\frac{1}{2\pi} \int_0^{2\pi} \mu \cos \theta \cdot f(a \sin \theta, a \cos \theta)\, d\theta = 0$$

Similarly, if we require $\dot{\phi}$ also to be periodic, this leads to the condition

15 $$\frac{1}{2\pi} \int_0^{2\pi} \frac{\mu}{a} \sin \theta \cdot f(a \sin \theta, a \cos \theta)\, d\theta = 0$$

Equations (14) and (15) can be used in the following way. Suppose we are interested in approximating the periodic solutions of (1) by functions of the form

16 $$y(t) = a \sin (1 + \delta)t$$

where a and δ are now *constants*. In this case, (14) and (15) simplify to

17 $$\int_0^{2\pi} \cos \theta \cdot f(a \sin \theta, a \cos \theta)\, d\theta = 0$$

18 $$\int_0^{2\pi} \sin \theta \cdot f(a \sin \theta, a \cos \theta)\, d\theta = 0$$

Because a is a constant, $f(a \sin \theta, a \cos \theta)$ is a periodic function of θ with period 2π and hence can be expanded in a Fourier series. Hence (17) and (18) state that in order for (16) to approximate (to the first order in μ) a periodic solution of (1), it is necessary for the first harmonic of the periodic function $f(a \sin \theta, a \cos \theta)$ to be zero. This requirement is sometimes called the *principle of harmonic balance*. We shall encounter the same reasoning again in Chap. 4 when we discuss the so-called describing function method.

Note that μ does not appear in (17) and (18), because when we study the periodic solutions of (1) we are in effect examining the *steady-state* oscillations of (1), and μ does not affect the steady-state solutions. However, μ is prominently present when we study the so-called "slowly varying" or transient solutions of (1). For this purpose, we make the approximations

$$\dot{a}(t) \simeq \frac{a(T) - a(0)}{T} \tag{19}$$

$$\dot{\phi}(t) \simeq \frac{\phi(T) - \phi(0)}{T} \tag{20}$$

where T is the period of the steady-state oscillations. However, as in studying the steady-state oscillations, we have

$$\frac{a(T) - a(0)}{T} = \frac{1}{2\pi}\int_0^{2\pi} \mu \cos\theta \cdot f(a\sin\theta, a\cos\theta)\, d\theta \tag{21}$$

$$\frac{\phi(T) - \phi(0)}{T} = \frac{1}{2\pi}\int_0^{2\pi} -\frac{\mu}{a}\sin\theta \cdot f(a\sin\theta, a\cos\theta)\, d\theta \tag{22}$$

Hence the approximate equations describing the slowly varying oscillations of (1) are

$$\dot{a} = \frac{1}{2\pi}\int_0^{2\pi} \mu \cos\theta \cdot f(a\sin\theta, a\cos\theta)\, d\theta \tag{23}$$

$$\dot{\phi} = \frac{1}{2\pi}\int_0^{2\pi} -\frac{\mu}{a}\sin\theta \cdot f(a\sin\theta, a\cos\theta)\, d\theta \tag{24}$$

25 **Example.** We apply the Krylov–Boguliubov method to Van der Pol's equation, which can be written in the form

$$\ddot{y}(t) + y(t) = \mu\dot{y}(t)[1 - y^2(t)] \tag{26}$$

This is of the form (1) with

$$f(y, \dot{y}) = \dot{y}(1 - y^2) \tag{27}$$

Hence

$$\begin{aligned} f(a\sin\theta, a\cos\theta) &= a\cos\theta(1 - a^2\sin^2\theta) \\ &= \left(a - \frac{a^3}{4}\right)\cos\theta + \frac{a^3}{4}\cos 3\theta \end{aligned} \tag{28}$$

so clearly

$$\frac{1}{2\pi}\int_0^{2\pi} \mu\cos\theta \cdot f(a\sin\theta, a\cos\theta)\, d\theta = \frac{\mu}{2}\left(a - \frac{a^3}{4}\right) \tag{29}$$

30 $$\frac{1}{2\pi}\int_0^{2\pi}\frac{\mu}{a}\sin\theta\cdot f(a\sin\theta, a\cos\theta)\,d\theta = 0$$

Thus the approximate equations governing slowly varying a and ϕ are given by (23) and (24) as

31 $$\dot{a} = \frac{\mu}{2}\left(a - \frac{a^3}{4}\right)$$

32 $$\dot{\phi} = 0$$

If we are interested in finding the steady-state periodic solutions of Van der Pol's equation, we set $\dot{a} = 0$, $\dot{\phi} = 0$, which gives $a = 2$. Hence, to first order in μ, the limit cycle of the Van der Pol oscillator is described by

33 $$y(t) = 2\sin(t + \phi_0)$$

To get the slowly varying solution, we solve (31) and (32), which results in

34 $$a(t) = 2\left[\frac{1}{1 + c\exp(-\mu t)}\right]^{1/2}$$

35 $$\phi(t) = \phi_0$$

where c is a constant determined by the initial conditions. Thus we see that even though the parameter μ does not affect the steady-state solution, it does affect the rate at which the transient solution approaches the steady-state solution.

2.5.2 Power Series Method

The power series method is applicable to autonomous second-order differential equations containing a "small" parameter μ and consists of attempting to expand the solution of the given equation as a power series in μ. We shall illustrate the procedure by means of an example.

Consider the differential equation

36 $$\ddot{y}(t) + \omega_0^2 y(t) + \mu[y(t)]^3 = 0$$

together with the simplified initial conditions

37 $$y(0) = a_0$$

38 $$\dot{y}(0) = 0$$

This equation represents the oscillations of a mass constrained by a nonlinear spring. If $\mu > 0$, the spring is said to be hard, whereas if $\mu < 0$, it is said to be soft.

Clearly, if $\mu = 0$, the solution of (36) satisfying (37)–(38) is

39 $$y_0(t) = a_0\cos\omega_0 t$$

If $\mu \neq 0$ but is "small," we can attempt to express the solution of (36)–(38) as a power series in μ, in the form

40 $$y(t) = y_0(t) + \mu y_1(t) + \mu^2 y_2(t) + \ldots$$

The idea is to substitute (40) into (36) and equate the coefficients of all powers of μ to zero. However, if this is done blindly, some of the $y_i(t)$ may contain *secular* terms, i.e., unbounded functions of time. To see this phenomenon, let us substitute (40) into (36) and set the coefficients of all powers of μ to zero. This gives

41 $$\ddot{y}_0(t) + \omega_0^2 y_0(t) = 0; \qquad y_0(0) = a_0,\ \dot{y}_0(0) = 0$$

42 $$\ddot{y}_1(t) + \omega_0^2 y_1^2(t) + y_0^3(t) = 0; \qquad y_1(0) = 0,\ \dot{y}_1(0) = 0$$

$$\vdots$$

Solving first of all for $y_0(\cdot)$, we get

43 $$y_0(t) = a_0 \cos \omega_0 t$$

This is to be expected, because $y_0(\cdot)$ is the solution of (36)–(38) corresponding to $\mu = 0$. Now the equation for $y_1(\cdot)$ becomes

44 $$\begin{aligned} \ddot{y}_1(t) + \omega_0^2 y_1(t) &= -y_0^3(t) = -a_0^3 \cos^3 \omega_0 t \\ &= -\tfrac{3}{4} a_0^3 \cos \omega_0 t - \tfrac{1}{4} a_0^3 \cos 3\omega_0 t \end{aligned}$$

The solution of this equation is

45 $$y_1(t) = -\frac{3a_0^3}{8\omega_0} t \sin \omega_0 t - \frac{a_0^3}{32\omega_0^2} \cos \omega_0 t + \frac{a_0^3}{32\omega_0^2} \cos 3\,\omega_0 t$$

The $t \cos \omega_0 t$ term on the right-hand side of (45) is the secular term, which arises because the forcing function of the equation for $y_1(\cdot)$ contains a component of angular frequency ω_0, while the system itself has a *resonant* frequency at ω_0. Combining the above expression for $y_0(\cdot)$ and $y_1(\cdot)$, the approximate solution of (36)–(38), which is good to the first order in μ, is obtained as

46 $$\begin{aligned} y(t) &\simeq y_0(t) + \mu y_1(t) \\ &= \left(a_0 - \frac{\mu a_0^3}{32\omega_0^2}\right) \cos \omega_0 t - \frac{3\mu a_0^3}{8\omega_0} t \sin \omega_0 t + \frac{\mu a_0^3}{32\omega_0^2} \cos 3\omega_0 t \end{aligned}$$

It is clear that this $y(t)$ is an unbounded function of t and hence is an unacceptable approximation.

The presence of the secular terms can be rationalized as follows: If $\mu = 0$, the solution of (36) is periodic with angular frequency ω_0. However, if $\mu \neq 0$, the angular frequency of the periodic solution of (36) is not necessarily ω_0 but is different in general. On the other hand, because $y_0(\cdot)$ [the so-called "generating solution" of the sequence of functions $y_0(\cdot), y_1(\cdot), \ldots$] has an angular frequency ω_0, it is clear that all functions $y_i(\cdot)$ consist of summations of terms of the type $\sin n\omega_0 t$ and $\cos n\omega_0 t$, where n is an integer. This attempt to express a function whose angular frequency is not ω_0 in terms of functions whose angular frequency is ω_0 leads to secular terms. As an example, suppose δ is "small," and let us express $\cos[(\omega_0 + \delta)t]$ as a power series in δ. This leads to

47 $$\cos[(\omega_0 + \delta)t] = \cos\omega_0 t - \delta t \sin\omega_0 t - \frac{\delta^2 t^2}{2}\cos\omega_0 t \ldots$$

This power series converges uniformly in t as t varies over any finite interval, but if the series is terminated after a finite number of terms, the resulting finite summation contains secular terms. Moreover, the periodicity and boundedness properties of the function $\cos[(\omega_0 + \delta)t]$ are not at all apparent from the power series expansion in δ.

To alleviate this difficulty, we shall suppose that the solution $y(\cdot)$ of (36)–(38) is periodic with an angular frequency ω, which itself is expressed as a power series in μ. In other words,

48 $$\omega^2 = \omega_0^2 + \mu\xi_1(a_0) + \mu^2\xi_2(a_0) + \ldots$$

which can be rewritten as

49 $$\omega_0^2 = \omega^2 - \mu\xi_1(a_0) - \mu^2\xi_2(a_0) - \ldots$$

In (48), we have explicitly identified the dependence of ω on a_0. Substituting (49) and (40) into (36) gives

50 $$\ddot{y}_0 + \mu\ddot{y}_1 + \ldots + \omega^2 y_0 - \mu\xi_1 y_0 + \mu\omega^2 y_1 + \ldots + \mu y_0^3 + \ldots = 0$$

Equating the coefficients of each power of μ to zero gives

51 $$\ddot{y}_0 + \omega^2 y_0 = 0; \qquad y_0(0) = a_0,\ \dot{y}_0(0) = 0$$

52 $$\ddot{y}_1 + \omega^2 y_1 = -y_0^3 + \xi_1 y_0; \qquad y_1(0) = 0,\ \dot{y}_1(0) = 0$$

$$\vdots$$

Hence

53 $$y_0(t) = a_0 \cos\omega t$$

54 $$\ddot{y}_1(t) + \omega^2 y_1(t) = -a_0^3\cos^3\omega t + \xi_1 a_0 \cos\omega t = -\tfrac{3}{4}a_0^3\cos\omega t - \tfrac{1}{4}a_0^3\cos 3\omega t + \xi_1 a_0\cos\omega t$$

For $y_1(\cdot)$ not to contain any secular terms, the forcing function on the right-hand side of (54) must not contain any terms of angular frequency ω. Hence we must have

55 $$\xi_1 = \tfrac{3}{4}a_0^2$$

With this condition, the solution for $y_1(\cdot)$ is obtained as

$$y_1(t) = -\frac{a_0^3}{32\omega^2}\cos\omega t + \frac{a_0^3}{32\omega^2}\cos 3\omega t \tag{56}$$

The overall solution of (36)–(38), which is accurate to the first order in μ, is

$$y(t) = a_0\cos\omega t - \frac{\mu a_0^3}{32\omega^2}\cos\omega t + \frac{\mu a_0^3}{32\omega^2}\cos 3\omega t \tag{57}$$

where

$$\omega^2 = \omega_0^2 + \tfrac{3}{4}\mu a_0^2 \tag{58}$$

59 **Example.** Consider the simple pendulum equation

$$\ddot{y} + \sin y = 0 \tag{60}$$

Equation (60) can be approximated by

$$\ddot{y} + y - \frac{y^3}{6} = 0 \tag{61}$$

which is of the form (36), with $\mu = -\frac{1}{6}$. Using the foregoing analysis, we see that the frequency of oscillation of the simple pendulum is related to the amplitude by

$$\omega^2 = 1 - \frac{a_0^2}{8} \tag{62}$$

Problem 2.15. Apply the Krylov–Boguliubov method to Rayleigh's equation

$$\ddot{y} + y = \mu\left(\dot{y} - \frac{\dot{y}^3}{3}\right)$$

optional: power series method

Solve the same equation using the perturbation method, and show that both methods give the same solution to the first order in μ.

Problem 2.16. Apply the perturbation method to the Van der Pol equation

$$\ddot{y} + y = \mu\dot{y}(1 - y^2)$$

Problem 2.17. Apply the Krylov–Boguliubov method to the pendulum equation

$$\ddot{y} + y - \frac{y^3}{6} = 0$$

Show that the expression derived for the frequency of oscillation is the same as (62).

Problem 2.18. Consider the second-order equation

$$\ddot{y} + y = \mu f(y, \dot{y})$$

$$y(0) = 0; \qquad \dot{y}(0) = a_0$$

where the function y is continuously differentiable with respect to its two arguments. Show that, to first order in μ, both the perturbation method and the Krylov–Boguliubov method give the same results.

3 Nonlinear differential equations

In this chapter, we shall undertake a systematic study of nonlinear ordinary differential equations (o.d.e.'s). As can be gathered from the examples given in Chap. 1, a nonlinear differential equation can in general exhibit very wild and unusual behavior. However, we shall show in this chapter that, for a practically significant class of nonlinear o.d.e.'s, the existence and uniqueness of solutions can be ascertained, as well as continuous dependence on initial conditions.

Except for very special cases, which are usually "cooked" in advance, it is not possible to obtain a closed-form expression for the solution of a nonlinear o.d.e. Hence it is necessary to devise alternative methods for analyzing the behavior of the solution of a given nonlinear o.d.e. without relying on being able to find a closed-form solution. In this chapter, we shall put forward two techniques for doing this:

1. A technique, known as Picard's iteration method, is given, which generates a sequence of functions that converge to the solution of a given equation. In this way, the solution of a given equation can be computed to an arbitarily high degree of accuracy.
2. A method is given for obtaining bounds on the solution of a given equation without actually solving the equation. Using this method, it is possible to determine, at each instant of time, a region in R^n within which the solution of the given equation must lie. Such a method is useful for two reasons:

(a) By obtaining bounds on the solution, one can draw conclusions on the qualitative behavior of the solution, and the labor involved is considerably less than that needed to actually solve the equation.
(b) The bounds obtained by this method serve as a check on solutions obtained by alternative methods, e.g., a numerical solution on a computer.

The study of nonlinear o.d.e.'s in general terms requires rather sophisticated mathematical tools. For this reason, we shall develop the necessary mathematical background in Secs. 3.1 and 3.2.

3.1 MATHEMATICAL PRELIMINARIES

3.1.1 Linear Vector Spaces

This subsection is devoted to an axiomatic development of linear vector spaces, both real and complex. In most practical applications, it is enough to deal with real vector spaces. However, it is sometimes necessary to deal with complex vector spaces in order to make the theory complete. For example, a polynomial of degree n has n zeroes only if one counts complex zeroes.

We note also that it is possible to define a linear vector space over an arbitrary field (e.g., the binary field, the field of rational functions, etc.). However, we do not need these concepts in this book.

1 *[definition]* A *real linear vector space* (resp. *complex linear vector space*) is a set V, together with two operations $+: V \times V \longrightarrow V$, termed addition, and $\cdot: R \times V \longrightarrow V$ (resp. $\cdot: C \times V \longrightarrow V$), termed scalar multiplication, such that the following axioms hold:

2 (V1) $x + y = y + x, \quad \forall\, x, y \in V$ (commutativity of addition)

3 (V2) $x + (y + z) = (x + y) + z, \quad \forall\, x, y \in V$ (associativity of addition)

(V3) There is an element in V, denoted by 0_V (or 0 if V is clear from the context), such that

4 $x + 0_V = 0_V + x = x, \quad \forall\, x \in V$ (existence of additive identity)

(V4) For each $x \in V$, there exists an element, denoted by $-x$, such that

5 $$x + (-x) = 0_V \quad \text{(existence of additive inverse)}$$

(V5) For each r_1, r_2 in R (resp. c_1, c_2 in C) and each $x \in V$,

6 $$r_1 \cdot (r_2 \cdot x) = (r_1 r_2) \cdot x \quad [\text{resp. } c_1 \cdot (c_2 \cdot x) = (c_1 c_2) \cdot x]$$

(V6) For each r in R (resp. c in C) and each $x_1, x_2 \in V$,

7 $$r \cdot (x_1 + x_2) = r \cdot x_1 + r \cdot x_2$$
$$[\text{resp. } c \cdot (x_1 + x_2) = c \cdot x_1 + c \cdot x_2]$$

(V7) For each r_1, r_2 in R (resp. c_1, c_2 in C) and each $x \in V$,

8 $$(r_1 + r_2) \cdot x = r_1 \cdot x + r_2 \cdot x$$

(V8) For each $x \in V$,

9 $$1 \cdot x = x$$

This axiomatic definition of a linear vector space is best illustrated by several examples.

10 **Example.** The set R^n, consisting of all ordered n-tuples of *real* numbers, becomes a *real* linear vector space if we define addition and scalar multiplication as follows: If $\mathbf{x} = (x_1, \ldots, x_n)$ and $\mathbf{y} = (y_1, \ldots, y_n)$ are vectors in R^n and r is a real number, then

11 $$\mathbf{x} + \mathbf{y} \triangleq (x_1 + y_1, \ldots, x_n + y_n)$$

12 $$r\mathbf{x} \triangleq (rx_1, \ldots, rx_n)$$

In other words, the *sum* of two n-tuples is obtained by componentwise addition, while the *product* of a real number and an n-tuple is obtained by componentwise multiplication.

As a limiting case, it is interesting to note that $R^1 = R$, the set of real numbers, is itself a real linear vector space.

The *complex* linear vector space C^n, consisting of all ordered n-tuples of *complex* numbers, is defined in a manner analogous to R^n.

13 REMARKS: It is important to realize that whether a linear vector space is real or complex is determined not by the nature of the elements of the space but by whether the associated set of scalars is the set of real numbers or the set of complex numbers. To bring out this distinction more clearly, note that C^n can be made into both a real linear vector space as well as a complex linear vector space.

14 **Example.** Let $F[a, b]$ denote the set of all real-valued functions defined over an interval $[a, b]$ in R. Therefore a typical element of $F[a, b]$ is a function $f(\cdot)$ mapping $[a, b]$ into R. The set $F[a, b]$ becomes a real

linear vector space if we make the following definitions: Let $x(\cdot)$ and $y(\cdot)$ be elements of $F[a, b]$. Then the sum of $x(\cdot)$ and $y(\cdot)$, denoted by $(x + y)(\cdot)$, is the *function* whose value at $t \in [a, b]$ is given by $x(t) + y(t)$. Symbolically,

15 $$(x + y)(t) \triangleq x(t) + y(t)$$

Similarly, the product of a real number r and a function $x(\cdot)$, denoted by $(rx)(\cdot)$, is the *function* whose value at $t \in [a, b]$ is $rx(t)$. Symbolically,

16 $$(rx)(t) \triangleq rx(t)$$

In other words, the sum of two functions in $F[a, b]$ is obtained by pointwise addition, and the product of a scalar and a function is obtained by pointwise multiplication.

17 **Example.** The set $F^n[a, b]$, consisting of all functions mapping an interval $[a, b]$ into R^n, is a real linear vector space if we make the following definitions: Let $\mathbf{x}(\cdot)$ and $\mathbf{y}(\cdot)$ map $[a, b]$ into R^n, and let r be a real number. Then, for all $t \in [a, b]$, set

18 $$(\mathbf{x} + \mathbf{y})(t) \triangleq \mathbf{x}(t) + \mathbf{y}(t)$$

19 $$(r\mathbf{x})(t) \triangleq r\mathbf{x}(t)$$

20 **Example.** Let S denote the set of all complex-valued sequences $(x_n)_{n=1}^{\infty}$. Then S can be made into a real or complex linear vector space, by appropriate choice of the associated set of scalars, if we define addition and scalar multiplication as follows: Let $(x_n)_1^{\infty}$ and $(y_n)_1^{\infty}$ be elements of S, and let α be a scalar (real or complex). Then the sum of $(x_n)_1^{\infty}$ and $(y_n)_1^{\infty}$ is defined as the sequence $(x_n + y_n)_1^{\infty}$, and the product of α and $(x_n)_1^{\infty}$ is defined as the sequence $(\alpha x_n)_1^{\infty}$.

Because we can think of a sequence as a function mapping the set of natural numbers into the set of real numbers, this example is basically the same as Example (14).

21 *[definition]* A subset M of a linear vector space V is called a *subspace* of V if

22 1. $x + y \in M$ whenever $x, y \in M$

23 2. $\alpha x \in M$ whenever $x \in M$ and α is a suitable scalar.

Loosely speaking, M is a subspace of V if it is a linear vector space in its own right.

24 **Example.** Let $F[a, b]$ be as in Example (14), let $t_0 \in [a, b]$, and let $F_{t_0}[a, b]$ denote the subset of $F[a, b]$ consisting of all $x(\cdot)$ in $F[a, b]$ such that $x(t_0) = 0$. In other words, $F_{t_0}[a, b]$ consists of all those functions in $F[a, b]$ that vanish at t_0. Then $F_{t_0}[a, b]$ is a subspace of $F[a, b]$.

3.1.2 Normed Linear Spaces

The concept of a linear vector space is a very useful one, because in that setting it is possible to define many of the standard engineering concepts, such as a linear operator, linear independence, etc. It is also possible to study the existence and uniqueness of solutions to linear equations. However, the limitation is that there is no notion of distance or proximity on a linear vector space. Hence it is not possible to discuss concepts such as convergence and continuity. This is the motivation for studying a normed linear space, which is basically a linear vector space with a measure of the "length" of a vector.

25 *[definition]* A normed linear space is an ordered pair $(X, \|\cdot\|)$, where X is a linear vector space and $\|\cdot\|$ is a real-valued function on X (called the norm function), such that

26 (N1) $\|x\| \geq 0, \quad \forall\, x \in X; \quad \|x\| = 0$ if and only if $x = 0_X$

27 (N2) $\|\alpha x\| = |\alpha| \cdot \|x\|, \quad \forall\, x \in X$, for all scalars α

28 (N3) $\|x + y\| \leq \|x\| + \|y\|, \quad \forall\, x, y \in X$ (triangle inequality)

The norm on a normed linear space is a natural generalization of the concept of the length of a vector in R^2 or R^3. Thus, given a vector x in a normed linear space $(X, \|\cdot\|)$, the nonnegative number $\|x\|$ can be thought of as the length of the vector x. Similarly, given two vectors x and y in $(X, \|\cdot\|)$, $\|x - y\|$ can be thought of as the distance between x and y. With the aid of this concept of distance or proximity, it is possible to define convergence and continuity in a normed linear space setting.

29 *[definition]* A sequence $(x_n)_1^\infty$ in a normed linear space $(X, \|\cdot\|)$ is said to *converge* to an element $x_0 \in X$ if $\|x_n - x_0\| \to 0$ as $n \to \infty$. Equivalently, $(x_n)_1^\infty$ converges to x_0 if, for every $\varepsilon > 0$, there exists an integer $N(\varepsilon)$ such that

30 $$\|x_n - x_0\| < \varepsilon \quad \text{whenever } n \geq N(\varepsilon).$$

This basic definition of convergence can be interpreted in many ways. Clearly the sequence of vectors $(x_n)_1^\infty$ converges to x_0 if and only if the sequence of *nonnegative numbers* $(\|x_n - x_0\|)_1^\infty$ converges to zero. Alternatively, let $B(x_0, \varepsilon)$ denote the ball in X defined by

31 $$B(x_0, \varepsilon) = \{x \in X: \|x - x_0\| < \varepsilon\}$$

Then the sequence $(x_n)_1^\infty$ converges to x_0 if, for any ε, the ball $B(x_0, \varepsilon)$ contains all except a finite number of vectors in the sequence $(x_n)_1^\infty$.

32 *[definition]* Let $(X, ||\cdot||_X)$ and $(Y, ||\cdot||_Y)$ be two normed linear spaces, and let f be a function mapping X into Y. We say that f is *continuous* at $x_0 \in X$ if, for every $\varepsilon > 0$, there exists a $\delta(\varepsilon, x_0) > 0$ such that

$$||f(x_0) - f(y)||_Y < \varepsilon \quad \text{whenever } ||x_0 - y||_X < \delta(\varepsilon, x_0) \tag{33}$$

f is *continuous* if it is continuous at every $x \in X$. Finally, f is *uniformly continuous* if (1) it is continuous and (2) for every $\varepsilon > 0$ there exists a $\delta(\varepsilon) > 0$ such that

$$||f(x) - f(y)||_Y < \varepsilon \quad \text{whenever } ||x - y||_X < \delta(\varepsilon) \tag{34}$$

We shall postpone examples of these concepts until we have given a few examples of normed linear spaces.

35 REMARKS:

1. The concept of a continuous function from one normed linear space to another is a natural extension of the concept of a continuous real-valued function of a real variable; in a general normed linear space setting the norm plays the same role as the absolute value does in the set of real numbers. Similar statements apply to uniform continuity.
2. The important difference between continuity and *uniform* continuity is that in that latter case δ depends only on ε and not on x or y.

A sequence $(x_n)_1^\infty$ in a normed linear space $(X, ||\cdot||)$ converges to x_0 if $||x_n - x_0||$ approaches zero as $n \to \infty$. However, in many cases, we generate a sequence $(x_n)_1^\infty$ without knowing beforehand what its limit is, or indeed whether it has a limit. (For example, this is the case whenever we try to iteratively solve a nonlinear equation.) Thus we need a characterization of sequences that does not involve the (possibly unknown) limit of the sequence. This is provided by the concept of a Cauchy sequence.

36 *[definition]* A sequence $(x_n)_1^\infty$ in a normed linear space $(X, ||\cdot||)$ is said to be a *Cauchy sequence* if, for every $\varepsilon > 0$, there exists an integer $N(\varepsilon)$ such that

$$||x_n - x_m|| < \varepsilon \quad \text{whenever } n, m \geq N(\varepsilon) \tag{37}$$

Thus, a sequence $(x_n)_1^\infty$ is *convergent* if its terms x_n approach arbitrarily closely a *fixed* vector x_0, while it is *Cauchy* if its terms approach *each other* arbitrarily closely as $n \to \infty$. The relationship between convergent sequences and Cauchy sequences is demonstrated next.

38 **fact** Every convergent sequence in a normed linear space is also a Cauchy sequence.

proof Suppose $(x_n)_1^\infty$ is a convergent sequence in a normed linear space $(X, ||\cdot||)$, and denote its limit by x_0. To prove that $(x_n)_1^\infty$ is also a Cauchy sequence, suppose $\varepsilon > 0$ is given; then pick an integer N such that

$$||x_n - x_0|| < \frac{\varepsilon}{2} \quad \text{whenever } n \geq N \tag{39}$$

It is always possible to find such an integer N because $(x_n)_1^\infty$ converges to x_0. Now, whenever $n, m \geq N$, we have, by the triangle inequality, that

$$||x_n - x_m|| \leq ||x_n - x_0|| + ||x_m - x_0|| < \frac{\varepsilon}{2} + \frac{\varepsilon}{2} = \varepsilon \tag{40}$$

where we have also used (39). Thus $(x_n)_1^\infty$ is a Cauchy sequence. ∎

The converse of fact (38) is not true in general—a sequence can be Cauchy without being convergent [see Example (82)]. However, some normed linear spaces have the special property that every Cauchy sequence in them is also convergent. This is clarified in the next definition.

41 *[definition]* A normed linear space $(X, ||\cdot||)$ is said to be a complete normed linear space, or a *Banach space*, if every Cauchy sequence in X converges (to an element of X).

Banach spaces are important for two reasons: (1) If $(X, ||\cdot||)$ is a Banach space, then (as stated in the above definition) every Cauchy sequence is convergent. (2) Even if $(X, ||\cdot||)$ is not a Banach space, it can be made into a Banach space by adding some extra elements to X Thus, in almost all applications, we can assume that we are dealing only with Banach spaces.

We shall now give some examples of normed linear spaces.

42 **Example.** Consider the linear vector space R^n, together with the function $||\cdot||_\infty : R^n \to R$ defined by

$$||\mathbf{x}||_\infty \triangleq \max_{1 \leq i \leq n} |x_i| \tag{43}$$

(The reason for the subscript ∞ will become clear later). The function $||\cdot||_\infty$ satisfies axioms (N1) through (N3), as can be easily verified. In fact, (N1) and (N2) can be verified by inspection. To verify (N3), suppose $\mathbf{x} = (x_1, \ldots, x_n)$ and $\mathbf{y} = (y_1, \ldots, y_n)$ are vectors in R^n. We know that, for each i, $|x_i + y_i| \leq |x_i| + |y_i|$. Therefore

$$\begin{aligned} ||\mathbf{x} + \mathbf{y}||_\infty &= \max_i |x_i + y_i| \leq \max_i \{|x_i| + |y_i|\} \\ &\leq \max_i |x_i| + \max_i |y_i| = ||\mathbf{x}||_\infty + ||\mathbf{y}||_\infty \end{aligned} \tag{44}$$

so that (N3) is satisfied. Thus the *ordered pair* $(R^n, \|\cdot\|_\infty)$ is a normed linear space. In fact, it can be shown that $(R^n, \|\cdot\|_\infty)$ is actually a Banach space.

45 **Example.** Consider once again the linear vector space R^n, but this time with the function $\|\cdot\|_1 \colon R^n \to R$ defined by

46 $$\|\mathbf{x}\|_1 \triangleq \sum_{i=1}^{n} |x_i|$$

Clearly $\|\cdot\|_1$ also satisfies axioms (N1) and (N2). To verify (N3), suppose $\mathbf{x}, \mathbf{y} \in R^n$. Then

47 $$\begin{aligned}\|\mathbf{x}+\mathbf{y}\|_1 &= \sum_{i=1}^{n} |x_i + y_i| \leq \sum_{i=1}^{n} \{|x_i| + |y_i|\} \\ &= \sum_{i=1}^{n} |x_i| + \sum_{i=1}^{n} |y_i| = \|\mathbf{x}\|_1 + \|\mathbf{y}\|_1\end{aligned}$$

Hence the *ordered pair* $(R^n, \|\cdot\|_1)$ is also a normed linear space and can in fact be shown to be a Banach space.

48 Remarks: It is important to note that the normed linear space $(R^n, \|\cdot\|_\infty)$ is a different entity from the normed linear space $(R^n, \|\cdot\|_1)$, even though the underlying linear vector space is the same in both cases, (namely R^n), because even though the linear vector space on which the two norm functions are defined is the same in both cases, the norm functions themselves are different.

49 **Example.** Consider once again the linear vector space R^n, together with the function $\|\cdot\|_p \colon R^n \to R$ defined by

50 $$\|\mathbf{x}\|_p = \left\{\sum_{i=1}^{n} |x_i|^p\right\}^{1/p}$$

where p ranges between 1 and ∞. If $p = 1$, $\|\cdot\|_p$ becomes the norm function of Example (45), while if $p = \infty$, $\|\cdot\|_p$ becomes the norm function of Example (42) (hence the subscripts 1 and ∞ on the norms in these examples). The function $\|\cdot\|_p$ clearly satisfies axioms (N1) and (N2) and can be shown to satisfy (N3) whenever $1 \leq p \leq \infty$. Thus the ordered pair $(R^n, \|\cdot\|_p)$ is a normed linear space whenever $1 \leq p \leq \infty$. (Of course, for different values of p, the corresponding normed linear spaces are different.)

In particular, if $p = 2$, we have

51 $$\|\mathbf{x}\|_2 = \left\{\sum_{i=1}^{n} |x_i|^2\right\}^{1/2}$$

which is generally known as the Euclidean norm or l_2-norm on R^n. The Euclidean norm is a particular example of an inner product norm, which is defined in Sec. 3.1.3.

Note that all the spaces $(R^n, \|\cdot\|_p)$ are Banach spaces.

Special Properties of R^n and C^n. Normed linear spaces whose underlying vector space is R^n or C^n have some special properties, which we shall now discuss. For ease of statement, we shall give the facts only for R^n, but they are equally true for C^n.

52 **fact** Let $\|\cdot\|_\alpha$ and $\|\cdot\|_\beta$ be *any* two norms on R^n. Then there exist finite positive constants k_1 and k_2 such that

53 $$k_1\|\mathbf{x}\|_\alpha \leq \|\mathbf{x}\|_\beta \leq k_2\|\mathbf{x}\|_\alpha, \qquad \forall\, \mathbf{x} \in R^n$$

We shall not prove this fact. Instead, by way of illustration, consider the two norms $\|\cdot\|_1$ and $\|\cdot\|_\infty$ on R^n. Then it is easy to verify that

54 $$\|\mathbf{x}\|_\infty \leq \|\mathbf{x}\|_1 \leq n\|\mathbf{x}\|_\infty, \qquad \forall\, \mathbf{x} \in R^n$$

Similarly, we have

55 $$\|\mathbf{x}\|_\infty \leq \|\mathbf{x}\|_2 \leq n^{1/2}\|\mathbf{x}\|_\infty, \qquad \forall\, \mathbf{x} \in R^n$$

Fact (52) merely states that a similar relationship (known as topological equivalence) exists between *any* two norms on R^n.

Now let us consider some consequences of fact (52). Let $\|\cdot\|_\alpha$ and $\|\cdot\|_\beta$ be two given norms on R^n; let $(\mathbf{x}_n)_1^\infty$ be a sequence in R^n, and suppose $\|\mathbf{x}_n - \mathbf{x}_0\|_\alpha \to 0$ as $n \to \infty$. Thus $(\mathbf{x}_n)_1^\infty$ converges to $\mathbf{x}_0$ in $(R^n, \|\cdot\|_\alpha)$; i.e., $(\mathbf{x}_n)_1^\infty$ converges to $\mathbf{x}_0$ in the sense of the norm $\|\cdot\|_\alpha$. Now, because (53) holds, we see that $\|\mathbf{x}_n - \mathbf{x}_0\|_\beta \leq k_2\|\mathbf{x}_n - \mathbf{x}_0\|_\alpha$, so that $\|\mathbf{x}_n - \mathbf{x}_0\|_\beta \to 0$ as $n \to \infty$. Hence $(\mathbf{x}_n)_1^\infty$ also converges to $\mathbf{x}_0$ in the sense of the norm $\|\cdot\|_\beta$. Conversely, suppose $\|\mathbf{x}_n - \mathbf{x}_0\|_\beta \to 0$ as $n \to \infty$. Then, because $\|\mathbf{x}_n - \mathbf{x}_0\|_\alpha \leq (1/k_1)\|\mathbf{x}_n - \mathbf{x}_0\|_\beta$, $\|\mathbf{x}_n - \mathbf{x}_0\|_\alpha$ also approaches zero as $n \to \infty$. Thus we see that $(\mathbf{x}_n)_1^\infty$ converges to $\mathbf{x}_0$ in the sense of $\|\cdot\|_\alpha$ if and only if it converges to $\mathbf{x}_0$ in the sense of $\|\cdot\|_\beta$. This means that convergence in R^n is independent of the norm chosen. Now, let us for the moment use the specific norm $\|\cdot\|_\infty$ to test convergence on R^n. Then one can readily see that $\|\mathbf{x}_n - \mathbf{x}_0\|_\infty \to 0$ as $n \to \infty$ if and only if, for each i between 1 and n, the component sequence of *real numbers* $(x_n^{(i)})_1^\infty$ converges to the real number $x_0^{(i)}$. Putting all these observations together, we can state the following.

56 **fact** Let $\|\cdot\|$ be *any* norm on R^n, let $(\mathbf{x}_n)_1^\infty$ be a sequence in R^n, and let $\mathbf{x}_0 \in R^n$. Then $\|\mathbf{x}_n - \mathbf{x}_0\| \to 0$ as $n \to \infty$ if and only if each component sequence $(x_n^{(i)})_1^\infty$ converges to $x_0^{(i)}$, for $i = 1, \ldots, n$.

Let us consider another implication of fact (52). Let $\mathbf{x}(\cdot)$ be a function mapping R into R^n. In accordance with definition (32), we would say that the function $\mathbf{x}(\cdot)$ is continuous at $t \in R$ in the sense of $\|\cdot\|_\alpha$ if, whenever $(t_n)_1^\infty$ is a sequence in $[a, b]$ such that $t_n \to t$ as $n \to \infty$, we have that $\|\mathbf{x}(t_n) - \mathbf{x}(t)\|_\alpha \longrightarrow 0$ as $n \longrightarrow \infty$. Thus if we view $x(\cdot)$ as a mapping from $(R, |\cdot|)$ into $(R^n, \|\cdot\|_\alpha)$, it would appear that whether or not $\mathbf{x}(\cdot)$ is continuous at $t \in R$ depends on the choice of the norm $\|\cdot\|_\alpha$ on R^n. However, in view of the above discussion, it follows that the continuity of $\mathbf{x}(\cdot)$ in one norm is precisely the same as the continuity of $\mathbf{x}(\cdot)$ in another norm. In particular, using fact (52), we can state the following.

57 **fact** Let $\|\cdot\|$ be *any* norm on R^n, and let $\mathbf{x}(\cdot)$ be a function mapping R into R^n. Then $\mathbf{x}(\cdot)$ is a continuous function from $(R, |\cdot|)$ into $(R^n, \|\cdot\|)$ if and only if each of the component functions $x_i(\cdot)$ is a continuous function on R.

Note that facts (52), (56), and (57) hold also for C^n. But they do *not* hold for an arbitrary normed linear space.

Further Examples of Normed Linear Spaces

58 **Example.** Let $[a, b]$ be a *bounded* interval in R, and let $C[a, b]$ denote the set of all continuous real-valued functions over $[a, b]$. Define a function $\|\cdot\|_C$ over $C[a, b]$ as follows: If $x(\cdot) \in C[a, b]$, then

59
$$\|x(\cdot)\|_C = \max_{t \in [a,b]} |x(t)|$$

It is easy to verify that $\|\cdot\|_C$ satisfies all the axioms of a norm—axioms (N1) and (N2) can be verified by inspection. To verify (N3), let $x(\cdot)$, $y(\cdot) \in C[a, b]$. Then

60
$$\|x(\cdot) + y(\cdot)\|_C = \max_t |x(t) + y(t)| \leq \max_t \left[|x(t)| + |y(t)|\right]$$
$$\leq \max_t |x(t)| + \max_t |y(t)| = \|x(\cdot)\|_C + \|y(\cdot)\|_C$$

where all maxima are taken over $[a, b]$. Thus the ordered pair $(C[a, b], \|\cdot\|_C)$ is a normed linear space. Notice that a sequence of functions $[x_n(\cdot)]_1^\infty$ in $C[a, b]$ converges to $x_0(\cdot)$ in $C[a, b]$ in the sense of $\|\cdot\|_C$ if and only if $[x_n(\cdot)]_1^\infty$ converges uniformly over $[a, b]$ to $x_0(\cdot)$. The norm $\|\cdot\|_C$ is usually known as the *sup* (for supremum) norm. The space $(C[a, b], \|\cdot\|_C)$ is in fact a Banach space.

61 **Example.** Let $\|\cdot\|$ be a given norm on R^n, and let $C^n[a, b]$ denote the set of all continuous functions mapping $[a, b]$ into R^n, where $[a, b]$ is a finite interval in R^n.[1] Define the function $\|\cdot\|_C$: $C^n[a, b] \longrightarrow R$ as

[1]The set $C^n[a, b]$ is *not* to be confused with $C^{(n)}[a, b]$, which is the set of all continuous *real-valued* functions over $[a, b]$ that are n-times continuously differentiable. We shall not have much occasion to use the latter symbol.

follows: Let $\underset{\sim}{x}(\cdot) \in C^n[a, b]$. Then

62 $$\|\underset{\sim}{x}(\cdot)\|_C = \max_{t \in [a,b]} \|x(t)\|$$

Once again, $\|\cdot\|_C$ constitutes a norm on $C^n[a, b]$, as can be readily verified. Axioms (N1) and (N2) are easy. To test (N3), let $x(\cdot), y(\cdot) \in C^n[a, b]$. Then

63 $$\begin{aligned}\|\mathbf{x}(\cdot) + \mathbf{y}(\cdot)\|_C &= \max_{t \in [a,b]} \|\mathbf{x}(t) + \mathbf{y}(t)\| \\ &\leq \max_{t \in [a,b]} [\|\mathbf{x}(t)\| + \|\mathbf{y}(t)\|] \quad \text{(by the triangle inequality on } R^n\text{)} \\ &\leq \max_{t \in [a,b]} \|\mathbf{x}(t)\| + \max_{t \in [a,b]} \|\mathbf{y}(t)\| \\ &= \|\mathbf{x}(\cdot)\|_C + \|\mathbf{y}(\cdot)\|_C\end{aligned}$$

The ordered pair $(C^n[a, b], \|\cdot\|_C)$ is in fact a Banach space.

In this example, it is essential to note the difference between $\|\cdot\|$ and $\|\cdot\|_C$. $\|\cdot\|$ is a norm on R^n, while $\|\cdot\|_C$ is a norm on $C^n[a, b]$. The former has an n-vector as its argument, while the latter has a vector-valued function as its argument. When we study nonlinear differential equations in Sec. 3.3 this distinction becomes crucial.

3.1.3 Inner Product Spaces

An inner product space is a special type of normed linear space in which, because of its special structure, it is possible to define geometrically appealing concepts such as orthogonality, Fourier series, etc. Axiomatically, an inner product space is defined as follows:

64 *[definition]* An *inner product space* is a linear vector space X with associated field $F = R$ or C, together with a function $\langle \cdot, \cdot \rangle$: $X \times X \longrightarrow F$, such that the following axioms are satisfied:

65 (I1) $\langle x, y \rangle = \langle y, x \rangle$ if $F = R$, $\langle x, y \rangle = \overline{\langle y, x \rangle}$ if $F = C$, $\forall\, x, y \in X$

66 (I2) $\langle x, y + z \rangle = \langle x, y \rangle + \langle x, z \rangle$, $\forall\, x, y, z \in X$

67 (I3) $\langle x, \alpha y \rangle = \alpha \langle x, y \rangle$, $\forall\, x, y \in X$, $\forall\, \alpha \in F$

68 (I4) $\langle x, x \rangle \geq 0$, $\forall\, x \in X$, $\langle x, x \rangle = 0$ if and only if $x = 0_X$

These four axioms present an abstraction of the familiar notion of the scalar product or *dot* product in R^2 or R^3.

The next result shows that an inner product space can be made into a normed linear space in a natural way.

69 ***Theorem*** Given an inner product space X with inner product $\langle \cdot, \cdot \rangle$, define $\|\cdot\|: X \longrightarrow R$ by

70 $$\|x\| = \langle x, x \rangle^{1/2}, \qquad \forall\, x \in X$$

Then $\|\cdot\|$ is a norm on X, so that the ordered pair $(X, \|\cdot\|)$ is a normed linear space.

proof Clearly $\|\cdot\|$ satisfies (N1) and (N2). To prove (N3), we need the following extremely useful inequality, known as Schwarz's inequality.

71 ***lemma*** Let x and y be arbitrary elements of an inner product space X with inner product $\langle \cdot, \cdot \rangle$. Then

72 $$|\langle x, y \rangle| \leq \|x\| \cdot \|y\|$$

73 $|\langle x, y \rangle| = \|x\| \cdot \|y\|$ if and only if $\alpha x + \beta y = 0_X$ for some scalars α, β, not both zero.

proof of lemma (71) We shall only prove the case where X is a real inner product space. The complex case is handled in much the same fashion.

Consider the function

74 $$f(\alpha, \beta) = \|\alpha x + \beta y\|^2 = \langle \alpha x + \beta y, \alpha x + \beta y \rangle$$
$$= \alpha^2 \|x\|^2 + 2\alpha\beta \langle x, y \rangle + \beta^2 \|y\|^2$$

By (I4), $f(\alpha, \beta) \geq 0$ for all α, β in R. Because f is a quadratic form in α and β, it follows that $f(\alpha, \beta) \geq 0 \;\forall\; \alpha, \beta \in R$ if and only if the discriminant of the quadratic form is nonpositive, i.e.,

75 $$\langle x, y \rangle^2 \leq \|x\|^2 \cdot \|y\|^2$$

Furthermore, suppose $\alpha x + \beta y \neq 0_X$ whenever either α or β is nonzero. Then $f(\alpha, \beta) > 0$ whenever either α or β is nonzero. This is the case if and only if the discriminant of (74) is strictly negative, i.e.,

76 $$\langle x, y \rangle^2 < \|x\|^2 \cdot \|y\|^2$$

Hence the lemma is proved. ∎

PROOF, Continuation of, for Theorem (69) For all x, y in X, we have

77 $$\|x + y\|^2 = \langle x + y, x + y \rangle = \|x\|^2 + \|y\|^2 + 2\langle x, y \rangle$$
$$\leq \|x\|^2 + \|y\|^2 + 2\|x\| \cdot \|y\| \quad \text{(by Schwarz's inequality)}$$
$$= (\|x\| + \|y\|)^2$$

Hence (N3) is satisfied, and $\|\cdot\|$ is a norm on X. ∎

Theorem (69) shows that every inner product space can be made into a normed linear space in a natural way. It is therefore possible to speak of the completeness of an inner product space.

78 *[definition]* An inner product space that is complete (in the sense of the norm induced by the inner product) is called a *Hilbert space*.

79 **Example.** Consider the linear vector space R^n, together with the function $\langle \cdot, \cdot \rangle_n: R^n \times R^n \longrightarrow R$ defined as follows: Let $\mathbf{x} = (x_1, \ldots, x_n)$ and $\mathbf{y} = (y_1, \ldots, y_n)$ belong to R^n. Then

80 $$\langle \mathbf{x}, \mathbf{y} \rangle_n = \sum_{i=1}^{n} x_i y_i$$

It is routine to verify that the function $\langle \cdot, \cdot \rangle_n$ satisfies axioms (I1) through (I4), so that the ordered pair $(R^n, \langle \cdot, \cdot \rangle_n)$ is a real inner product space. It is in fact a Hilbert space. The norm on R^n corresponding to this inner product is

81 $$\|\mathbf{x}\| = \left\{ \sum_{i=1}^{n} x_i^2 \right\}^{1/2}$$

which is the Euclidean norm discussed earlier.

82 **Example.** Let $C^n[a, b]$ be the linear space of Example (61), and define $\langle \cdot, \cdot \rangle_C: C^n[a, b] \times C^n[a, b] \longrightarrow R$ as follows: If $\mathbf{x}(\cdot)$ and $\mathbf{y}(\cdot)$ belong to $C^n[a, b]$, then

83 $$\langle \mathbf{x}(\cdot), \mathbf{y}(\cdot) \rangle_C = \int_a^b \langle \mathbf{x}(t), \mathbf{y}(t) \rangle_n \, dt$$

In this case, the ordered pair $(C^n[a, b], \langle \cdot, \cdot \rangle_C)$ is a real inner product space. It is *not*, however, a Hilbert space. [Contrast with $(C^n[a, b], \|\cdot\|_C)$, which *is* a Banach space.] The completion of $(C^n[a, b] \langle \cdot, \cdot \rangle_C)$ is a space denoted by $L_2^n[a, b]$, which is the space of all square-integrable Lebesgue-measurable functions. The inner product on $L_2^n[a, b]$ is also given by (83), except that the integral now has to be interpreted as a Lebesgue integral (see also Chap. 6).

We shall close this section by giving two examples of continuous functions.

84 **fact** Let $(X, \|\cdot\|)$ be a normed linear space. Then the norm function $\|\cdot\|$ is uniformly continuous on X.

proof Use definition (32) of uniform continuity, and let $\delta(\varepsilon) = \varepsilon$, for each ε. Because, for each $x, y \in X$, we have

85 $$\big| \|x\| - \|y\| \big| \leq \|x - y\|$$

it follows that

86 $$\big|\,\|x\| - \|y\|\,\big| < \varepsilon \quad \text{whenever } \|x - y\| < \varepsilon$$

Thus $\|\cdot\|: X \longrightarrow R$ is uniformly continuous on X. ∎

87 ***corollary*** Suppose that $(X, \|\cdot\|)$ is a normed linear space and that $(x_n)_1^\infty$ is a sequence in X converging to x_0. Then the sequence of *real numbers* $(\|x_n\|)_1^\infty$ converges to $\|x_0\|$.

88 **fact** Suppose $(X, \langle\cdot, \cdot\rangle)$ is an inner product space. Then, for each $y \in X$, the function $x \mapsto \langle x, y\rangle$ is uniformly continuous.

proof Use definition (32) of uniform continuity, and let $\delta(\varepsilon) = \varepsilon/\|y\|$, for each ε. Now, suppose $\|x - z\| < \delta(\varepsilon)$. By Schwarz's inequality, we have

89 $$|\langle x, y\rangle - \langle z, y\rangle| = |\langle x - z, y\rangle| \le \|x - z\|\cdot\|y\| < \delta(\varepsilon)\|y\| = \varepsilon$$

Hence the function $x \mapsto \langle x, y\rangle$ is uniformly continuous. ∎

Problem 3.1. Show that the zero element of a linear vector space is unique. [*Hint:* Assume that a linear vector space V has two zero elements, 0_1 and 0_2, and use axiom (V3).]

Problem 3.2. Show that, in a linear vector space, the additive inverse of an element is unique.

Problem 3.3. Let S be the sequence space of Example (20), and let S_r be the subset of S consisting of sequences converging to r. For what values of r is S_r a subspace of S?

Problem 3.4. Sketch the unit spheres, i.e., the sets

$$\{\mathbf{x} \in R^n: \|\mathbf{x}\|_p = 1\}$$

for (a) $p = 1$, (b) $p = 1.2$, (c) $p = 2$, (d) $p = 3$, (e) $p = \infty$.

Problem 3.5. (a) Using the triangle inequality, show that

$$\left\|\sum_{i=1}^{m} \mathbf{x}_i\right\| \le \sum_{i=1}^{m} \|\mathbf{x}_i\|$$

where m is any finite number, $\mathbf{x}_i \in R^n$ for $i = 1, \ldots, m$, and $\|\cdot\|$ is any norm on R^n.

(b) Let $C^n[a, b]$ be as in Example (61). Using the Riemannian approximation for the integral, show that

$$\left\|\int_a^b \mathbf{x}(t)\, dt\right\| \le \int_a^b \|\mathbf{x}(t)\|\, dt$$

Problem 3.6. Prove Schwarz's inequality for complex inner product spaces. Show that if $(x_n)_1^\infty$ and $(y_n)_1^\infty$ are sequences in X converging to x_0 and

y_0, respectively, then the sequence of real numbers $(\langle x_n, y_n \rangle)_1^\infty$ converges to $\langle x_0, y_0 \rangle$ (*Hint:* Write $\langle x_n, y_n \rangle - \langle x, y \rangle$ as

$$\langle x_n, y_n \rangle - \langle x, y_n \rangle + \langle x, y_n \rangle - \langle x, y \rangle$$

and use Schwarz's inequality.)

3.2 INDUCED NORMS AND MATRIX MEASURES

In this section, we shall introduce the concepts of the induced norm of a matrix and the measure of a matrix. These concepts are vitally needed in Sec. 3.5, where we shall study methods for estimating the solutions to nonlinear differential equations without actually solving them.

3.2.1 Induced Norms

Let $C^{n \times n}(R^{n \times n})$ denote the set of all $n \times n$ matrices with complex (real) elements. Then $C^{n \times n}$ can be made into a linear vector space if addition and scalar multiplication are done componentwise. Moreover, each element $\mathbf{A} \in C^{n \times n}$ defines a corresponding linear mapping $\boldsymbol{\alpha}$ from C^n into C^n, according to the rule

1 $$\boldsymbol{\alpha}(\mathbf{x}) = \mathbf{A}\mathbf{x}, \qquad \forall\, \mathbf{x} \in C^n$$

Conversely, every linear mapping $\boldsymbol{\beta}$ from C^n into itself can be associated with an $n \times n$ matrix $\mathbf{B}$, i.e., an element $\mathbf{B}$ of $C^{n \times n}$. This can be shown as follows: Let $\mathbf{e}_j$ be the vector in C^n which has all zero components, except for the jth component, which is equal to 1. Now, $\boldsymbol{\beta}(\mathbf{e}_j)$ is an element of C^n, because $\boldsymbol{\beta}$ maps C^n into itself. Accordingly, let $(b_{ij}, i = 1, \ldots, n)$ be the components of the vector $\boldsymbol{\beta}(\mathbf{e}_j)$, so that

2 $$\boldsymbol{\beta}(\mathbf{e}_j) = [b_{1j}, b_{2j}, \ldots, b_{nj}]' \triangleq \mathbf{b}_j$$

Define $\mathbf{b}_j$, as above, to be the element of C^n whose ith component is b_{ij}. Finally, if we form the $n \times n$ matrix

3 $$\mathbf{B} = [\mathbf{b}_1 \,|\, \mathbf{b}_2 \,|\, \ldots \,|\, \mathbf{b}_n]$$

then one can easily verify that

4 $$\boldsymbol{\beta}(\mathbf{x}) = \mathbf{B}\mathbf{x}, \qquad \forall\, \mathbf{x} \in C^n$$

Therefore, there is a one-to-one correspondence between elements of $C^{n \times n}$ and linear mappings from C^n into C^n.[2] We do not in general dis-

[2]The matrix $\mathbf{B}$ is known as a *matrix representation* of the mapping $\boldsymbol{\beta}$ *with respect to the basis* $(\mathbf{e}_1, \ldots, \mathbf{e}_n)$, which is sometimes known as the *natural basis*. It is possible for the same linear mapping to have different matrix representations with respect to different bases. However, in this book we shall not explore such subtleties of linear algebra.

tinguish between an element $\mathbf{A}$ in $C^{n\times n}$ and the corresponding linear mapping $\mathbf{A}$ from C^n into C^n. However, this one-to-one correspondence is important for the notions of induced matrix norm and matrix measure.

5 *[definition]* Let $||\cdot||$ be a given norm on C^n. Then for each matrix $\mathbf{A} \in C^{n\times n}$, the quantity $||\mathbf{A}||_i$ defined by

6 $$||\mathbf{A}||_i = \sup_{\substack{\mathbf{x}\neq\mathbf{0}\\ \mathbf{x}\in C^n}} \frac{||\mathbf{Ax}||}{||\mathbf{x}||} = \sup_{||\mathbf{x}||=1} ||\mathbf{Ax}|| = \sup_{||\mathbf{x}||\leq 1} ||\mathbf{Ax}||$$

is called the *induced (matrix) norm* of $\mathbf{A}$ corresponding to the vector norm $||\cdot||$.

It should be noted that there are two distinct functions involved in definition (5): One is the *norm* function $||\cdot||$ mapping C^n into R, and the other is the *induced norm* function $||\cdot||_i$ mapping $C^{n\times n}$ into R.

The induced norm of a matrix $\mathbf{A}$ (or, what is the same, the induced norm of a linear mapping $\mathbf{A}$) can be given a simple geometric interpretation. Equation (4) shows that $||\mathbf{A}||_i$ is the least upper bound of the ratio $||\mathbf{Ax}||/||\mathbf{x}||$ as $\mathbf{x}$ varies over C^n. In this sense, $||\mathbf{A}||_i$ can be thought of as the maximum "gain" of the mapping $\mathbf{A}$. Alternatively, let B be the unit ball in C^n; i.e., let

7 $$B = \{\mathbf{x} \in C^n : ||\mathbf{x}|| \leq 1\}$$

Now, suppose we distort B by replacing each $\mathbf{x}$ in B by $\mathbf{Ax}$, i.e., by its image under the mapping $\mathbf{A}$. Then what results is the image of the set B under the mapping $\mathbf{A}$. In this setting, the induced norm of $\mathbf{A}$, $||\mathbf{A}||_i$, can be thought of as the radius of the smallest ball in C^n that completely covers the image of B under $\mathbf{A}$.

Lemma (8) shows that the induced norm function actually constitutes a valid norm on $C^{n\times n}$.

8 ***lemma*** For each norm $||\cdot||$ on C^n, the corresponding induced norm function $||\cdot||_i$ maps $C^{n\times n}$ into $[0, \infty)$, satisfies axioms (N1) through (N3), and is therefore a norm on $C^{n\times n}$.

proof Clearly, $||\mathbf{A}||_i \geq 0 \;\forall\; \mathbf{A} \in C^{n\times n}$, and axioms (N1) and (N2) can be verified by inspection. To verify (N3), suppose $\mathbf{A}, \mathbf{B} \in C^{n\times n}$. Then

9 $$\begin{aligned} ||\mathbf{A}+\mathbf{B}||_i &= \sup_{||\mathbf{x}||=1} ||(\mathbf{A}+\mathbf{B})\mathbf{x}|| = \sup_{||\mathbf{x}||=1} ||\mathbf{Ax}+\mathbf{Bx}|| \\ &\leq \sup_{||\mathbf{x}||=1} [||\mathbf{Ax}|| + ||\mathbf{Bx}||] \quad \text{(by the triangle inequality on } C^n) \\ &\leq \sup_{||\mathbf{x}||=1} ||\mathbf{Ax}|| + \sup_{||\mathbf{x}||=1} ||\mathbf{Bx}|| = ||\mathbf{A}||_i + ||\mathbf{B}||_i \end{aligned}$$

Hence (N3) is also satisfied, and thus $||\cdot||_i$ is a norm on $C^{n\times n}$. ∎

In view of lemma (8), it is clear that, corresponding to each norm on C^n, there is an induced norm on $C^{n \times n}$. However, the converse is not true. Consider the function $\|\cdot\|_s\colon C^{n \times n} \longrightarrow [0, \infty)$ defined by

$$\|\mathbf{A}\|_s = \max_{i,j} |\mathbf{a}_{ij}| \tag{10}$$

Then one can easily verify that $\|\cdot\|_s$ defines a norm on $C^{n \times n}$. However, no norm $\|\cdot\|$ on C^n exists such that $\|\cdot\|_s$ is the corresponding induced norm. Hereafter, when we say that $\|\cdot\|_i$ is an induced norm on $C^{n \times n}$, we mean that there exists a norm $\|\cdot\|$ on C^n such that $\|\cdot\|$ induces $\|\cdot\|_i$. (Note that some authors use the term *bound norm* instead of *induced norm*.)

In fact [3.1(52)], it is stated that any two norms on C^n are topologically equivalent. By the same token, any two norms on $C^{n \times n}$ are also equivalent. In particular, an induced norm is equivalent to another "noninduced" norm. But a special property of induced norms is bought out below.

11 ***lemma*** Let $\|\cdot\|_i$ be an induced norm on $C^{n \times n}$ corresponding to the norm $\|\cdot\|$ on C^n. Then

$$\|\mathbf{AB}\|_i \leq \|\mathbf{A}\|_i \|\mathbf{B}\|_i, \qquad \forall\ \mathbf{A}, \mathbf{B} \in C^{n \times n} \tag{12}$$

proof We have

$$\|\mathbf{AB}\|_i = \sup_{\|\mathbf{x}\|=1} \|\mathbf{ABx}\| \tag{13}$$

However, by definition,

$$\|\mathbf{Ay}\| \leq \|\mathbf{A}\|_i \|\mathbf{y}\|, \qquad \forall\ \mathbf{y} \in C^n \tag{14}$$

So, in particular,

$$\|\mathbf{ABx}\| \leq \|\mathbf{A}\|_i \|\mathbf{Bx}\| \tag{15}$$

Similarly,

$$\|\mathbf{Bx}\| \leq \|\mathbf{B}\|_i \|\mathbf{x}\| \tag{16}$$

As a result, we get

$$\|\mathbf{ABx}\| \leq \|\mathbf{A}\|_i \|\mathbf{B}\|_i \|\mathbf{x}\| \tag{17}$$

from which (11) follows immediately. ∎

Thus induced norms have the special feature that they are *submultiplicative*; i.e., the induced norm of the product of two matrices **A** and **B** is less than or equal to the product of the induced norms of **A** and **B**. It can be readily verified by example that the norm $\|\cdot\|_s$ defined in (8) does not have this property (and hence cannot be an induced norm).

In general, if we are given a specific norm on C^n (say, for instance, the norm $\|\cdot\|_p$ defined in Example [3.1(49)]), it is not always easy to find an *explicit* expression for the corresponding induced norm on $C^{n\times n}$—the equations (6) serve more as definitions and do not always result in explicit expressions. However, the induced matrix norms corresponding to the vector norms $\|\cdot\|_\infty$, $\|\cdot\|_1$, and $\|\cdot\|_2$, respectively (as defined in Examples [3.1(42)], [3.1(45)], and [3.1(49)], respectively), are known and are displayed in Table 3.1.

TABLE 3.1

Norm on C^n	*Induced Norm on $C^{n\times n}$*
$\|\mathbf{x}\|_\infty = \max_i \|\mathbf{x}_i\|$	$\|\mathbf{A}\|_{i\infty} = \max_i \sum_j \|a_{ij}\|$ (row sum)
$\|\mathbf{x}\|_1 = \sum_i \|\mathbf{x}_i\|$	$\|\mathbf{A}\|_{i1} = \max_j \sum_i \|a_{ij}\|$ (column sum)
$\|\mathbf{x}\|_2 = (\sum_i \|x_i\|^2)^{1/2}$	$\|\mathbf{A}\|_{i2} = [\lambda_{\max}(\mathbf{A}^*\mathbf{A})]^{1/2}$, where $\lambda_{\max}(\mathbf{A}^*\mathbf{A})$ = maximum eigenvalue of $\mathbf{A}^*\mathbf{A}$

3.2.2 Matrix Measures

Let $\|\cdot\|_i$ be an induced matrix norm on $C^{n\times n}$. Then the corresponding *matrix measure*[3] is the function $\mu_i\colon C^{n\times n} \to R$ defined by

18 $$\mu_i(\mathbf{A}) = \lim_{\varepsilon\to 0^+} \frac{\|\mathbf{I}+\varepsilon\mathbf{A}\|_i - 1}{\varepsilon}$$

From a purely mathematical point of view, the measure $\mu_i(\mathbf{A})$ of a matrix $\mathbf{A}$ can be thought of as the directional derivative of the norm function $\|\cdot\|_i$, as evaluated *at* $\mathbf{I}$ *in the direction* $\mathbf{A}$. However, the measure function has several useful properties, as stated below.

19 ***Theorem*** Whenever $\|\cdot\|_i$ is an induced norm on $C^{n\times n}$, the corresponding measure $\mu_i(\cdot)$ has the following properties:

(i) For each $\mathbf{A} \in C^{n\times n}$, the limit indicated in (18) exists and is well defined.

(ii) $-\|\mathbf{A}\|_i \le -\mu_i(-\mathbf{A}) \le \mu_i(\mathbf{A}) \le \|A\|_i, \quad \forall\, \mathbf{A} \in C^{n\times n}$

(iii) $\mu_i(\alpha\mathbf{A}) = \alpha\mu_i(\mathbf{A}), \quad \forall\, \alpha > 0, \quad \forall\, \mathbf{A} \in C^{n\times n}$

(iv) $\max\,[\mu_i(\mathbf{A}) - \mu_i(-\mathbf{B}), -\mu_i(-\mathbf{A}) + \mu_i(\mathbf{B})] \le \mu_i(\mathbf{A}+\mathbf{B}) \le \mu_i(\mathbf{A}) + \mu_i(\mathbf{B})$

---

[3] Some authors also use the term *logarithmic derivative.*

(v) $\mu_i(\cdot)$ is a convex function on $C^{n\times n}$; i.e.,

$$\mu_i[\alpha\mathbf{A} + (1-\alpha)\mathbf{B}] \leq \alpha\mu_i(\mathbf{A}) + (1-\alpha)\mu_i(\mathbf{B}), \quad \forall\ \alpha \in [0, 1], \quad \forall\ \mathbf{A}, \mathbf{B} \in C^{n\times n}$$

(vi) $-\mu_i(-\mathbf{A}) \leq \mathrm{Re}\ \lambda \leq \mu_i(\mathbf{A})$ whenever λ is an eigenvalue of $\mathbf{A}$.

proof

(i) The function $\|\cdot\|_i$ is a convex function on $C^{n\times n}$; i.e., it satisfies the property

20
$$\|\alpha\mathbf{A} + (1-\alpha)\mathbf{B}\|_i \leq \alpha\|\mathbf{A}\|_i + (1-\alpha)\|\mathbf{B}\|_i, \quad \forall\ \alpha \in [0, 1], \quad \forall\ \mathbf{A}, \mathbf{B} \in C^{n\times n}$$

Because $\|\cdot\|_i$ is a convex function, it can be shown ([11] Eggleston) that it has a directional derivative at *every* point in $C^{n\times n}$ in *every* direction. However, we shall also give a constructive proof that is applicable to the present case.

Let us define

21
$$f(\varepsilon; \mathbf{A}) = \frac{\|\mathbf{I} + \varepsilon\mathbf{A}\|_i - 1}{\varepsilon}$$

We first show that the function $\varepsilon \mapsto f(\varepsilon; A)$ is nonincreasing as $\varepsilon \longrightarrow 0^+$; i.e.,

22
$$f(\varepsilon_1; \mathbf{A}) \leq f(\varepsilon_2; \mathbf{A}) \quad \text{whenever } 0 < \varepsilon_1 \leq \varepsilon_2$$

By definition, we have

23
$$\begin{aligned}
\varepsilon_1 f(\varepsilon_1; \mathbf{A}) &= \|\mathbf{I} + \varepsilon_1\mathbf{A}\|_i - 1 \\
&= \left\|\frac{\varepsilon_1}{\varepsilon_2}(\mathbf{I} + \varepsilon_2\mathbf{A}) + \left(1 - \frac{\varepsilon_1}{\varepsilon_2}\right)\mathbf{I}\right\|_i - 1 \\
&\leq \frac{\varepsilon_1}{\varepsilon_2}\|\mathbf{I} + \varepsilon_2\mathbf{A}\|_i + 1 - \frac{\varepsilon_1}{\varepsilon_2} - 1 \\
&= \frac{\varepsilon_1}{\varepsilon_2}(\|\mathbf{I} + \varepsilon_2\mathbf{A}\|_i - 1) \\
&= \varepsilon_1 f(\varepsilon_2; \mathbf{A})
\end{aligned}$$

Dividing both sides of the inequality (23) by ε_1 yields (22). Also, because

24
$$1 - \varepsilon\|\mathbf{A}\|_i \leq \|I + \varepsilon\mathbf{A}\|_i \leq 1 + \varepsilon\|\mathbf{A}\|_i$$

it follows that

25
$$-\|\mathbf{A}\|_i \leq f(\varepsilon; \mathbf{A}) \leq \|\mathbf{A}\|_i, \quad \forall\ \varepsilon > 0$$

Thus the function $\varepsilon \mapsto f(\varepsilon; \mathbf{A})$ is nonincreasing and bounded below as ε decreases to zero and therefore has a well-defined limit as $\varepsilon \longrightarrow 0^+$.

(ii) The inequality (25) shows that

26
$$-\|\mathbf{A}\|_i \leq \mu_i(\mathbf{A}) \leq \|\mathbf{A}\|_i$$

Thus, to prove (ii), it only remains to show that

$$-\mu_i(-\mathbf{A}) \leq \mu_i(\mathbf{A}) \tag{27}$$

Now, clearly,

$$1 = \|\mathbf{I} + \varepsilon(\mathbf{A} - \mathbf{A})\|_i \leq \frac{\|\mathbf{I} + 2\varepsilon\mathbf{A}\|_i}{2} + \frac{\|\mathbf{I} - 2\varepsilon\mathbf{A}\|_i}{2} \tag{28}$$

Therefore

$$0 \leq \frac{\|\mathbf{I} + 2\varepsilon\mathbf{A}\|_i - 1}{2\varepsilon} + \frac{\|\mathbf{I} - 2\varepsilon\mathbf{A}\|_i - 1}{2\varepsilon} \tag{29}$$

Taking the limit as $\varepsilon \longrightarrow 0^+$ yields

$$0 \leq \mu_i(\mathbf{A}) + \mu_i(-\mathbf{A}) \tag{30}$$

which is clearly equivalent to (ii).

(iii) It is a simple consequence of the definition of $f(\varepsilon; \mathbf{A})$ that

$$f(\varepsilon; \alpha\mathbf{A}) = \alpha f(\alpha\varepsilon; \mathbf{A}), \qquad \forall\ \alpha > 0 \tag{31}$$

Therefore,

$$\mu_i(\alpha\mathbf{A}) = \lim_{\varepsilon \to 0^+} f(\varepsilon; \alpha\mathbf{A}) = \alpha\mu_i(\mathbf{A}), \qquad \forall\ \alpha > 0 \tag{32}$$

This proves the validity of (iii) in the case $\alpha > 0$. If $\alpha = 0$, (iii) follows trivially because the measure of the zero matrix is zero.

(iv) We first show that

$$\mu_i(\mathbf{A} + \mathbf{B}) \leq \mu_i(\mathbf{A}) + \mu_i(\mathbf{B}) \tag{33}$$

First, we have

$$\begin{aligned} \frac{\|\mathbf{I} + \varepsilon(\mathbf{A} + \mathbf{B})\|_i - 1}{\varepsilon} &= \frac{1}{2\varepsilon}(\|\mathbf{I} + 2\varepsilon\mathbf{A} + \mathbf{I} + 2\varepsilon\mathbf{B}\|_i - 2) \\ &\leq \frac{\|\mathbf{I} + 2\varepsilon\mathbf{A}\|_i - 1}{2\varepsilon} + \frac{\|\mathbf{I} + 2\varepsilon\mathbf{B}\|_i - 1}{2\varepsilon} \end{aligned} \tag{34}$$

Taking the limits in (34) as $\varepsilon \longrightarrow 0^+$ proves (33) and establishes the right-hand inequality in (iv). To prove the left-hand inequality in (iv), we simply make use of (33). Clearly,

$$\mu_i(\mathbf{A}) = \mu_i(\mathbf{A} + \mathbf{B} - \mathbf{B}) \leq \mu_i(\mathbf{A} + \mathbf{B}) + \mu_i(-\mathbf{B}) \tag{35}$$

so that

$$\mu_i(\mathbf{A}) - \mu_i(-\mathbf{B}) \leq \mu_i(\mathbf{A} + \mathbf{B}) \tag{36}$$

By symmetry, we must also have

$$\mu_i(\mathbf{B}) - \mu_i(-\mathbf{A}) \leq \mu_i(\mathbf{A} + \mathbf{B}) \tag{37}$$

The inequalities (36) and (37) together establish the left-hand inequality in (iv).

(v) From (iii) and (iv), we have

38
$$\mu_i[\alpha\mathbf{A} + (1-\alpha)\mathbf{B}] \leq \mu_i(\alpha\mathbf{A}) + \mu_i[(1-\alpha)\mathbf{B}] \quad \text{by (iv)}$$
$$= \alpha\mu_i(\mathbf{A}) + (1-\alpha)\mu_i(\mathbf{B}) \qquad \forall\, \alpha \in [0,1] \quad \text{by (iii)}$$

This proves (v).

(vi) Let λ be an eigenvalue of $\mathbf{A}$, and let $\mathbf{v}$ be an associated eigenvector such that $\|\mathbf{v}\| = 1$ (where $\|\cdot\|$ is the *vector* norm that induces the *matrix* norm $\|\cdot\|_i$). Then

39
$$(\mathbf{I} + \varepsilon\mathbf{A})\mathbf{v} = (1 + \varepsilon\lambda)\mathbf{v}$$

Therefore,

40
$$\|\mathbf{I} + \varepsilon\mathbf{A}\|_i = \sup_{\|\mathbf{x}\|=1} \|(\mathbf{I} + \varepsilon\mathbf{A})\mathbf{x}\|$$
$$\geq \|(\mathbf{I} + \varepsilon\mathbf{A})\mathbf{v}\| \quad \text{(because } \mathbf{v} \text{ is a specific unit vector)}$$
$$= |1 + \varepsilon\lambda|$$

This implies that

41
$$\frac{\|\mathbf{I} + \varepsilon\mathbf{A}\|_i - 1}{\varepsilon} \geq \frac{|1 + \varepsilon\lambda| - 1}{\varepsilon}$$

As $\varepsilon \longrightarrow 0^+$, the left-hand side of (41) converges to $\mu_i(\mathbf{A})$, while the right-hand side of (41) (as can be easily verified) converges to Re λ.

Thus, (41) reduces to the right-hand inequality in (vi). The proof of the left-hand inequality in (vi) is similar and is left as an exercise. ∎

Summarizing the properties of the matrix measure function, we see that although the measure of a matrix is a convex function as is a norm, the similarity almost ends there. The measure can have positive as well as negative values, whereas a norm can assume only nonnegative values. The measure is *sign-sensitive* in that $\mu_i(-\mathbf{A})$ is in general different from $\mu_i(\mathbf{A})$, whereas $\|-\mathbf{A}\|_i = \|\mathbf{A}\|_i$. Because of these special properties, the measure function is useful in obtaining "tight" upper and lower bounds on the norms of solutions of vector differential equations.

Theorem (19) lists only some of the many interesting properties of the measure function. A more complete discussion can be found in [10] Desoer and Vidyasagar.

In defining the measure of a matrix in $C^{n\times n}$, we have assumed that the norm used in (18) is an *induced* norm. It is possible, given *any* norm $\|\cdot\|$ on $C^{n\times n}$ (induced or not), to define a corresponding measure function $\mu(\cdot)$ mapping $C^{n\times n}$ into R. In this case, all the properties of Theorem (19) still hold except for (vi). However, for the purpose of estimating solutions of *vector* differential equations, only measure functions corresponding to *induced* norms prove useful.

Given a particular vector norm $\|\cdot\|$ on C^n, it is in general a very difficult task to obtain an explicit expression for the corresponding induced norm (as mentioned earlier), and therefore it is more difficult to obtain an explicit expression for the corresponding measure function. However, the measure functions corresponding to the norms $\|\cdot\|_\infty$, $\|\cdot\|_1$, and $\|\cdot\|_2$ can be calculated explicitly and are displayed in Table 3.2.

TABLE 3.2

Norm on C^n	*Matrix Measure on* $C^{n \times n}$
$\|\mathbf{x}\|_\infty = \max_i \|x_i\|$	$\mu_{i\infty}(\mathbf{A}) = \max_i \{a_{ii} + \sum_{j \neq i} \|a_{ij}\|\}$
$\|\mathbf{x}\|_1 = \sum_{i=1}^{n} \|x_i\|$	$\mu_{i1}(\mathbf{A}) = \max_j \{a_{jj} + \sum_{i \neq j} \|a_{ij}\|\}$
$\|\mathbf{x}\|_2 = \left(\sum_{i=1}^{n} \|x_i\|^2\right)^{1/2}$	$\mu_{i2}(\mathbf{A}) = \lambda_{\max}[(\mathbf{A}^* + \mathbf{A})]/2$

42 **Example.** Let

$$\mathbf{A} = \begin{bmatrix} -2 & 1 \\ -1 & -3 \end{bmatrix}$$

Using the formulas given in Table 3.2, one can verify that

$$\mu_{i1}(\mathbf{A}) = -1; \qquad \mu_{i1}(-\mathbf{A}) = 4$$
$$\mu_{i2}(\mathbf{A}) = -2; \qquad \mu_{i2}(-\mathbf{A}) = 3$$
$$\mu_{i\infty}(\mathbf{A}) = -1; \qquad \mu_{i\infty}(-\mathbf{A}) = 4$$

This clearly shows that the measure of a given matrix depends on the particular norm used to define the measure function. The eigenvalues of **A** are

$$\lambda_{1,2} = -\frac{5}{2} \pm j\frac{\sqrt{5}}{2}$$

If we use property (vi) of Theorem (19) to obtain bounds on the real parts of the eigenvalues of **A**, we get

$$-4 \leq \operatorname{Re} \lambda_{1,2} \leq -1 \qquad \text{using } \mu_{i1} \text{ and } \mu_{i\infty} \text{ measure}$$
$$-3 \leq \operatorname{Re} \lambda_{1,2} \leq -2 \qquad \text{using } \mu_{i2} \text{ measure}$$

Thus for the particular matrix **A** studied in this example, the μ_{i2} measure gives better bounds on the eigenvalues of **A**. However, this is very much dependent on the specific matrix **A**, as shown by the following example.

43 **Example.** Let

$$\mathbf{A} = \begin{bmatrix} 2 & -1 \\ 0 & 1 \end{bmatrix}$$

The eigenvalues of **A** are 2 and 1, so that the smallest interval containing the real parts of both eigenvalues of **A** is [1, 2]. Now, if we use the measure functions μ_{i1}, μ_{i2}, and $\mu_{i\infty}$ to estimate this interval, we get [1, 3] using $\mu_{i\infty}$,

$$\left[\frac{3}{2} - \frac{1}{\sqrt{2}}, \frac{3}{2} + \frac{1}{\sqrt{2}}\right]$$

or approximately [0.793, 2.207] using μ_{i2} and [0, 2] using μ_{i1}. Therefore $\mu_{i\infty}$ gives an exact lower bound, μ_{i1} gives an exact upper bound, while μ_{i2} gives the interval of the smallest "width."

In practice, to estimate the range of the real parts of the eigenvalues of a given matrix, one would utilize various measure functions to obtain various intervals on the real line. Then, because property (vi) of Theorem (19) is valid for *all* measure functions corresponding to induced norms, one would take the intersection of *all* intervals in order to obtain the best estimate. In Example (43), this intersection is [1, 2], which happens to be an exact estimate.

Before we close this section, some comments are in order regarding the use of the matrix measure in connection with norms on R^n and the corresponding induced norms on $R^{n \times n}$. First, it is easy to see that every element in $R^{n \times n}$ corresponds to a linear mapping from R^n into itself and vice versa. Next, suppose $\|\cdot\|$ is a norm on R^n. Then, given any $\mathbf{A} \in R^{n \times n}$, we can define its *induced norm* $\|\mathbf{A}\|_i$ in a manner entirely analogous to definition (5) as well as a corresponding *measure function* $\mu_i(\mathbf{A})$ in a manner analogous to (18). The question is, What properties does this measure function have? Properties (i)–(v) of Theorem (19) hold, and the proof for the complex case can be used as is. However, it can be noted that in proving property (vi) of Theorem (19) (namely, the bounds on the real parts of the eigenvalues of **A**) essential use was made of the fact that **A** maps C^n into itself, because λ might be a complex eigenvalue of **A**, with a corresponding eigenvector having some complex components. In spite of this fact, however, it can be shown that, given any norm $\|\cdot\|$ on R^n and any $\mathbf{A} \in R^{n \times n}$, it is possible to "extend" these, respectively, to a norm on C^n and to a mapping from C^n into itself. The details are omitted here, but the end result can be summarized as follows.

44 ***Theorem*** Let $\|\cdot\|$ be a norm on R^n, and let $\|\cdot\|_i$: $R^{n \times n} \longrightarrow R$ and $\mu_i(\cdot)$: $R^{n \times n} \longrightarrow R$ be defined in a manner analogous with (5) and (18), respectively. Then $\mu_i(\cdot)$ satisfies properties (i) through (vi) of Theorem (19).

Problem 3.7. Calculate $\|\mathbf{A}\|_{i1}$, $\|\mathbf{A}\|_{i2}$, and $\|\mathbf{A}\|_{i\infty}$, when

$$\text{(i) } \mathbf{A} = \begin{bmatrix} -2 & 1 \\ 2 & -3 \end{bmatrix}; \qquad \text{(ii) } \mathbf{A} = \begin{bmatrix} 0 & 1 \\ -1 & 0 \end{bmatrix}$$

$$\text{(iii) } \mathbf{A} = \begin{bmatrix} -4 & 1 & 2 \\ 0 & -2 & 1 \\ -1 & 1 & -3 \end{bmatrix}$$

Problem 3.8. Calculate $\mu_{i1}(\mathbf{A})$, $\mu_{i2}(\mathbf{A})$, $\mu_{i\infty}(\mathbf{A})$, and $\mu_{i1}(-\mathbf{A})$, $\mu_{i2}(-\mathbf{A})$, $\mu_{i\infty}(-\mathbf{A})$, for the three matrices above. Find an interval in the real line that contains the real parts of all eigenvalues of $\mathbf{A}$.

3.3 CONTRACTION MAPPING THEOREM

In this section, we shall state and prove a very important theorem, which we shall use in Sec. 3.4 to derive the existence and uniqueness of solutions to a class of nonlinear vector differential equations.

The theorem that we shall prove is generally known as the contraction mapping theorem (and sometimes as the Banach fixed point theorem) and is usually given in two forms: the *global* version and the *local* version. The local theorem assumes a weaker hypothesis than the global theorem and obtains correspondingly weaker conclusions. We shall first give the global version.

Note that, hereafter, we use *mapping*, *function*, and *operator* interchangeably.

3.3.1 Global Contractions

1 ***Theorem*** Let $(X, \|\cdot\|)$ be a Banach space, and let $T: X \longrightarrow X$ be a mapping for which there exists a fixed constant $\rho < 1$ such that

2 $$\|Tx - Ty\| \leq \rho\|x - y\|, \qquad \forall\, x, y \in X$$

[where we write Tx instead of $T(x)$ in the interests of brevity]. Then

(i) There exists exactly one $x^* \in X$ such that $Tx^* = x^*$.

(ii) For *any* $x \in X$, the sequence $(x_n)_1^\infty$ in X defined by

3 $$x_{n+1} = Tx_n; \qquad x_0 = x$$

converges to x^*. Moreover,

4 $$\|x^* - x_n\| \leq \frac{\rho^n}{1-\rho}\|x_1 - x_0\| = \frac{\rho^n}{1-\rho}\|Tx_0 - x_0\|$$

REMARKS: An operator T satisfying the condition (2) is known as a *contraction*, because the images of any two elements x and y are closer together than x and y are. Furthermore, T is a *global* contraction because (2) holds for all x, y in the *entire* space X. An element $x^* \in X$ such that $Tx^* = x^*$ is called a fixed point of the operator T, because x^* remains fixed when we apply T to X. Theorem (1) asserts that every contraction has exactly one fixed point in X; furthermore, this fixed point can be determined by taking *any* arbitrary starting point $x \in X$ and repeatedly applying the operator T to it; finally, (4) provides an estimate of the rate of convergence of this sequence to the fixed point.

proof Let $x \in X$ be arbitrary. We first show that $(x_n)_1^\infty$ is a Cauchy sequence. For each $n \geq 0$, we have

$$\|x_{n+1} - x_n\| \leq \rho\|x_n - x_{n-1}\| \leq \ldots \leq \rho^n\|x_1 - x_0\| \tag{5}$$

Now suppose $m = n + r$, $r \geq 0$. Then

$$\begin{aligned} \|x_m - x_n\| &= \|x_{n+r} - x_n\| \leq \sum_{i=0}^{r-1} \|x_{n+i+1} - x_{n+i}\| \\ &\leq \sum_{i=0}^{r-1} \rho^{n+i}\|x_1 - x_0\| \leq \sum_{i=0}^{\infty} \rho^{n+i}\|x_1 - x_0\| \\ &= \frac{\rho^n}{1-\rho}\|x_1 - x_0\| \end{aligned} \tag{6}$$

It is clear from (6) that $\|x_n - x_m\|$ can be made arbitrarily small by choosing n large enough. Hence $(x_n)_1^\infty$ is a Cauchy sequence, and because X is complete, $(x_n)_1^\infty$ converges (to an element of X). Let us denote this limit as x^*. Now, using definition [3.1(32)] of uniform continuity, one can easily show from (2) that T is uniformly continuous. Therefore,

$$Tx^* = T(\lim_{n\to\infty} x_n) = \lim_{n\to\infty} Tx_n = \lim_{n\to\infty} x_{n+1} = x^* \tag{7}$$

Hence x^* remains invariant under T. To show that x^* is the only element of X satisfying (7), suppose that $x \in X$ and that $Tx = x$. Then, by (2),

$$\|x^* - x\| = \|Tx^* - Tx\| \leq \rho\|x^* - x\| \tag{8}$$

which can hold only if $\|x^* - x\| = 0$, i.e., $x^* = x$ (recall that $\rho < 1$). Finally, to prove the estimate (4), consider the inequality (6). Because the norm function $\|\cdot\|: X \longrightarrow R$ is continuous, we have

$$\begin{aligned} \|x^* - x_n\| &= \|(\lim_{m\to\infty} x_m) - x_n\| \\ &= \lim_{m\to\infty} \|x_m - x_n\| \leq \frac{\rho^n}{1-\rho}\|x_1 - x_0\| \end{aligned} \tag{9}$$

where we have used the fact that the right-hand side of (6) is independent of m. ∎

Note that in general it is *not* possible to replace (2) by the weaker condition

10 $$\|Tx - Ty\| < \|x - y\| \quad \forall\, x, y \in X \text{ such that } x \neq y$$

It is easy to show that any mapping satisfying (10) has *at most* one fixed point, but quite possibly it may not have any at all. As a simple example, let $X = R$, and let $f\colon R \longrightarrow R$ be defined by

11 $$f(x) = x + \frac{\pi}{2} - \arctan x$$

Then

12 $$f'(x) = 1 - \frac{1}{1 + x^2} = \frac{x^2}{1 + x^2} < 1 \qquad \text{for all } x \in R$$

Thus by the mean value theorem,

13 $$f(x) - f(y) = f'(\xi)(x - y) \qquad \text{for some } \xi \in [x, y]$$

and hence f satisfies (10). However, a quick sketch of f reveals that f has no fixed points in R.

14 **Example.** Let $f\colon R \longrightarrow R$ be a continuously differentiable function, and suppose

15 $$\sup_{x \in R} |f'(x)| \triangleq \rho < 1$$

Then Theorem (1) tells us that there is a unique number $x^* \in R$ such that $f(x^*) = x^*$; furthermore, this number x^* can be determined as the limit of the sequence $(x_i)_1^\infty$ obtained by choosing *any* arbitrary $x_0 \in R$ as the starting point and then setting $x_{i+1} = f(x_i)\ \forall\, i \geq 0$. The sequence of points so obtained is depicted in Fig. 3.1.

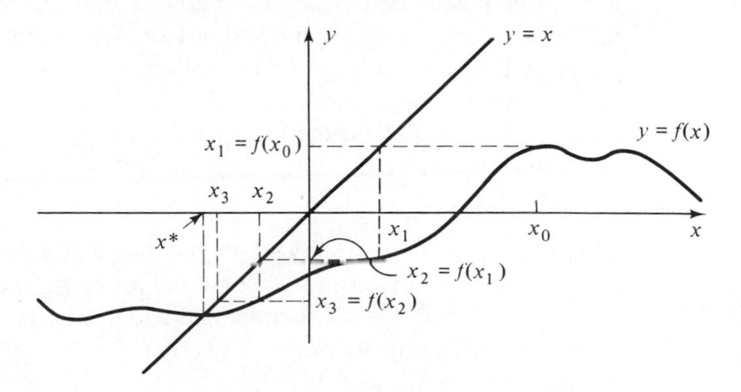

FIG. 3.1

3.3.2 Local Contractions

The application of Theorem (1) is limited by the fact that the operator T is required to satisfy (2) for *all* x, y in X. In other words, T has to be a *global* contraction. In Theorem (16), we examine the case where T satisfies (2) only over some region M in X, i.e., the case where T is a *local* contraction, and we derive correspondingly weaker results.

16 ***Theorem*** Let $(X, \|\cdot\|)$ be a Banach space, let M be a subset of X, and let $T: x \longrightarrow x$. Suppose there exists a constant $\rho \leq 1$ such that

17 $$\|Tx - Ty\| \leq \rho \|x - y\|, \qquad \forall\, x, y \in M$$

and suppose one can find an element $x_0 \in X$ such that the ball

18 $$B = \left\{x \in X: \|x - x_0\| \leq \frac{\|Tx_0 - x_0\|}{1 - \rho}\right\}$$

is entirely contained within M. Then

(i) T has exactly one fixed point in M (call it x^*).

(ii) The sequence defined by

19 $$x_{i+1} = Tx_i, \qquad i \geq 0$$

and x_0 is the element defined in (18), converges to x^*. Further,

20 $$\|x_n - x^*\| < \frac{\rho^n}{1 - \rho} \|Tx_0 - x_0\|$$

21 REMARKS:

1. The significance of Theorem (16) lies in the fact that *it* is only required to be a contraction over the set M, and not over all of X. The price we pay is that the conclusions of Theorem (16) are also weaker than those of Theorem (1).
2. Everything is contingent on finding a suitable element x_0 in M such that the ball B defined in (18) is contained in M. In effect, this means that we must be able to find an element x_0 in M such that repeated applications of T to x_0 result in a sequence that is entirely contained within M. Even if T satisfies (17), it may not be possible to find such an x_0. For example, let $X = R$, and let T be the function defined by

22 $$Tx = \begin{cases} 2 & \text{if } |x| \leq 1 \\ 0 & \text{if } |x| > 1 \end{cases}$$

   If we choose M as the interval $[-1, 1]$, then T is a contraction on M. However, it is not possible to find an $x_0 \in M$ such that the ball B defined by (18) is contained within M. Accordingly, T has no fixed point in M.
3. Suppose we do succeed in finding an $x_0 \in M$ such that the hypothesis of Theorem (16) holds. Then the *particular* sequence

$(x_0, x_1, \ldots)$ converges to x^* (the unique fixed point of T in M). However, if y is *some other* element of M, the sequence $(y, Ty, T^2y, \ldots)$ may or may not converge to x^*. In contrast, if T is a global contraction, then the sequence $(y, Ty, T^2y, \ldots)$ converges to x^* for any *arbitrary* starting point y.

4. Theorem (16) does *not* rule out the possibility that T has some fixed points outside M—it states only that T has exactly one fixed point in M.

proof of Theorem (16) First, it is clear that T has *at most* one fixed point in M, because of (17). If $x_0 \in M$ is chosen in such a way that the ball B defined in (18) is contained within M, then clearly the sequence $(x_i)_1^\infty$ is contained within M [apply the inequality (6) with $n = 0$]. Because the contraction condition (17) holds in B, one can show, as in the proof of theorem (1), that (x_i) is a Cauchy sequence and therefore converges, say, to x^*. Because of the continuity of the norm function, $x^* \in B$. The rest of the proof exactly follows that of Theorem (1). ■

23 **Example.** Consider once again the case where $X = R$, and let $f: R \longrightarrow R$ be continuously differentiable. Suppose

24
$$\sup_{x \in [-1, 1]} |f'(x)| \triangleq < 1$$

and that there exists an $x_0 \in [-1, 1]$ such that the interval

$$B = \left[x_0 - \frac{|f(x_0) - x_0|}{1 - \rho},\ x_0 + \frac{|f(x_0) - x_0|}{1 - \rho}\right] \triangleq [a, b]$$

is a subset of $[-1, 1]$. Then Theorem (16) states that there is a unique $x^* \subset [-1, 1]$ such that $f(x^*) = x^*$ and that x^* is the limit of the sequence $\{x_0, f(x_0), f[f(x_0)], \ldots\}$. The situation is depicted in Fig. 3.2.

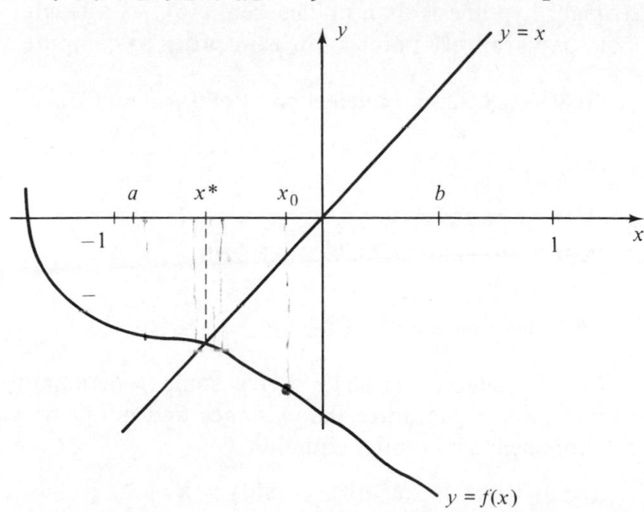

FIG. 3.2

We shall close this section by giving an alternative version of the local contraction mapping theorem, which assumes a somewhat stronger hypothesis than Theorem (16) but which is more convenient for later applications.

25 ***Theorem*** Let X be a Banach space, and let B be a closed ball in X, i.e., a set of the form

$$B = \{x: \|x - z\| \leq r\} \tag{26}$$

for some $z \in X$ and some $r \leq \infty$. Let P: $X \longrightarrow X$ be an operator satisfying the following conditions:

(i) P maps B into itself, i.e., $Px \in B$ whenever $x \in B$.

(ii) There is a constant $\rho < 1$ such that

$$\|Px - Py\| \leq \rho\|x - y\|, \qquad \forall\, x, y \in B \tag{27}$$

Then

(i) P has exactly one fixed point in B (call it x^*).

(ii) For *any* $x_0 \in B$, the sequence $(x_n)_1^\infty$ defined by

$$x_{n+1} = Px_n, \qquad n \geq 0 \tag{28}$$

converges to x^*. Moreover,

$$\|x_n - x^*\| \leq \frac{\rho^n}{1 - \rho}\|Px_0 - x_0\| \tag{29}$$

proof Obvious from Theorem (16). ∎

30 REMARKS: The difference between Theorems (16) and (25) is that in the present case P is assumed to map B into itself, which is a stronger hypothesis than in Theorem (16). As a result, we can start from *any* arbitrary point x_0 in B in order to compute x^*.

Problem 3.8. Give a detailed proof of Theorem (25).

3.4 NONLINEAR DIFFERENTIAL EQUATIONS

3.4.1 Introduction

In this section, we shall derive some general and very useful conditions which guarantee the existence and uniqueness of solutions to the nonlinear differential equation

$$\dot{\mathbf{x}}(t) = \mathbf{f}[t, \mathbf{x}(t)], \quad t \geq 0; \qquad \mathbf{x}(0) = \mathbf{x}_0 \tag{1}$$

where $\mathbf{x}(t)$: R^n and $\mathbf{f}$: $R_+ \times R^n \longrightarrow R^n$. As shown in Chap. 1, the existence and uniqueness of solutions to (1) is not guaranteed unless some restrictions are placed on the nature of $\mathbf{f}$. That is the subject of this section. By a *solution* of (1) over an interval $[0, T]$, we mean an element $\mathbf{x}(\cdot)$ of $C^n[0, T]$ such that (i) $\mathbf{x}(\cdot)$ is differentiable almost everywhere {i.e., $\dot{\mathbf{x}}(t)$ is defined for all $t \in [0, T]$, with the possible exception of a countable number of points} and (ii) equation (1) holds at all t where $\dot{\mathbf{x}}(t)$ is defined.

We shall first establish some conditions under which (1) has exactly one solution over every finite interval $[0, \delta]$ for δ sufficiently small, i.e., conditions for *local* existence and uniqueness. We shall then obtain stronger conditions for *global* existence and uniqueness, i.e., conditions under which (1) has exactly one solution over $[0, \infty)$.

One small point is to be cleared up before we proceed to the theorems. First, if $\mathbf{x}(\cdot)$ is a solution of (1) over $[0, T]$, then $\mathbf{x}(\cdot)$ also satisfies

$$\mathbf{x}(t) = \mathbf{x}_0 + \int_0^t \mathbf{f}[\tau, \mathbf{x}(\tau)]\, d\tau \tag{2}$$

On the other hand, if $\mathbf{x}(\cdot) \in C^n[0, T]$ satisfies (2), then clearly $\mathbf{x}(\cdot)$ is actually differentiable and satisfies (1). Thus (1) and (2) are equivalent in the sense that every solution of (1) is also a solution of (2) and vice versa.

3.4.2 Local Existence and Uniqueness

3 ***Theorem*** Suppose the function $\mathbf{f}$ in (1) is continuous in t and $\mathbf{x}$ and satisfies the following conditions: There exist finite constants T, r, h, k such that

$$\|\mathbf{f}(t, \mathbf{x}) - \mathbf{f}(t, \mathbf{y})\| \leq k\|\mathbf{x} - \mathbf{y}\|, \qquad \forall\, \mathbf{x}, \mathbf{y} \in B, \quad \forall\, t \in [0, T] \tag{4}$$

$$\|\mathbf{f}(t, \mathbf{x}_0)\| \leq h, \qquad \forall\, t \in [0, T] \tag{5}$$

where B is a ball in R^n of the form

$$B = \{\mathbf{x} \in R^n : \|\mathbf{x} - \mathbf{x}_0\| \leq r\} \tag{6}$$

Then (1) has exactly one solution over $[0, \delta]$ whenever

$$h\delta \exp(k\delta) \leq r \tag{7}$$

and

$$\delta \leq \min\left(T, \frac{\rho}{k}, \frac{r}{h + kr}\right) \tag{8}$$

for some constant $\rho < 1$.

9 REMARKS:

1. While following the proof of Theorem (3), it is important to keep in mind the distinction between $||\cdot||$ (which is a norm on R^n), and $||\cdot||_C$ (which is a norm on $C^n[0, \delta]$). Also, it should be noted that B is a ball in R^n, while S defined in (10) is a ball in $C^n[0, \delta]$.
2. The condition (4) is known as a *Lipschitz condition*, and the constant k is known as a *Lipschitz constant*. Notice that we say *a* Lipschitz constant, because if k is a Lipschitz constant for the function $\mathbf{f}$, so is any constant larger than k. Some authors reserve the term Lipschitz constant for the smallest number k for which (4) holds. Also, (4) is a *local* Lipschitz condition, because it holds only for all $\mathbf{x}$, $\mathbf{y}$ in some ball around $\mathbf{x}_0$, and for $t \in [0, T]$. Accordingly, theorem (3) is a local existence and uniqueness theorem, because it guarantees only existence and uniqueness of solutions over a sufficiently small interval $[0, \delta]$. Note that, given k, b, a, and h, (7) and (8) can always be satisfied by choosing δ sufficiently small.

***proof of Theorem** (3)* By a slight abuse of notation, we use $\mathbf{x}_0(\cdot)$ to denote the function in $C^n[0, \delta]$ whose value is $\mathbf{x}_0$ for all $t \in [0, \delta]$. Suppose δ satisfies (7) and (8), and let S be the ball in $C^n[0, \delta]$ defined by

10 $$S = \{\mathbf{x}(\cdot) \in C^n[0, \delta]: ||\mathbf{x}(\cdot) - \mathbf{x}_0(\cdot)||_C \leq r\}$$

Let P denote the mapping of $C^n[0, \delta]$ into itself defined by

11 $$(P\mathbf{x})(t) = \mathbf{x}_0 + \int_0^t \mathbf{f}[\tau, \mathbf{x}(\tau)]\, d\tau \qquad \forall\ t \in [0, \delta]$$

Clearly $\mathbf{x}(\cdot)$ is a solution of (2) over $[0, \delta]$ if and only if $(P\mathbf{x})(\cdot) = \mathbf{x}(\cdot)$; i.e., $\mathbf{x}(\cdot)$ is a fixed point of the mapping P.

We first show that P is a contraction on S. Let $\mathbf{x}(\cdot)$ and $\mathbf{y}(\cdot)$ be arbitrary elements of S; then $\mathbf{x}(t)$ and $\mathbf{y}(t)$ lie in the ball B, $\forall\ t \in [0, \delta]$. Thus

12 $$(P\mathbf{x})(t) - (P\mathbf{y})(t) = \int_0^t \{\mathbf{f}(\tau, \mathbf{x}(\tau)] - \mathbf{f}[\tau, \mathbf{y}(\tau)]\}\, d\tau$$

13 $$\begin{aligned} ||(P\mathbf{x})(t) - (P\mathbf{y})(t)|| &\leq \int_0^t ||\mathbf{f}[\tau, \mathbf{x}(\tau) - \mathbf{f}[\tau, \mathbf{y}(\tau)]||\, d\tau \\ &\leq \int_0^t k||\mathbf{x}(\tau) - \mathbf{y}(\tau)||\, d\tau \\ &\leq kt||\mathbf{x}(\cdot) - \mathbf{y}(\cdot)||_C \\ &\leq \rho||\mathbf{x}(\cdot) - \mathbf{y}(\cdot)||_C \end{aligned}$$

because $kt \leq k\delta \leq \rho$ by (8). Because the last term on the right-hand side of (13) is independent of t, it follows that

14 $$\|(P\mathbf{x})(\cdot) - (P\mathbf{y})(\cdot)\|_C = \sup_{t\in[0,\delta]} \|(P\mathbf{x})(t) - (P\mathbf{y})(t)\| \leq \rho\|\mathbf{x}(\cdot) - \mathbf{y}(\cdot)\|_C$$

so that P is a contraction on S.

Next, we show that P maps S into itself. Let $\mathbf{x}(\cdot) \in S$. Then

15 $$(P\mathbf{x})(t) - \mathbf{x}_0 = \int_0^t \mathbf{f}[\tau, \mathbf{x}(\tau)]\, d\tau = \int_0^t \{(\mathbf{f}[\tau, \mathbf{x}(\tau)] - \mathbf{f}[\tau, \mathbf{x}_0)] + \mathbf{f}[\tau, \mathbf{x}_0)]\}\, d\tau$$

$$\|(P\mathbf{x})(t) - \mathbf{x}_0\| \leq \int_0^t \{\|\mathbf{f}[\tau, \mathbf{x}(\tau)] - \mathbf{f}(\tau, \mathbf{x}_0)\| + \|\mathbf{f}(\tau, \mathbf{x}_0)\|\}\, d\tau \leq k\rho\delta + h\delta \leq r$$

by (7). Hence

16 $$\|(P\mathbf{x})(\cdot) - \mathbf{x}_0(\cdot)\|_C = \sup_{t\in[0,\delta]} \|(P\mathbf{x})(t) - \mathbf{x}_0\| \leq r$$

so that $(P\mathbf{x})(\cdot)$ belongs to S.

Now, because P maps S into itself and is a contraction on S, it has exactly one fixed point in S, by Theorem [3.3(25)]. Our objective, however, is to show that P has exactly one fixed point in $C^n[0, \delta]$, not just in S (the point being that S is a proper subset of $C^n[0, \delta]$). Thus the proof is complete if we show that any fixed point of P in $C^n[0, \delta]$ must in fact be in S. Accordingly, suppose $\mathbf{x}(\cdot) \in C^n[0, \delta]$ satisfies (2). Then we have

17 $$\mathbf{x}(t) - \mathbf{x}_0 = \int_0^t \mathbf{f}[\tau, \mathbf{x}(\tau)]\, d\tau = \int_0^t \{\mathbf{f}[\tau, \mathbf{x}(\tau)] - \mathbf{f}(\tau, \mathbf{x}_0) + \mathbf{f}(\tau, \mathbf{x}_0)\}\, d\tau$$

18 $$\|\mathbf{x}(t) - \mathbf{x}_0\| \leq \int_0^t k\|\mathbf{x}(\tau) - \mathbf{x}(0)\|\, d\tau + ht \leq h\delta + \int_0^t k\|\mathbf{x}(\tau) - \mathbf{x}(0)\|\, d\tau$$

Applying the Bellman–Gronwall inequality (see Appendix I) to (18), we get

19 $$\|\mathbf{x}(t) - \mathbf{x}_0\| \leq h\delta \exp(kt) \leq h\delta \exp(k\delta), \qquad \forall\, t \in [0, \delta].$$

From (8) and (19), it follows that $\mathbf{x}(\cdot) \in S$. Thus we have shown that any fixed point of P in $C^n[0, \delta]$ must in fact be in S. Because P has exactly one fixed point in S, it follows that P has exactly one fixed point in $C^n[0, \delta]$. By the manner in which P is defined, we conclude that (2) has exactly one solution over $[0, \delta]$. ∎

The following result is actually a corollary to Theorem (3) but is in a form that can be readily applied.

20 ***corollary*** Consider the differential equation (1). Suppose that in some neighborhood of $(0, \mathbf{x}_0)$ the function $\mathbf{f}$ has continuous partial derivatives

with respect to its second argument ($\mathbf{x}$) and continuous one-sided partial derivatives with respect to its first argument (t). Then (1) has exactly one solution over $[0, \delta]$, provided δ is sufficiently small.

proof The differentiability properties assumed on $\mathbf{f}$ ensure that $\mathbf{f}$ satisfies (4) and (5) for some set of finite constants a, b, k, and h. ∎

21 **Example.** Consider the Van der Pol oscillator, which can be described by the set of two first-order equations

22 $$\dot{x}_1 = x_2$$

23 $$\dot{x}_2 = -x_1 + \mu(x_1^2 - 1)x_2$$

This set of equations is of the form (1), with $\mathbf{f}: R^2 \longrightarrow R^2$ defined by

24 $$\mathbf{f}(\mathbf{x}) = [x_2 \quad -x_1 + \mu(x_1^2 - 1)x_2]'$$

Because each component of $\mathbf{f}$ is continuously differentiable, it follows by corollary (20) that, starting from any arbitrary initial condition $[x_{10} \quad x_{20}]'$, (22)–(23) has a unique solution over $[0, \delta]$ for some sufficiently small δ.

3.4.3 Global Existence and Uniqueness

In this subsection, we shall show that (loosely speaking) if $\mathbf{f}$ satisfies a global Lipschitz condition, then (1) has a unique solution over $[0, \infty)$.

25 ***Theorem*** Suppose that for each $T \in [0, \infty)$ there exist finite constants k_T and h_T such that

26 $$\|\mathbf{f}(t, \mathbf{x}) - \mathbf{f}(t, \mathbf{y})\| \le k_T\|\mathbf{x} - \mathbf{y}\|, \qquad \forall\ \mathbf{x}, \mathbf{y} \in R^n, \forall\ t \in [0, T]$$

27 $$\|\mathbf{f}(t, \mathbf{x}_0\| \le h_T, \qquad \forall\ t \in [0, T]$$

Then (1) has exactly one solution over $[0, T]$, $\forall\ T \in [0, \infty)$.

proof We shall give two alternative proofs.

no. 1 Let $T < \infty$ be specified, and let k_T and h_T be finite constants such that (26)–(27) hold. Then the hypotheses of Theorem (3) are satisfied, with $a = b = \infty$. Thus, by Theorem (3), it follows that (1) has a unique solution over $[0, \delta]$ whenever δ satisfies

28 $$\delta \le \frac{\rho}{k_T}$$

for some constant $\rho < 1$. Suppose a positive number δ satisfying (28) is chosen. If $T \le \delta$, the theorem is proved, so suppose $T > \delta$. Now (1) has a unique solution over $[0, \delta]$. Denote this solution by $\mathbf{y}_1(\cdot)$, and consider the differential equation

29 $$\dot{\mathbf{x}}(t) = \mathbf{f}_1[t, \mathbf{x}(t)]; \qquad \mathbf{x}(0) = \mathbf{y}_1(\delta)$$

where

30 $$\mathbf{f}_1(t, \mathbf{x}) = \mathbf{f}(t + \delta, \mathbf{x})$$

Then $\mathbf{f}_1$ also satisfies (26) and (27); therefore, once again by Theorem (3), (29) has a unique solution over $[0, \delta]$, *where δ is the same as before*. Denote this solution by $\mathbf{y}_2(\cdot)$. It is easy to verify that the function $\mathbf{x}_2(\cdot)$ defined by

31 $$\mathbf{x}_2(t) = \begin{cases} \mathbf{y}_1(t), & 0 \leq t \leq \delta \\ \mathbf{y}_2(t - \delta), & \delta < t \leq 2\delta \end{cases}$$

is the unique solution of (1) over the interval $[0, 2\delta]$. Proceeding by induction, let $\mathbf{x}_m(\cdot)$ denote the unique solution of (1) over the interval $[0, m\delta]$, and consider the differential equation

32 $$\dot{\mathbf{x}}(t) = \mathbf{f}_m[t, \mathbf{x}(t)]; \qquad \mathbf{x}(0) = \mathbf{x}_m(m\delta)$$

where

33 $$\mathbf{f}_m(t, \mathbf{x}) = \mathbf{f}(t + m\delta, \mathbf{x})$$

Let $\mathbf{y}_{m+1}(\cdot)$ denote the unique solution of (32) over the interval $[0, \delta]$ (the same δ as before). Then the function $\mathbf{x}_{m+1}(\cdot)$ defined by

34 $$\mathbf{x}_{m+1}(t) = \begin{cases} \mathbf{x}_m(t), & 0 \leq t \leq m\delta \\ \mathbf{y}_{m+1}(t), & m\delta < t \leq m\delta + \delta \end{cases}$$

is the unique solution of (1) over the interval $[0, m\delta + \delta]$. In this manner, the unique solution can be extended to all of $[0, T]$.

no. 2 Let $T < \infty$ be given, let $P: C^n[0, T] \longrightarrow C^n[0, T]$ be given by (11), and let $\mathbf{x}_0(\cdot)$ denote (as before) the element of $C^n[0, T]$ whose value is $\mathbf{x}_0 \ \forall \ t \in [0, T]$. We show first that the sequence $[P^m\mathbf{x}_0(\cdot)]_{m=1}^{\infty}$ is a Cauchy sequence in $C^n[0, T]$ and that it converges to a solution of (2).

Let $\mathbf{x}_m(\cdot) = (P^m\mathbf{x}_0)(\cdot)$. Then we have, first,

35 $$\mathbf{x}_1(t) - \mathbf{x}_0(t) = \int_0^t \mathbf{f}(\tau, \mathbf{x}_0)\, d\tau$$

36 $$\|\mathbf{x}_1(t) - \mathbf{x}_0(t)\| \leq \int_0^t \|\mathbf{f}(\tau, \mathbf{x}_0\|\, d\tau \leq h_T t$$

In general, for $m \geq 1$, we have

37 $$\|\mathbf{x}_{m+1}(t) - \mathbf{x}_m(t)\| \leq \int_0^t \|\mathbf{f}[\tau, \mathbf{x}_m(\tau)] - \mathbf{f}[\tau, \mathbf{x}_{m-1}(\tau)\|\, d\tau$$
$$\leq k_T \int_0^t \|\mathbf{x}_m(\tau) - \mathbf{x}_{m-1}(\tau)\|\, d\tau$$

Substituting (36) into (37) and proceeding by induction, we get

38 $$\|\mathbf{x}_m(t) - \mathbf{x}_{m-1}(t)\| \leq k_T^{m-1} h \frac{t^m}{m!}$$

Thus for any integer $p \geq 0$ we have

$$\|\mathbf{x}_{m+p}(t) - \mathbf{x}_m(t)\| \leq \sum_{i=0}^{p-1} \|\mathbf{x}_{m+i+1}(t) - \mathbf{x}_{m+i}(t)\| \leq \sum_{i=0}^{p-1} h_T k_T^{m+i} \frac{t^{m+i+1}}{(m+i+1)!} = \sum_{i=m+1}^{m+p} h_T k_T^{i-1} \frac{t^i}{i!} \tag{39}$$

$$\|\mathbf{x}_{m+p}(\cdot) - \mathbf{x}_m(\cdot)\|_C = \sup_{t\in[0,T]} \|\mathbf{x}_{m+p}(t) - \mathbf{x}_m(t)\| \leq \sum_{i=m+1}^{m+p} h_T k_T^{i-1} \frac{T^i}{i!} \leq \sum_{i=m+1}^{\infty} h_T k_T^{i-1} \frac{T^i}{i!} \tag{40}$$

Now consider the sequence

$$\left(\sum_{i=0}^{m} h_T k_T^{i-1} \frac{T^i}{i!}\right)_1^{\infty} \tag{41}$$

As $m \longrightarrow \infty$, this sequence converges to $(h_T/k_T) \exp(k_T T)$. Moreover, the last term in (40) is the difference between $\sum_{i=0}^{m} h_T k_T^{i-1}(T^i/i!)$ and $(h_T/k_T) \exp(k_T T)$ and can therefore be made arbitrarily small by choosing m sufficiently large. Hence $[\mathbf{x}_m(\cdot)]_1^{\infty}$ is a Cauchy sequence in $C^n[0, T]$, and because $C^n[0, T]$ is a Banach space, $[\mathbf{x}_m(\cdot)]_1^{\infty}$ converges to a limit in $C^n[0, T]$. Denote this limit by $\mathbf{x}^*(\cdot)$.

Whenever $\mathbf{z}_1(\cdot)$ and $\mathbf{z}_2(\cdot)$ are two elements in $C^n[0, T]$, we have

$$(P\mathbf{z}_1)(t) - (P\mathbf{z}_2)(t) = \int_0^t \{\mathbf{f}[\tau, \mathbf{z}_1(\tau)] - \mathbf{f}[\tau, \mathbf{z}_2(\tau)]\}\, d\tau \tag{42}$$

$$\|(P\mathbf{z}_1)(t) - (P\mathbf{z}_2)(t)\| \leq \int_0^t \|\mathbf{f}[\tau, \mathbf{z}_1(\tau)] - \mathbf{f}[\tau, \mathbf{z}_2(\tau)]\|\, d\tau \leq k_T T \|\mathbf{z}_1(\cdot) - \mathbf{z}_2(\cdot)\|_C \tag{43}$$

$$\|(P\mathbf{z}_1)(\cdot) - (P\mathbf{z}_2)(\cdot)\|_C = \sup_{t\in[0,T]} \|(P\mathbf{z}_1)(t) - (P\mathbf{z}_2)(t)\| \leq k_T T \|\mathbf{z}_1(\cdot) - \mathbf{z}_2(\cdot)\|_C \tag{44}$$

Because $k_T T$ is a finite constant, it follows that P is uniformly continuous on $C^n[0, T]$. Hence, if $[\mathbf{x}_m(\cdot)]_1^{\infty}$ converges to $\mathbf{x}^*(\cdot)$, we have

$$(P\mathbf{x}^*)(\cdot) = \lim_{m\to\infty} (P\mathbf{x}_m)(\cdot) = \lim_{m\to\infty} \mathbf{x}_{m+1}(\cdot) = \mathbf{x}^*(\cdot) \tag{45}$$

Thus $\mathbf{x}^*(\cdot)$ is a solution of (2).

Next, to show that $\mathbf{x}^*(\cdot)$ is the *only* solution of (2), suppose $\mathbf{y}(\cdot)$ also satisfies (2). Then

$$\mathbf{y}(t) - \mathbf{x}^*(t) = \int_0^t \{\mathbf{f}[\tau, \mathbf{y}(t)] - \mathbf{f}[\tau, \mathbf{x}^*(\tau)]\}\, d\tau \tag{46}$$

$$\|\mathbf{y}(t) - \mathbf{x}^*(t)\| \leq k \int_0^t \|\mathbf{y}(\tau) - \mathbf{x}^*(\tau)\|\, d\tau \tag{47}$$

Applying the Bellman–Gronwall inequality to (47), we get

48 $$\|\mathbf{y}(t) - \mathbf{x}^*(t)\| = 0 \qquad \forall\, t \in [0, T]$$

i.e., $\mathbf{y}(\cdot) = \mathbf{x}^*(\cdot)$. This shows that $\mathbf{x}^*(\cdot)$ is the unique solution of (2). ∎

49 REMARKS:

1. The sequence $[P^m \mathbf{x}_0(\cdot)]$ that converges to the solution $\mathbf{x}^*(\cdot)$ of (2) is known as the sequence of Picard's iterations, and this method of generating a solution to (2) is known as Picard's method. Actually, it is easy to show that Picard's iterations converge starting from *any* arbitrary starting function in $C^n[0, T]$ and not just $\mathbf{x}_0(\cdot)$.
2. Note that some authors assume that $\mathbf{f}(t, \mathbf{0}) = \mathbf{0}$. This assumption, together with (4), implies (5), because then $\|\mathbf{f}(t, \mathbf{x}_0)\| \leq k\|\mathbf{x}_0\|$. However, in "forced" nonlinear systems, it is not necessarily true that $\mathbf{f}(t, \mathbf{0}) = \mathbf{0}$. The present development does not require this assumption.

We shall next prove two theorems regarding the solutions of (2). In effect, Theorem (25) states that (2) has a unique solution corresponding to each initial condition. Theorem (50) shows that, at any given time, there is exactly one solution trajectory of (2) passing through each point in R^n. Theorem (58) shows that the solution of (2) depends continuously on the initial condition.

50 ***Theorem*** Let $\mathbf{f}$ satisfy the hypotheses of Theorem (25). Then for each $\mathbf{z} \in R^n$ and each $T \in [0, \infty)$ there exists exactly one element $\mathbf{z}_0 \in R^n$ such that the unique solution over $[0, T]$ of the differential equation

51 $$\dot{\mathbf{x}}(t) = \mathbf{f}[t, \mathbf{x}(t)]; \qquad \mathbf{x}(0) = \mathbf{z}_0$$

satisfies

52 $$\mathbf{x}(T) = \mathbf{z}$$

proof Consider the equation

53 $$\dot{\mathbf{x}}(t) = \mathbf{f}_s[t, \mathbf{x}(t)]; \qquad \mathbf{x}(0) = \mathbf{z}$$

54 $$\mathbf{f}_s(t, \mathbf{x}) = -\mathbf{f}(T - t, \mathbf{x}), \qquad \forall\, t \in [0, T]$$

Then $\mathbf{f}_s$ also satisfies the hypotheses of Theorem (25), so that (53) has a unique solution over $[0, T]$. Denote this solution by $\mathbf{y}(\cdot)$, and define $\mathbf{z}_0 = \mathbf{y}(T)$. Then one can easily verify that the functions $\mathbf{y}_1(\cdot)$ defined by

55 $$\mathbf{y}_1(t) = \mathbf{y}(T - t), \qquad \forall\, t \in [0, T]$$

satisfies (51) and also satisfies (52). To prove the uniqueness of the element $\mathbf{z}_0$ corresponding to a particular $\mathbf{z}$, assume by way of contradiction that there exist two functions $\mathbf{y}_1(\cdot)$ and $\mathbf{y}_2(\cdot)$ in $C^n[0, T]$ that satisfy (51) and (52). Let $\mathbf{y}_1(0) = \mathbf{z}_1$, $\mathbf{y}_2(0) = \mathbf{z}_2$. Then the functions $\mathbf{y}_a(\cdot)$ and $\mathbf{y}_b(\cdot)$ defined by

56 $$\mathbf{y}_a(t) = \mathbf{y}_1(T - t)$$

57 $$\mathbf{y}_b(t) = \mathbf{y}_2(T - t)$$

must *both* satisfy (53). However, because the solution to (53) is unique, it follows that $\mathbf{y}_1(\cdot) = \mathbf{y}_2(\cdot)$. Thus $\mathbf{z}_1 = \mathbf{z}_2$. ∎

58 ***Theorem*** Let $\mathbf{f}$ satisfy the hypotheses of Theorem (25). Let $T \in [0, \infty)$ be specified, and suppose $\mathbf{x}(\cdot)$ and $\mathbf{y}(\cdot)$ are two functions in $C^n[0, T]$ satisfying

59 $$\dot{\mathbf{x}}(t) = \mathbf{f}[t, \mathbf{x}(t)]; \qquad \mathbf{x}(0) = \mathbf{x}_0$$

60 $$\dot{\mathbf{y}}(t) = \mathbf{f}[t, \mathbf{y}(t)]; \qquad \mathbf{y}(0) = \mathbf{y}_0$$

Then for each $\epsilon > 0$, there exists a $\delta(\epsilon, T) > 0$ such that

61 $$\|\mathbf{x}(\cdot) - \mathbf{y}(\cdot)\|_C < \epsilon \quad \text{whenever } \|\mathbf{x}_0 - \mathbf{y}_0\| < \delta(\epsilon, T)$$

proof The functions $\mathbf{x}(\cdot)$ and $\mathbf{y}(\cdot)$ also satisfy

62 $$\mathbf{x}(t) = \mathbf{x}_0 + \int_0^t \mathbf{f}[\tau, \mathbf{x}(\tau)]\, d\tau$$

63 $$\mathbf{y}(t) = \mathbf{y}_0 + \int_0^t \mathbf{f}[\tau, \mathbf{y}(\tau)]\, d\tau$$

Subtracting, we get

64 $$\mathbf{x}(t) - \mathbf{y}(t) = \mathbf{x}_0 - \mathbf{y}_0 + \int_0^t \{\mathbf{f}[\tau, \mathbf{x}(\tau)] - \mathbf{f}[\tau, \mathbf{y}(\tau)]\}\, d\tau$$

65 $$\|\mathbf{x}(t) - \mathbf{y}(t)\| \leq \|\mathbf{x}_0 - \mathbf{y}_0\| + k_T \int_0^t \|\mathbf{x}(\tau) - \mathbf{y}(\tau)\|\, d\tau$$

Applying the Bellman–Gronwall inequality to (65), we get

66 $$\|\mathbf{x}(t) - \mathbf{y}(t)\| \leq \|\mathbf{x}_0 - \mathbf{y}_0\| \exp(k_T t)$$

Hence

67 $$\|\mathbf{x}(\cdot) - \mathbf{y}(\cdot)\|_C \leq \|x_0 - \mathbf{y}_0\| \exp(k_T T)$$

Thus given $\epsilon > 0$, (61) is satisfied if we choose $\delta(\epsilon, T) = \epsilon/\exp(k_T T)$. ∎

REMARKS: The results contained in Theorems (50) and (58) can be given a simple geometric interpretation in terms of certain mappings being continuous. Let $\phi\colon R^n \longrightarrow C^n[0, T]$ be the mapping that associates, with each initial condition $\mathbf{x}_0$ in R^n, the corresponding unique solution of (2). Then Theorem (58) states that ϕ is uniformly continuous on R^n. In the same vein, let $\psi: R^n \longrightarrow R^n$ be the mapping that associates, with each initial condition $\mathbf{x}_0 \in R^n$, the value at time T of the corresponding unique solution of (2). Then Theorem (50) states that ψ is one-to-one [i.e., given $\psi(\mathbf{x})$, one can uniquely determine $\mathbf{x}$] and onto (i.e., the range of ψ is all of R^n). Furthermore, Theorem (58) shows that both ψ and its inverse map ψ^{-1} are continuous.

68 **Example.** Consider the scalar differential equation

69 $$\dot{x}(t) = \tanh x(t) \triangleq f[x(t)]; \qquad x(0) = x_0$$

Since the function $x \longrightarrow \tanh x$ is everywhere continuously differentiable, and since this derivative is everywhere bounded (in magnitude) by 1, it is easy to verify that $f(\cdot)$ satisfies a global Lipschitz condition of the form (26) with $k_T = 1$ for all T (see also Problem 3.10 infra). Also, for every x_0, there exists a finite h_T such that (27) holds. Hence, by Theorem (25), (69) has a unique solution over $[0, \infty)$ corresponding to each x_0; moreover, this solution depends continuously on x_0.

70 **Example.** Consider the *linear* vector differential equation

71 $$\dot{\mathbf{x}}(t) = \mathbf{A}(t)\mathbf{x}(t); \qquad \mathbf{x}(0) = \mathbf{x}_0$$

where $\mathbf{A}(\cdot)$ is piecewise-continuous. Let $\|\cdot\|$ be a given norm on R^n. Since $\mathbf{A}(\cdot)$ is piecewise-continuous, for every finite T, there exists a finite constant k_T such that

72 $$\|\mathbf{A}(t)\|_i \leq k_T, \qquad \forall\, t \in [0, T]$$

Hence we have

73 $$\|\mathbf{A}(t)\mathbf{x} - \mathbf{A}(t)\mathbf{y}\| \leq k_T \|\mathbf{x} - \mathbf{y}\|, \qquad \forall\, \mathbf{x}, \mathbf{y} \in R^n; \qquad \forall\, t \in [0, T]$$

74 $$\|\mathbf{A}(t)\mathbf{x}_0\| \leq k_T \|\mathbf{x}_0\|, \qquad \forall\, t \in [0, T]$$

So (26) is satisfied with k_T as above, and (27) is satisfied with $h_T = k_T \|\mathbf{x}_0\|$. Therefore, (71) has a unique solution over each finite interval $[0, T]$ corresponding to each initial condition $\mathbf{x}_0$; moreover, this solution depends continuously on $\mathbf{x}_0$.

In conclusion, in this section we have derived some conditions that are sufficient to ensure that a given nonlinear vector differential equation has a unique solution over some interval. It is easy to construct counterexamples to show that the conditions derived here are by no means necessary for the existence and uniqueness of solutions. For instance, consider the scalar differential equation

$$\dot{x}(t) = -x^2; \qquad x(0) = 1$$

This equation has a unique solution over $[0, \infty)$ [namely, $x(t) = 1/(t+1)$], even though the function $f(x) = x^2$ is not globally Lipschitz-continuous.

Problem 3.9. Let $\|\cdot\|_a$ and $\|\cdot\|_b$ be two given norms on R^n. Show that for each finite T there exists a finite k_{aT} such that

$$\|\mathbf{f}(t, \mathbf{x}) - \mathbf{f}(t, \mathbf{y})\|_a \leq k_{aT}\|\mathbf{x} - \mathbf{y}\|_a, \qquad \forall\, \mathbf{x}, \mathbf{y} \in R^n, \qquad \forall\, t \in [0, T]$$

if and only if, for each finite T there exists a finite k_{bT} such that

$$\|\mathbf{f}(t, \mathbf{x}) - \mathbf{f}(t, \mathbf{y})\|_b \leq k_{bT}\|\mathbf{x} - \mathbf{y}\|_b, \qquad \forall\, \mathbf{x}, \mathbf{y} \in R^n, \qquad \forall\, t \in [0, T]$$

In other words, show that, in verifying whether a given function $\mathbf{f}$ satisfies (26), the norm used on R^n is immaterial. [*Hint:* Use fact (3.1.52)].

Problem 3.10. (a) Let $f\colon R_+ \times R \longrightarrow R$ be continuous, and continuously differentiable in the second argument. Show that f satisfies (26) if and only if, for each finite T there exists a finite k_T, such that

$$\left|\frac{\partial f(t, x)}{\partial x}\right| \leq k_T, \qquad \forall\, x \in R, \qquad \forall\, t \in [0, T]$$

i.e., $\partial f(t, x)/\partial x$ is bounded independently of x over each finite interval $[0, T]$. (Hint: Use the mean value theorem).

(b) Let $f\colon R_+ \times R^n \longrightarrow R^n$ be continuous, and continuously differentiable in the second argument. Show that $\mathbf{f}$ satisfies (26) if and only if, for each finite T there exists a finite k_T such that

$$\left|\frac{\partial f_i(t, \mathbf{x})}{\partial x_j}\right| \leq k_T, \qquad \forall\, i, j, \qquad \forall\, \mathbf{x} \in R^n, \qquad \forall\, t \in [0, T]$$

(Hint: Use the results of Problem 3.9, above).

(c) Determine whether the following functions satisfy a global Lipschitz condition:

(i) $\mathbf{f}(\mathbf{x}) = [x_1^2 - x_1x_2 \quad 2x_1 - x_2^2]'$
(ii) $\mathbf{f}(\mathbf{x}) = [x_1 \exp(-x_2^2) \quad x_2 \exp(-x_1^2)]'$

3.5 SOLUTION ESTIMATES FOR LINEAR EQUATIONS

In this section, we shall give a method for obtaining both upper and lower bounds on the norm of the solution of a given differential equation. The Bellman–Gronwall inequality (Appendix I) does give an easily applicable upper bound on the norm of the solution of a linear differential equation, and a similar inequality known as Langenhop's inequality provides a lower bound. However, both of these bounds suffer from the deficiency of being *sign-insensitive*; i.e., they give exactly the same estimates for

$$\dot{\mathbf{x}}(t) = \mathbf{A}(t)\mathbf{x}(t)$$

and

$$\dot{\mathbf{x}}(t) = -\mathbf{A}(t)\mathbf{x}(t)$$

They do so because both the Bellman–Gronwall inequality and Langenhop's inequality utilize $\|\mathbf{A}(t)\|$, which is, of course, sign-insensitive. In contrast, the method given here utilizes the concept of the matrix measure, which is sign-sensitive. As a result, the bounds derived in this section are always "tighter" than (or the same as) those given by the Bellman–Gronwall and Langenhop's inequalities.

1 ***Theorem*** Consider the differential equation

$$\dot{\mathbf{x}}(t) = \mathbf{A}(t)\mathbf{x}(t), \qquad t \geq 0 \tag{2}$$

where $\mathbf{x}(t) \in R^n$, $\mathbf{A}(t) \in R^{n \times n}$, and $\mathbf{A}(\cdot)$ is piecewise-continuous.[4] Let $\|\cdot\|$ be a norm on R^n, and let $\|\cdot\|_i$ and μ_i denote, respectively, the corresponding induced norm and matrix measure on $R^{n \times n}$. Then, whenever $t \geq t_0 \geq 0$, we have

$$\|\mathbf{x}(t_0)\| \exp\left\{\int_{t_0}^{t} -\mu_i[-\mathbf{A}(\tau)]\, d\tau\right\} \leq \|\mathbf{x}(t)\| \leq \|\mathbf{x}(t_0)\| \exp\left\{\int_{t_0}^{t} \mu_i[\mathbf{A}(\tau)]\, d\tau\right\} \tag{3}$$

proof Clearly, the right-hand side of (2) is globally Lipschitz-continuous in $\mathbf{x}$, so that by Theorem [3.4(25)], (2) has a unique solution over all bounded intervals of the form $[0, T]$. To prove the inequalities (3), we begin by observing that, from the integral form of (2),

$$\mathbf{x}(t + \delta) = \mathbf{x}(t) + \delta\mathbf{A}(t)\mathbf{x}(t) + \mathbf{o}(\delta), \qquad \forall\, \delta > 0 \tag{4}$$

where $\mathbf{o}(\delta)$ is used to denote an error term with the property that

$$\lim_{\delta \to 0^+} \frac{\|\mathbf{o}(\delta)\|}{\delta} = 0 \tag{5}$$

Rearranging (4), we get, successively,

$$\mathbf{x}(t + \delta) = [\mathbf{I} + \delta\mathbf{A}(t)]\mathbf{x}(t) + \mathbf{o}(\delta) \tag{6}$$

$$\|\mathbf{x}(t + \delta)\| \leq \|\mathbf{I} + \delta\mathbf{A}(t)\|_i\, \|\mathbf{x}(t)\| + \|o(\delta)\| \tag{7}$$

$$\|\mathbf{x}(t + \delta)\| - \|\mathbf{x}(t)\| \leq (\|\mathbf{I} + \delta\mathbf{A}(t)\|_i - 1)\|\mathbf{x}(t)\| + o(\delta) \tag{8}$$

---

[4]By this we mean that, over each finite interval, $\mathbf{A}(\cdot)$ is continuous at all except a finite number of points; at each point of discontinuity, $\mathbf{A}(\cdot)$ has well-defined left and right limits; and the norm of the difference of these limits is finite.

9 $$\frac{d^+}{dt}\|\mathbf{x}(t)\| = \lim_{\delta\to 0^+}\frac{\|\mathbf{x}(t+\delta)\| - \|\mathbf{x}(t)\|}{\delta} \le \mu_i[\mathbf{A}(t)]\|\mathbf{x}(t)\|$$

where $d^+/dt[\cdot]$ denotes the right-hand derivative. If we multiply both sides of inequality (9) by the integrating factor

10 $$\exp\left\{-\int_{t_0}^{t}\mu_i[\mathbf{A}(\tau)]\,d\tau\right\}$$

we get the right-hand inequality in (3). The proof of the left-hand inequality in (3) is entirely similar, starting with

11 $$\mathbf{x}(t-\delta) = \mathbf{x}(t) - \delta\mathbf{A}(t)\mathbf{x}(t) + \mathbf{o}(\delta)$$

The completion of the proof is left as an exercise. ■

Theorem (1) provides both upper and lower bounds for the norm of the solution of the unforced linear equation (2). In applying the bounds (3), it is important to remember that the norm being used and the measure must correspond to each other. The application of Theorem (1) is illustrated by the following examples.

12 **Example.** Consider equation (2), with $n = 2$ and

13 $$\mathbf{A}(t) = \begin{bmatrix} -2t & 1 \\ -1 & -t \end{bmatrix}$$

14 $$\mathbf{x}(0) = [1 \quad 0]'$$

First, we calculate the measures μ_{i1}, μ_{i2}, and $\mu_{i\infty}$ of $\mathbf{A}(t)$ as well as $-\mathbf{A}(t)$. This gives

15 $$\mu_{i1}[\mathbf{A}(t)] = \mu_{i\infty}[\mathbf{A}(t)] = -t + 1$$

16 $$\mu_{i1}[-\mathbf{A}(t)] = \mu_{i\infty}[-\mathbf{A}(t)] = 2t + 1$$

17 $$\mu_{i2}[\mathbf{A}(t)] = -t$$

18 $$\mu_{i2}[-\mathbf{A}(t)] = 2t$$

Thus, applying inequalities (3) with each of the above measures gives

19 $$\exp(-t - t^2) \le |x_1(t)| + |x_2(t)| \le \exp\frac{t - t^2}{2} \qquad \text{for } \|\cdot\|_1,\ \mu_{i1}$$

20 $$\exp(-t - t^2) \le |x_1(t)|, |x_2(t)| \le \exp\frac{t - t^2}{2} \qquad \text{for } \|\cdot\|_\infty,\ \mu_{i\infty}$$

21 $$\exp(-t^2) \le \{|x_1(t)|^2 + |x_2(t)|^2\}^{1/2} \le \exp\frac{-t^2}{2} \qquad \text{for } \|\cdot\|_2,\ \mu_{i2}$$

Thus the same two inequalities (3), when applied with different measures, yield different estimates for the vector $\mathbf{x}(t)$. By way of illustrating the bounds obtained above, the regions to which the vector $\mathbf{x}(1)$ is confined by each of these bounds are shown in Figs. 3.3, 3.4, and 3.5.

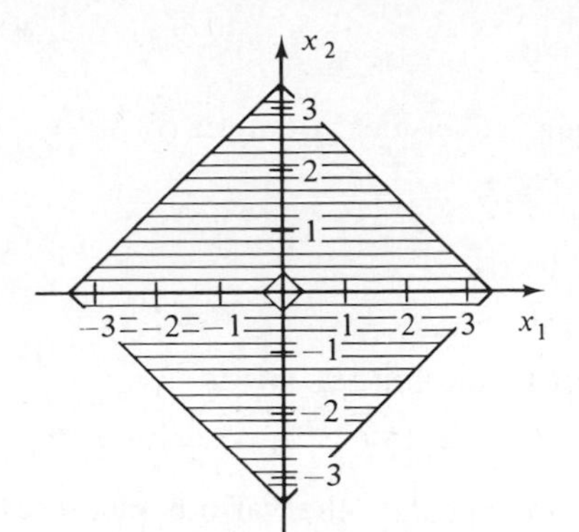

FIG. 3.3

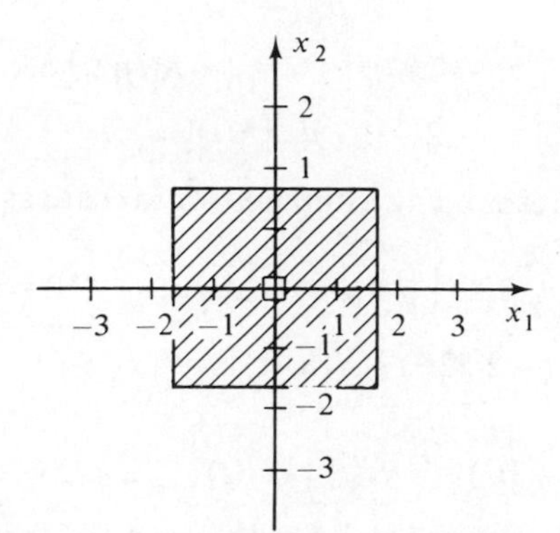

FIG. 3.4

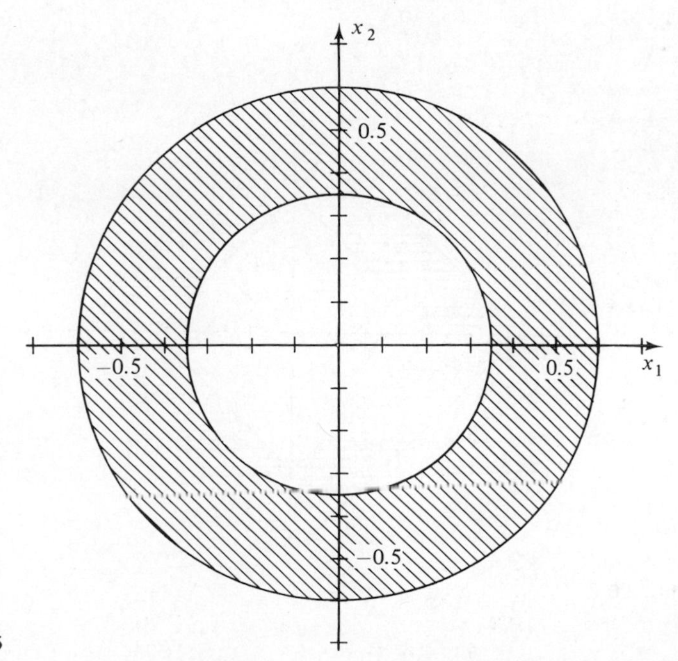

FIG. 3.5

22 **Example.** Consider equation (2) with $n = 2$ and

$$\mathbf{A}(t) = \begin{bmatrix} -3t & t \\ 2t & -4t \end{bmatrix} \tag{23}$$

$$\mathbf{x}(0) = [1 \quad 2]' \tag{24}$$

Then the actual solution for $\mathbf{x}(t)$ is

$$\mathbf{x}(t) = [\tfrac{4}{3}e^{-t^2} - \tfrac{1}{3}e^{-\frac{5}{2}t^2} \quad \tfrac{4}{3}e^{-t^2} + \tfrac{2}{3}e^{-\frac{5}{2}t^2}]' \tag{25}$$

However, if we calculate the various measures of $\mathbf{A}(t)$, we get

$$\mu_{i1}[\mathbf{A}(t)] = -t; \qquad \mu_{i1}[-\mathbf{A}(t)] = 5t \tag{26}$$

$$\mu_{i2}[\mathbf{A}(t)] \simeq -2.97t; \qquad \mu_{i2}[-\mathbf{A}(t)] \simeq 5.03t \tag{27}$$

$$\mu_{i\infty}[\mathbf{A}(t)] = -2t; \qquad \mu_{i\infty}[\mathbf{A}(t)] = 6t \tag{28}$$

Thus the corresponding estimates for $\mathbf{x}(t)$ are as follows:

$$3 \exp \frac{-5t^2}{2} \leq |x_1(t)| + |x_2(t)| \leq 3 \exp \frac{-t^2}{2} \qquad \text{for } \|x(t)\|_1 \tag{29}$$

$$\sqrt{5} \exp(-2.52t^2) \leq \{|x_1(t)|^2 + |x_2(t)|^2\}^{1/2} \leq \sqrt{5} \exp(-1.49t^2) \qquad \text{for } \|x(t)\|_2 \tag{30}$$

$$2 \exp(-3t^2) \leq [|x_1(t)|, |x_2(t)|] \leq 2 \exp(-t^2) \qquad \text{for } \|\mathbf{x}(t)\|_\infty \tag{31}$$

These bounds are depicted for the case $t = 0.5$ in Figs. 3.6, 3.7, and 3.8.

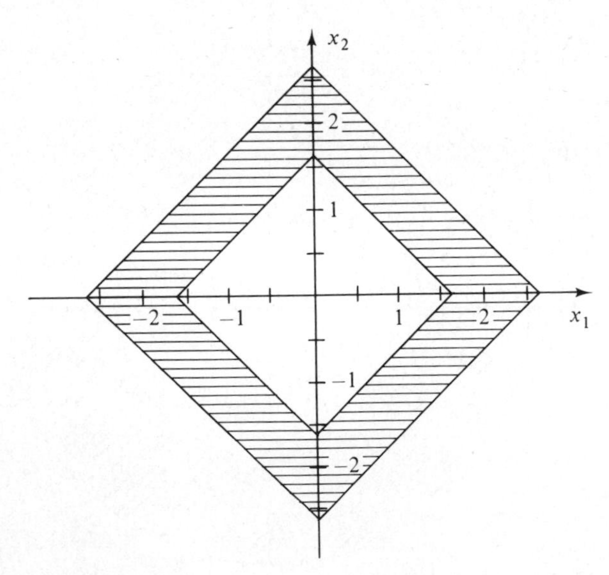

FIG. 3.6

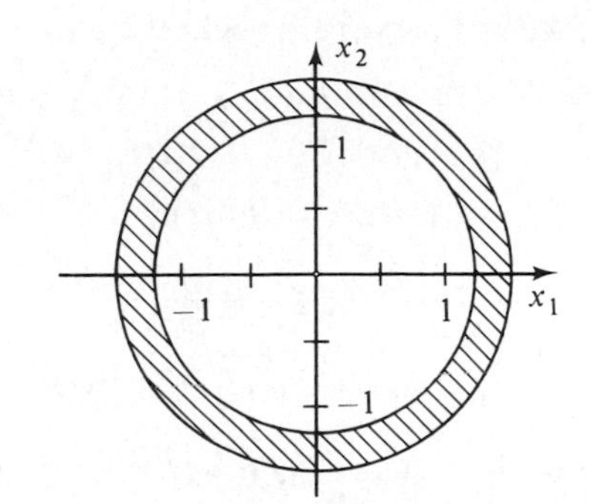

FIG. 3.7

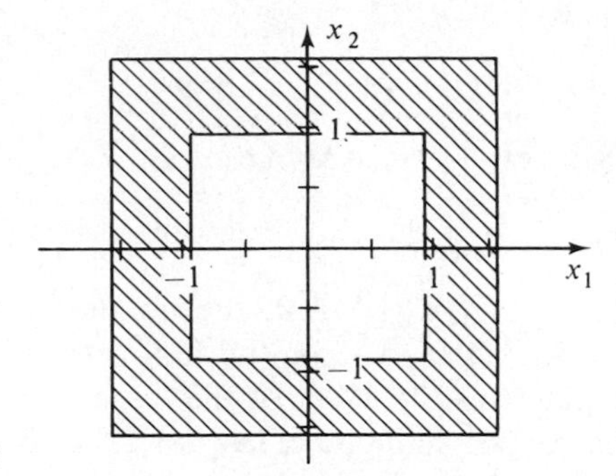

FIG. 3.8

We turn now to "forced" linear equations.

32 ***Theorem*** Consider the linear differential equation

33 $$\dot{\mathbf{x}}(t) = \mathbf{A}(t)\mathbf{x}(t) + \mathbf{v}(t), \qquad \forall\, t > 0$$

where $\mathbf{x}(t), \mathbf{v}(t) \in R^n$, $\mathbf{A}(t) \in R^{n \times n}\ \forall\, t \geq 0$, and $\mathbf{v}(\cdot), \mathbf{A}(\cdot)$ are piecewise-continuous. Let $\|\cdot\|$ be a norm on R^n, and let $\|\cdot\|_i$, $\mu_i(\cdot)$ denote, respectively, the corresponding induced norm and matrix measure on $R^{n \times n}$. Then, whenever $t \geq t_0 \geq 0$, we have

34 $$\|\mathbf{x}(t)\| \leq \exp\left\{\int_{t_0}^{t} \mu_i[\mathbf{A}(\tau)]\, d\tau\right\}\Bigg[\|\mathbf{x}(t_0)\| + \int_{t_0}^{t} \exp\left\{\int_{\tau}^{t_0} \mu_i[\mathbf{A}(s)]\, ds\right\}\|\mathbf{v}(\tau)\|\, d\tau\Bigg]$$

35 $$\|\mathbf{x}(t)\| \geq \exp\left\{\int_{t_0}^{t} -\mu_i[-\mathbf{A}(\tau)]\, d\tau\right\}\Bigg[\|\mathbf{x}(t_0)\| - \int_{t_0}^{t} \exp\left\{\int_{\tau}^{t_0} -\mu_i[-\mathbf{A}(s)]\, ds\right\}\|\mathbf{v}(\tau)\|\, d\tau\Bigg]$$

proof We first prove (35). As in the proof of Theorem (1), we have

36 $$\mathbf{x}(t-\delta) = \mathbf{x}(t) - \delta\mathbf{A}(t)\mathbf{x}(t) - \delta\mathbf{v}(t) + \mathbf{o}(\delta) = [\mathbf{I} - \delta\mathbf{A}(t)]\mathbf{x}(t) - \delta\mathbf{v}(t) + \mathbf{o}(\delta)$$

37 $$\|\mathbf{x}(t-\delta)\| \leq \|\mathbf{I} - \delta\mathbf{A}(t)\|_i \|\mathbf{x}(t)\| + \delta\|\mathbf{v}(t)\| + o(\delta)$$

38 $$\frac{\|\mathbf{x}(t)\| - \|\mathbf{x}(t-\delta)\|}{\delta} \geq \frac{1 - \|\mathbf{I} - \delta\mathbf{A}(t)\|_i}{\delta}\|\mathbf{x}(t)\| - \|\mathbf{v}(t)\| + \frac{o(\delta)}{\delta}$$

39 $$\frac{d^-}{dt}[\|\mathbf{x}(t)\|] \geq -\mu_i[-\mathbf{A}(t)]\|\mathbf{x}(t)\| - \|\mathbf{v}(t)\|$$

If we multiply both sides of (39) by the integrating factor

40 $$\exp\left\{\int_{t_0}^t \mu_i[-\mathbf{A}(\tau)]\,d\tau\right\}$$

and regroup terms, we get

41 $$\frac{d^-}{dt}\left(\|\mathbf{x}(t)\| \exp\left\{\int_{t_0}^t \mu_i[-\mathbf{A}(\tau)]\,d\tau\right\}\right) \geq -\|\mathbf{v}(t)\| \exp\left\{\int_{t_0}^t \mu_i[-\mathbf{A}(\tau)]\,d\tau\right\}$$

Integration of (41) yields (35). Inequality (34) is proved similarly. ∎

It is clear that if $\mathbf{v}(t) \equiv \mathbf{0}$, the bounds (34) and (35) reduce to those in (3). In (34) and (35), the first term on the right-hand side can be thought of as the effect of the initial condition $\mathbf{x}(t_0)$, while the second term can be thought of as the effect of the forcing function $\mathbf{v}(\cdot)$.

We close this subsection with an observation on systems of the type (2), i.e., unforced linear systems. Suppose $\mathbf{x}(t_0) \neq \mathbf{0}$, so that $\|\mathbf{x}(t_0)\| > 0$. Then for all finite t, the extreme left-hand side of (3) is always strictly positive, while the extreme right-hand side is always finite. Physically speaking, this means that a system of type (2) cannot have a *finite settling time* (i.e., it cannot go from a nonzero initial state to a zero state in a finite amount of time), or a *finite escape time* [i.e., $\mathbf{x}(t)$ cannot "blow up" in a finite amount of time]. In contrast, a linear *discrete-time* system described by

42 $$\mathbf{x}_{k+1} = \mathbf{A}_k\mathbf{x}_k$$

can exhibit finite settling time (though not finite escape time). For example, if $\mathbf{A}_k$ is singular and $\mathbf{x}_k$ belongs to the null space of $\mathbf{A}_k$, then $\mathbf{x}_i = \mathbf{0}\ \forall\ i \geq k+1$. As a more extreme example, if $\mathbf{A}_k = \mathbf{A}\ \forall\ k$, where $\mathbf{A}$ is a nilpotent $n \times n$ matrix, then *every* initial state is reduced to zero in at most n sampling intervals.

Problem 3.11. Calculate upper and lower bounds for $\|\mathbf{x}(t)\|_1$, $\|\mathbf{x}(t)\|_2$, $\|\mathbf{x}(t)\|_\infty$, given

(i) $$\begin{bmatrix} \dot{x}_1(t) \\ \dot{x}_2(t) \end{bmatrix} = \begin{bmatrix} -2t & 1 \\ -3 & t \end{bmatrix} \begin{bmatrix} x_1(t) \\ x_2(t) \end{bmatrix}; \begin{bmatrix} x_1(0) \\ x_2(0) \end{bmatrix} = \begin{bmatrix} 1 \\ -1 \end{bmatrix}$$

(ii) $$\begin{bmatrix} \dot{x}_1(t) \\ \dot{x}_2(t) \end{bmatrix} = \begin{bmatrix} -2 + 2\sin t & -1 \\ 2 & -3 - 2\cos t \end{bmatrix} \begin{bmatrix} x_1(t) \\ x_2(t) \end{bmatrix}; \begin{bmatrix} x_1(0) \\ x_2(0) \end{bmatrix} = \begin{bmatrix} 2 \\ 1 \end{bmatrix}$$

Problem 3.12. Calculate upper and lower bounds for $||\mathbf{x}(t)||_1$, $||\mathbf{x}(t)||_2$, $||\mathbf{x}(t)||_\infty$, given

(i) $$\begin{bmatrix} \dot{x}_1(t) \\ \dot{x}_2(t) \end{bmatrix} = \begin{bmatrix} -t + 2\sin t & 0 \\ -1 & -2t - \cos t \end{bmatrix} \begin{bmatrix} x_1(t) \\ x_2(t) \end{bmatrix} + \begin{bmatrix} 1 \\ e^{-t} \end{bmatrix};$$

$$\begin{bmatrix} x_1(0) \\ x_2(0) \end{bmatrix} = \begin{bmatrix} 1 \\ 1 \end{bmatrix}$$

(ii) $$\begin{bmatrix} \dot{x}_1(t) \\ \dot{x}_2(t) \end{bmatrix} = \begin{bmatrix} 3 - 2t & 1 \\ 2 & 4 - t \end{bmatrix} \begin{bmatrix} x_1(t) \\ x_2(t) \end{bmatrix} + \begin{bmatrix} e^{-t} \\ e^{t} \end{bmatrix}; \begin{bmatrix} x_1(0) \\ x_2(0) \end{bmatrix} = \begin{bmatrix} 1 \\ 2 \end{bmatrix}$$

4 Approximate analysis methods

In this chapter, we shall present several methods for *approximately* analyzing a given nonlinear system. Because a closed-form analytic solution of a nonlinear differential equation is usually impossible to obtain (unless the equation has been "cooked" in advance), methods for carrying out an approximate analysis are very useful in practice. The methods presented here fall into three categories:

1. *Describing function methods* consist of replacing a nonlinear element within the system by a linear element and carrying out further analysis. The utility of these methods is in predicting the existence and stability of limit cycles, in predicting jump resonance, etc.
2. *Numerical solution methods* are specifically aimed at carrying out a numerical solution of a given nonlinear differential equation using a computer.
3. *Singular perturbation methods* are especially well suited for the analysis of systems where the inclusion or exclusion of a particular component changes the order of the system. (For example, in an amplifier, the inclusion of a stray capacitance in the system model increases the order of the dynamic model by one.)

It should be emphasized that the above three types of methods are only some of the many varieties of techniques that are available for the approximate analysis of nonlinear systems. Moreover, even with regard to the three subject areas mentioned above, the presentation here only scratches the surface, and references are given, at appropriate places, to works that treat the subjects more thoroughly.

4.1 DESCRIBING FUNCTIONS

4.1.1 Optimal Quasilinearization

Consider a nonlinear element N, which can be thought of as an operator from $C[0, \infty)$ into itself, where $C[0, \infty)$ denotes the linear space of continuous real-valued functions over $[0, \infty)$. In other words, given any continuous function $x(\cdot)$ over $[0, \infty)$, the nonlinear element N associates, with $x(\cdot)$, another function $(Nx)(\cdot)$ in $C[0, \infty)$. It is natural to think of $(Nx)(\cdot)$ as the output of N corresponding to the input $x(\cdot)$. A commonly encountered type of nonlinearity is the so-called *memoryless nonlinearity*, where $(Nx)(\cdot)$ is of the form

$$(Nx)(t) = n[t, x(t)] \tag{1}$$

where $n(t, \cdot): R \rightarrow R$. The rationale for calling an operator of the form (1) *memoryless* is that the output at time t [namely $(Nx)(t)$] depends solely on the input value at the same time t [namely $x(t)$]—and not on the past or future values of $x(\cdot)$.

The problem studied in this subsection is the following: Suppose that a particular function $x_0(\cdot)$ in $C[0, \infty)$, which might be called the *reference signal*, is given. The objective is to approximate the function $(Nx_0)(\cdot)$ by means of another function $(Wx_0)(\cdot)$ of the form

$$(Wx_0)(t) = \int_0^t w_0(t - \tau)x_0(\tau)\, d\tau \tag{2}$$

← key pt. to functional analysis.

Note that $(Wx_0)(\cdot)$ can be interpreted as the output of a linear time-invariant system with input $x_0(\cdot)$ and *impulse response* $w_0(\cdot)$[1]. Thus the objective is to approximate the output of the nonlinear operator [namely $(Nx_0)(\cdot)$] by means of the output of a *linear* system [namely $(Wx_0)(\cdot)$]. Furthermore, we would like the approximation to be the "best possible," in a sense to be made precise next. We would like to choose the impulse response $w_0(\cdot)$ in such a way as to minimize the error criterion

$$e(w) = \lim_{T \to \infty} \frac{1}{T} \int_0^T [(Nx_0)(t) - (Wx_0)(t)]^2\, dt \tag{3}$$

[1]See Sec. 6.1 for a definition of some of these terms.

assuming, of course, that the indicated limit exists and is finite. The error criterion e is recognized as a measure of the average mean squared deviation between $(Nx_0)(\cdot)$ and $(Wx_0)(\cdot)$.

In the sequel, we make the following assumption regarding the nonlinearity N: For each finite constant b, there exists a corresponding finite constant $m_0(b)$ such that

$$|(Nx)(t)| \leq m_0(b) \text{ whenever } x(\cdot) \in C[0, \infty) \text{ and } |x(t)| \leq b \tag{4}$$

Assumption (4) is not overly restrictive and places the mathematical manipulations that follow on a firm mathematical footing. For the same reason, we restrict the impulse response $w(\cdot)$ by requiring that[2]

$$\int_0^\infty |w(t)|\, dt < \infty \tag{5}$$

These two assumptions, together with the restriction that the reference input $x_0(\cdot)$ is bounded, are enough to assure that all the integrals encountered in the following arguments exist and are finite.

To determine the optimal choice for the impulse response $w(\cdot)$, we use a standard variational argument. Assume that, for the problem at hand, an optimum choice of $w(\cdot)$ exists, and denote it by $w_0(\cdot)$. Suppose $w(\cdot)$ is any other impulse response; then, because we are assuming that $w_0(\cdot)$ is the best possible choice, we must have

$$e(w) \geq e(w_0) \tag{6}$$

or, in other words,

$$e(w) - e(w_0) \geq 0, \qquad \text{for all } w(\cdot) \tag{7}$$

Now, by definition,

$$e(w_0) = \lim_{T \to \infty} \frac{1}{T} \int_0^T [(Nx_0)(t) - (W_0x_0)(t)]^2\, dt \tag{8}$$

Hence

$$\begin{aligned} e(w) - e(w_0) &= \lim_{T \to \infty} \frac{1}{T} \int_0^T [(Nx_0 - Wx_0)^2(t) \\ &\qquad - (Nx_0 - W_0x_0)^2(t)]\, dt \\ &= \lim_{T \to \infty} \frac{1}{T} \int_0^T [(Dx_0)^2(t) + 2(Dx_0)(t) \cdot \\ &\qquad (W_0x_0 - Nx_0)(t)]\, dt \end{aligned} \tag{9}$$

where we define

$$(Dx_0)(t) = (Wx_0)(t) - (W_0x_0)(t) \tag{10}$$

---

[2]Assumption (5) implies that the linear system represented by W is bounded-input/bounded-output stable. See Sec. 6.3.

Next, for the right-hand side of (9) to be nonnegative for all Dx_0, the necessary and sufficient condition is that the linear term in Dx_0 should be identically zero; i.e., we must have

$$\lim_{T\to\infty} \frac{1}{T} \int_0^T \left\{ \int_0^t [w(t-\tau) - w_0(t-\tau)] x_0(\tau)\, d\tau \right\} \cdot (W_0 x_0 - N x_0)(t)\, dt = 0 \tag{11}$$

Next, it is easy to verify that

$$\int_0^t [w(t-\tau) - w_0(t-\tau)] x_0(\tau)\, d\tau = \int_0^t [w(\tau) - w_0(\tau)] x_0(t-\tau)\, d\tau \tag{12}$$

Substituting from (12) into (11) and interchanging the order of integration, we obtain

$$\lim_{T\to\infty} \frac{1}{T} \int_0^T [w(\tau) - w_0(\tau)] \cdot \left\{ \int_\tau^T x_0(t-\tau)(W_0 x_0 - N x_0)(t)\, dt \right\} d\tau = 0 \tag{13}$$

For (13) to hold for *all* functions $w(\cdot) - w_0(\cdot)$, we must have (after interchanging limiting and integration with respect to τ)

$$\lim_{T\to\infty} \frac{1}{T} \int_\tau^T x_0(t-\tau)(W_0 x_0 - N x_0)(t)\, dt = 0 \qquad \forall\, \tau \geq 0 \tag{14}$$

Equation (14) can be rewritten more compactly. Define

$$\phi_{W_0}(\tau) = \lim_{T\to\infty} \frac{1}{T} \int_\tau^T x_0(t-\tau)(W_0 x_0)(t)\, dt \tag{15}$$

$$\phi_N(\tau) = \lim_{T\to\infty} \frac{1}{T} \int_\tau^T x_0(t-\tau)(N x_0)(t)\, dt \tag{16}$$

The function $\phi_{W_0}(\cdot)$ is the *cross-correlation function* between the functions $x_0(\cdot)$ and $(W_0 x_0)(\cdot)$, and similarly for $\phi_N(\cdot)$. With these definitions, (14) reduces simply to

$$\phi_{W_0}(\tau) = \phi_N(\tau), \qquad \forall\, \tau \geq 0 \tag{17}$$

Equation (17) represents an important principle, namely: A necessary and sufficient condition for an impulse response $w_0(\cdot)$ to be optimal [in the sense of minimizing the error criterion in (3)] is that the cross-correlation function between $x_0(\cdot)$ and $(W_0 x_0)(\cdot)$ is the same as the cross-correlation function between $x_0(\cdot)$ and $(N x_0)(\cdot)$.

A few comments are in order regarding the above result: Given a particular reference input $x_0(\cdot)$, if some $w_0(\cdot)$ satisfies (14), this can be taken to mean that a linear system with impulse response $w_0(\cdot)$ is the best possible linear approximation to the nonlinear operator N, *with the reference input* $x_0(\cdot)$. However, (1) for a given $x_0(\cdot)$, there is in general more than one $w_0(\cdot)$ that satisfies (14), and (2) a function $w_0(\cdot)$ that satisfies (14) for one choice of reference input $x_0(\cdot)$ need not do so for some other choice of $x_0(\cdot)$. Keeping these points in mind, we refer to any $w_0(\cdot)$ that satisfies (14) as an *optimal quasilinearization* of N with reference input $x_0(\cdot)$.

In the preceding derivation, we assumed that the reference input $x_0(\cdot)$ is deterministic. However, it is also possible to carry out an optimal quasilinearization of N with respect to *random* reference input. Also, the above development can be extended to *multi-input/multi-output* nonlinearities, with no essential changes. For a detailed discussion, see [12] Gelb and Vander Velde.

4.1.2 Equivalent Linearization and Harmonic Balance

In this subsection, we shall specialize the results of Sec. 4.1.1 to obtain the optimal quasilinearization of a memoryless time-invariant nonlinearity in two special cases: (1) when the reference input is a constant or *bias* signal and (2) when the reference input is sinusoidal.

Constant Reference Input

Suppose the reference input $x_0(\cdot)$ is a constant, i.e.,

18 $$x_0(t) \equiv k, \qquad \forall\, t \geq 0$$

for some real number k. Suppose also that the nonlinearity N is memoryless and time-invariant; i.e., suppose that

19 $$(Nx)(t) = n[x(t)], \qquad \forall\, t \geq 0, \qquad \forall\, x(\cdot) \in c[0, \infty)$$

where $n\colon R \longrightarrow R$. Then the cross-correlation function $\phi_N(\tau)$ becomes

20 $$\phi_N(\tau) = kn(k), \qquad \forall\, \tau \geq 0$$

Thus the optimality condition (17) becomes

21 $$\phi_{W_0}(\tau) = kn(k), \qquad \forall\, \tau \geq 0$$

There are several possible solutions to (21). However, if $k \neq 0$, the simplest choice for $w_0(\cdot)$ that satisfies (21) is

22 $$w_0(t) = \frac{n(k)}{k}\,\delta(t)$$

A system whose impulse response is given by (22) is nothing but a constant gain of value $[n(k)/k]$. Thus we have shown the following.

Conclusion. If N is a memoryless time-invariant nonlinearity of the form (15), then an optimal quasilinearization of N with respect to a constant reference input $k \neq 0$ is a constant gain of value $[n(k)/k]$. The latter constant gain is sometimes referred to as the *equivalent linearization* of the nonlinearity N with respect to the constant reference input k.

Sinusoidal Reference Input

Next, suppose $x_0(\cdot)$ is a sinusoidal function with period $2\pi/\omega$ and amplitude a; i.e., suppose $x_0(\cdot)$ is of the form

23 $$x_0(t) = a \sin \omega t$$

where we take $a \geq 0$ without loss of generality. In (23), we do not include a phase angle [i.e., we do not choose $x_0(t) = a \sin(\omega t + \phi)$] because, as will be evident, ϕ can be taken to be zero.

If N is of the form (19), then $(Nx_0)(\cdot)$ is a periodic function with period $2\pi/\omega$. Hence $(Nx_0)(\cdot)$ can be expanded in a Fourier series, of the form

24 $$(Nx_0)(t) = c_0 + \sum_{i=1}^{\infty} (c_i \cos i\omega t + d_i \sin i\omega t)$$

In general, there is no reason to assume that the *d.c. bias* term c_0 in (24) is zero. However, if the nonlinear element N, in addition to satisfying (19), is also *odd*, i.e.,

25 $$n(-\sigma) = -n(\sigma), \qquad \forall\, \sigma \in R$$

then not only is c_0 zero, but all the c_i's in (24) are zero. In this case, (24) simplifies to

26 $$(Nx_0)(t) = \sum_{i=1}^{\infty} d_i \sin i\omega t$$

Next, when we apply the optimality condition (14), it is easy to show that

27 $$\phi_N(\tau) = \lim_{T \to \infty} \frac{1}{T} \int_{\tau}^{T} a \sin \omega(t - \tau)\, d_1 \sin \omega t\, dt, \qquad \forall\, \tau \geq 0$$

because of the orthogonality property of the family of functions $(\sin i\omega t)_{i=1}^{\infty}$, namely,

28 $$\int_0^{2\pi/\omega} \sin i\omega t \sin j\omega t\, dt = \delta_{ij} \frac{\pi}{\omega}$$

where δ_{ij} is the Knonecker delta defined by

29 $$\delta_{ij} = \begin{cases} 1 & \text{if } i = j \\ 0 & \text{if } i \neq j \end{cases}$$

In other words, $\phi_N(\cdot)$ is the same as the cross-correlation between the functions $a \sin \omega t$ and $d_1 \sin \omega t$. Thus (17) is satisfied if $w_0(\cdot)$ is chosen such that

30 $$(W_0 x_0)(t) = d_1 \sin \omega t$$

In turn, (30) is satisfied if $w_0(\cdot)$ is chosen as

31 $$w_0(t) = \frac{d_1}{a} \delta(t)$$

i.e., if $w_0(\cdot)$ corresponds to a constant gain of value d_1/a.

Thus we have the

Conclusion: In the case of a sinusoidal reference input of the form (23), the optimal quasilinearization is once again a constant gain.

An interpretation of the *equivalent* gain d_1/a is the following: If the nonlinear element N is replaced by a constant gain of value γ, the resulting output of the constant gain element in response to an input $x_0(\cdot)$ is

32 $$\gamma a \sin \omega t$$

Comparing the function (32) with the output (26) of the nonlinear element N, we see that if we choose $\gamma = d_1/a$, then the first harmonic of the output (26) is exactly matched by (32). This is sometimes referred to as the *principle of harmonic balance.*

Next, we shall give an alternative derivation of this principle—one which, incidentally, removes the unnatural assumption (25). As shown previously, if the reference input $x_0(\cdot)$ is chosen according to (23), and if the operator N satisfies (19), then $(Nx_0)(\cdot)$ is of the form (24). Now, if we calculate $(W_0x_0)(\cdot)$, keeping in mind that W_0 represents a linear time-invariant system, then, *in the steady state*, $(W_0x_0)(\cdot)$ is of the form

33 $$(W_0x_0)(t) = \gamma_1 a \sin \omega t + \gamma_2 a \cos \omega t$$

where

34 $$\gamma_1 + j\gamma_2 = \hat{w}_0(j\omega)$$

[In other words, if we calculate the Laplace transform of $w_0(\cdot)$ and denote it by $\hat{w}_0(\cdot)$, then γ_1 is the real part of $\hat{w}_0(j\omega)$ and γ_2 is the imaginary part of $\hat{w}_0(j\omega)$.] Our objective, as before, is to choose $w_0(\cdot)$ in such a way that the function (33) is the best possible approximation, in the mean least-squares sense, to the function (24). However, in view of the orthogonality property (28), we have that (33) is the best possible approximation to (24), provided

35 $$\gamma_1 a = d_1$$

36 $$\gamma_2 a = c_1$$

The conditions can be satisfied if we choose the operator W_0 as

37 $$(W_0x)(t) = \frac{d_1}{a}x(t) + \frac{c_1}{a\omega}\dot{x}(t)$$

Such an operator W_0 is not, strictly speaking, within the class of operators satisfying (5). However, we choose to overlook this fact, and for the purposes of sinusoidal steady-state analysis at the frequency ω, we treat W_0 as a complex gain of value $(d_1/a) + j(c_1/a)$. The equations (35) and (36) once again state that in order to have the best possible linear time-invariant approximation to the nonlinear element N, the first harmonic of the nonlinear element output $(Nx_0)(\cdot)$ should be exactly matched by the linear element output $(W_0x_0)(\cdot)$.

38 *[definition]* Given a nonlinearity N satisfying (19), the complex number

39 $$\eta(a;\omega) = \frac{d_1}{a} + j\frac{c_1}{a}$$

is called the *equivalent gain* of the nonlinearity N; the function $a \mapsto \eta(a;\omega)$ is called the *describing function* of N.

REMARKS: At this point, we have designated η as $\eta(a;\omega)$ because, in general, the describing function may depend on both the amplitude and frequency of the reference input.

40 **fact** If the nonlinear characteristic $n(\cdot)$ is memoryless and time-invariant, then $\eta(a;\omega)$ is independent of ω.

proof Consider two reference inputs

41 $$x_0^{(1)}(t) = a \sin \omega_1 t$$

42 $$x_0^{(2)}(t) = a \sin \omega_2 t$$

both of which have the same amplitude but different frequencies. If both ω_1 and ω_2 are nonzero, we can write

43 $$x_0^{(2)}(t) = x_0^{(1)}\left(\frac{\omega_2 t}{\omega_1}\right)$$

i.e., $x_0^{(2)}(\cdot)$ can be obtained from $x_0^{(1)}(\cdot)$ by *time scaling*. Because N is memoryless, we have

44 $$(Nx_0^{(2)})(t) = (Nx_0^{(1)})\left(\frac{\omega_2 t}{\omega_1}\right)$$

Hence, if

45 $$(Nx_0^{(1)})(t) = c_0 + \sum_{i=1}^{\infty} (c_i \cos i\omega_1 t + d_i \sin i\omega_1 t)$$

then

46 $$(Nx_0^{(2)})(t) = c_0 + \sum_{i=1}^{\infty} (c_i \cos i\omega_2 t + d_i \sin i\omega_2 t)$$

This shows that

47 $$\eta(a;\omega_1) = \eta(a;\omega_2)$$

Because ω_1 and ω_2 are arbitrary, (47) shows that $\eta(a;\omega)$ is independent of ω. ∎

48 **fact** If $n(\cdot)$ is odd in addition to being memoryless and time-invariant, then $\eta(a)$ is a real number.

proof If N satisfies (25) and $x_0(\cdot)$ is of the form (23), then $(Nx_0)(\cdot)$ is of the form (26). This shows that $\eta(a)$ is real. ∎

49 **Example.** Consider the *sign* nonlinearity $\eta_1(\cdot)$ shown in Fig. 4.1. If we apply an input $x_0(\cdot)$ of the form (23) to this nonlinearity, the resulting output is a square wave of amplitude 1, regardless of what a

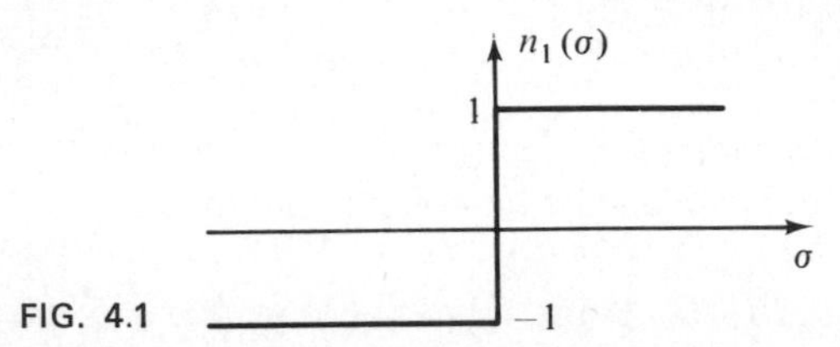

FIG. 4.1

is (as long as $a \neq 0$). The first harmonic of a square wave of amplitude 1 has an amplitude of $4/\pi$, so that the describing function of this nonlinearity $\eta_1(\cdot)$ is given by

50 $$\eta_1(a) = \frac{4}{\pi a}$$

51 **Example.** Consider an element $n(\cdot)$ which is *piecewise-linear*, as shown in Fig. 4.2. For $|\sigma| \leq \delta$, $n(\cdot)$ acts like a linear gain of value m_1, whereas for $|\sigma| > \delta$, $n(\cdot)$ acts (for small perturbations in σ) like a linear

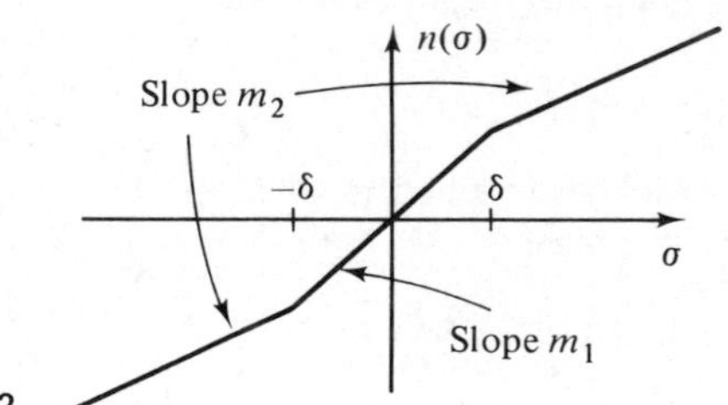

FIG. 4.2

gain of value m_2. Therefore, if we apply a sinusoidal input of amplitude a, and if $a \leq \delta$, then the output is another sinusoid of amplitude $m_1 a$, so that

52 $$\eta(a) = m_1 \quad \text{if } 0 \leq a \leq \delta$$

However, if $a > \delta$, the output of the nonlinearity is of the form shown in Fig. 4.3. In this case, it can be verified through laborious but straightforward calculations that

53 $$\eta(a) = \frac{2(m_1 - m_2)}{\pi}\left[\sin^{-1}\frac{\delta}{a} + \frac{\delta}{a}\left(1 - \frac{\delta^2}{a^2}\right)^{1/2}\right] + m_2, \qquad a > \delta$$

Figure 4.2 depicts the nonlinearity $n(\cdot)$ in the case where $m_1 > m_2$.

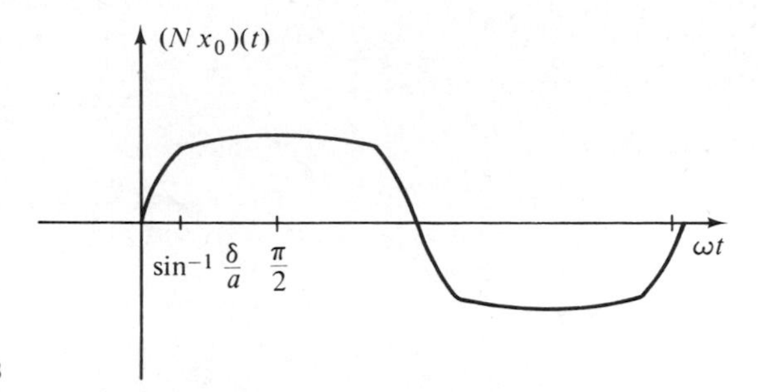

FIG. 4.3

However, expression (53) for $\eta(a)$ is valid for any m_1, m_2. (See also Problem 4.5.) Thus (52) and (53) completely characterize the describing function $\eta(\cdot)$.

By selecting various specific values for m_1 and m_2, we can now derive the expression for the describing functions of several commonly encountered forms of nonlinearities. For example, if we let $m_1 = 0$, we

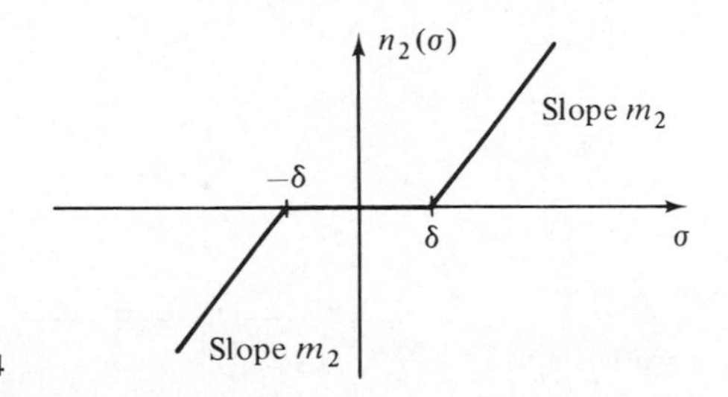

FIG. 4.4

get the dead-zone nonlinearity $n_2(\cdot)$ shown in Fig. 4.4. The corresponding describing function $\eta_2(\cdot)$ is obtained from (52) and (53) as

54
$$\eta_2(a) = \begin{cases} 0, & 0 \le a \le \delta \\ m_2 - \dfrac{2m_2}{\pi}\left[\sin^{-1}\dfrac{\delta}{a} + \dfrac{\delta}{a}\left(1 - \dfrac{\delta^2}{a^2}\right)^{1/2}\right], & a > \delta \end{cases}$$

Similarly, if we let $m_2 = 0$, we get the limiter nonlinearity $n_3(\cdot)$ shown in Fig. 4.5. The describing function $\eta_3(\cdot)$ of $n_3(\cdot)$ is

55
$$\eta_3(a) = \begin{cases} m_1, & 0 \le a \le \delta \\ \dfrac{2m_1}{\pi}\left[\sin^{-1}\dfrac{\delta}{a} + \dfrac{\delta}{a}\left(1 - \dfrac{\delta^2}{a^2}\right)^{1/2}\right], & a > \delta \end{cases}$$

There is one function that occurs repeatedly in the above expressions, namely,

56
$$f(x) \triangleq \begin{cases} 1, & \text{for } x \ge 1 \\ \dfrac{2}{\pi}[\sin^{-1} x + x(1 - x^2)^{1/2}] & \text{for } 0 \le x \le 1 \end{cases}$$

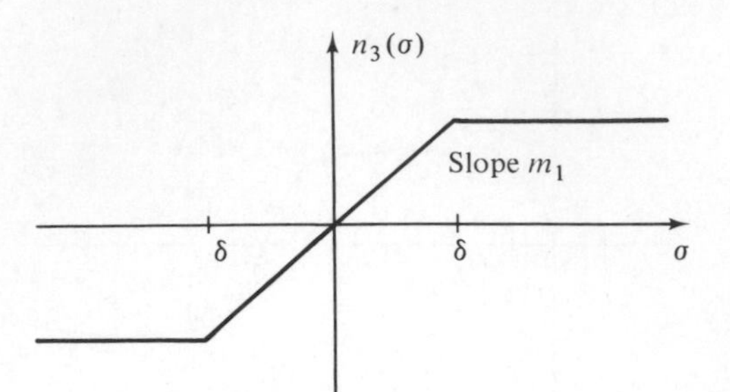

FIG. 4.5

A sketch of $f(x)$ is found in Fig. 4.6. With the above definition, $\eta_2(a)$ can be expressed compactly as

$$\eta_2(a) = m_2\left[1 - f\left(\frac{\delta}{a}\right)\right] \tag{57}$$

while $\eta_3(a)$ can be written as

$$\eta_3(a) = m_1 f\left(\frac{\delta}{a}\right) \tag{58}$$

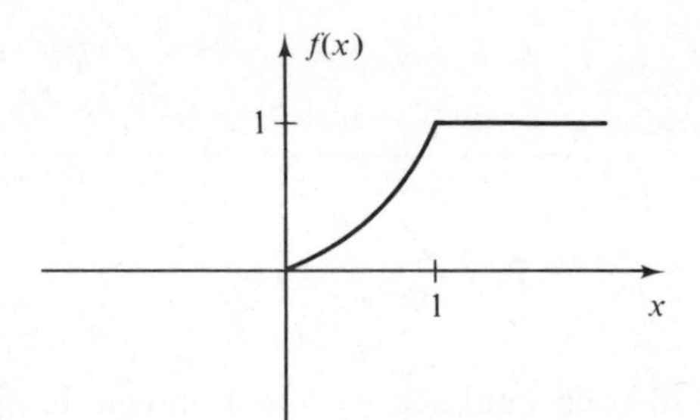

FIG. 4.6

While (54) and (57) give the precise expressions for $\eta_2(a)$, one can readily derive the approximate shape of $\eta_2(a)$ just by using common sense. If the input amplitude a is less than δ, the output is zero, so that $\eta_2(a)$ is also zero. On the other hand, as a becomes larger, the effect of the dead zone becomes smaller, so that $\eta_2(a)$ is a monotonically increasing function of a. As a becomes extremely large compared with δ, the effect of the dead zone all but disappears, and $n_2(\cdot)$ begins to look like a linear element of gain m_2. Hence $\eta_2(a) \longrightarrow m_2$ as $a \longrightarrow \infty$. The characteristic shape of $\eta_3(a)$ can be rationalized in the same way.

We shall close this subsection by deriving some bounds on the describing function. Suppose the function $n(\cdot)$ in (19) satisfies a condition of the form

$$k_1\sigma^2 \leq \sigma n(\sigma) \leq k_2\sigma^2, \qquad \forall\, \sigma \in R \tag{59}$$

for some real numbers k_1 and k_2. (We say that $n(\cdot)$ *lies in the sector*

$[k_1, k_2]$.) If, in addition, $n(\cdot)$ also satisfies (25) [i.e., $n(\cdot)$ is odd], we show below that

60 $$k_1 \le \eta(a) \le k_2, \qquad \forall\, a \in R$$

In other words, if $n(\cdot)$ lies in the sector $[k_1, k_2]$, its describing function also lies between the limits k_1 and k_2.

This sector condition (59) has a simple graphical interpretation, as shown in Fig. 4.7. It means that the graph of the nonlinearity lies between two straight lines, having slopes k_1 and k_2, respectively. Thus the above result states that if the graph of a nonlinear element lies between two straight lines of slopes k_1 and k_2, its describing function takes values only between k_1 and k_2.

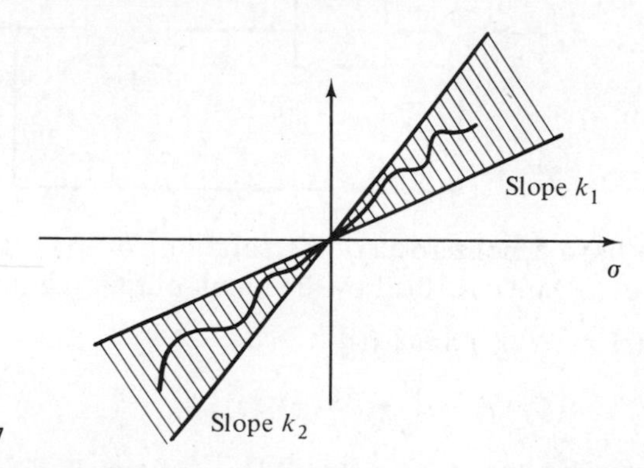

FIG. 4.7

The bound (60) is rather easy to prove. Because $n(\cdot)$ is odd, $\eta(a)$ is real. Moreover,

$$\begin{aligned}\eta(a) &= \frac{\omega}{\pi a}\int_0^{2\pi/\omega} n(a \sin \omega t) \sin \omega t \, dt \\ &= \frac{1}{\pi a}\int_0^{2\pi} n(a \sin \theta) \sin \theta \, d\theta \quad (\text{letting } \theta = \omega t) \\ &\ge \frac{1}{\pi a^2}\int_0^{2\pi} k_1 (a \sin \theta)^2 \, d\theta \quad [\text{by (59)}] \\ &= k_1\end{aligned}$$

The upper bound in (60) is proved similarly.

For a generalization of this result, see Problem 4.6.

4.1.3 Existence of Periodic Solutions

In this section, we shall present one of the main applications of describing functions, namely, predicting the existence of limit cycles (periodic solutions). It should be borne in mind that the analysis based on describing functions is only approximate. For example, there are instances where the describing function analysis predicts the existence of periodic solutions, but the actual system does not exhibit any, and other instances where the situation is reversed. Hence it is perhaps more accurate to say that describing function analysis predicts the likelihood of limit cycles. However, in spite of these shortcomings, describing functions are widely used in practice, because of the ease of analyzing nonlinear systems (albeit approximately) by using them.

Consider the nonlinear feedback system shown in Fig. 4.8. The objective is to determine whether, in the absence of an input, it is pos-

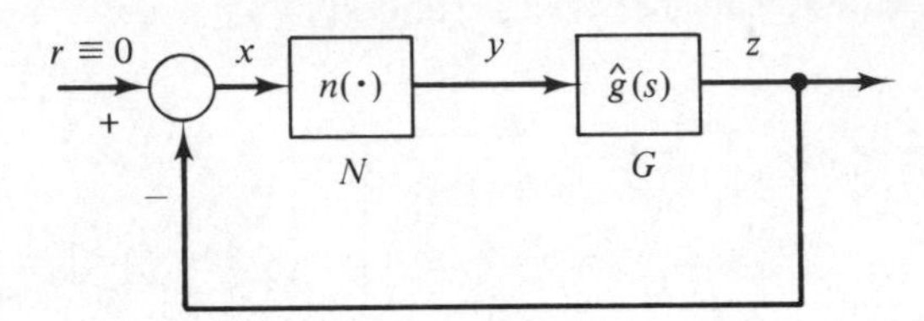

FIG. 4.8

sible to have a nonzero periodic solution for $x(\cdot)$. If the two blocks "N" and "G" are described by the input–output relationships

61 $$y(t) = (Nx)(t) = n[x(t)]$$

62 $$z(t) = (Gy)(t) = \int_0^t g(t - \tau)y(\tau)\,d\tau$$

then, with zero input, the closed-loop system is described by the integral equation

63 $$x(t) = -(GNx)(t) = -\int_0^t g(t - \tau)n[x(\tau)]\,d\tau$$

The problem is to determine whether (63) has any nonzero solutions for $x(\cdot)$.

We approach the problem by *assuming* that (63) has a periodic solution of the form

64 $$x(t) = a \sin \omega t$$

where a and ω are to be determined. As pointed out in Sec. 4.1.2, whenever $x(\cdot)$ is of the form (64), $(Nx)(\cdot)$ is also periodic with period $2\pi/\omega$, and the fundamental harmonic of $(Nx)(\cdot)$ is

65 $$\gamma_1 a \sin \omega t + \gamma_2 a \cos \omega t$$

where

66 $$\eta(a; \omega) = \gamma_1 + j\gamma_2$$

is the describing function of $n(\cdot)$. Next, let

67 $$\hat{g}(j\omega) = \hat{g}_r(\omega) + j\hat{g}_i(\omega)$$

be the Fourier transform of the impulse response $g(\cdot)$. Because (62) describes a linear time-invariant operation, and because $y(\cdot)$ is periodic with period $2\pi/\omega$, it follows that in the steady state $z(\cdot)$ is also periodic with period $2\pi/\omega$ and that the fundamental harmonic of $z(\cdot)$ is

68 $$ua \sin \omega t + va \cos \omega t$$

where

69 $$u + jv \triangleq g(j\omega)\eta(a; \omega)$$

Next, for (63) to hold, the first harmonics of both sides of (3) must be equal. Now, if $x(\cdot)$ is of the form (64), equating the first harmonics of both sides of (63) gives

70 $$a \sin \omega t = -(ua \sin \omega t + va \cos \omega t)$$

or, equivalently,

71 $$1 + u + jv = 0$$

This procedure is also referred to by some authors as the principle of harmonic balance. Note that it is a very approximate procedure because even if (63) has a periodic solution, it is highly unlikely that this periodic solution is a pure sinusoid of the form (64). However, the rationale is that (64) is a good approximation to this periodic solution. (These points are discussed in greater detail later on.)

The essence of the *method* for predicting the existence of periodic solutions using describing functions can now be stated: Given a system description of the form (63), formulate the harmonic balance equation (71), and solve for a and ω (note that u and v are functions of a and ω). Now, if a_0 and ω_0 satisfy (71), i.e., if

72 $$1 + \hat{g}(j\omega_0)\eta(a_0; \omega_0) = 0$$

then we consider it likely that (63) has a periodic solution with an amplitude of approximately a_0 and a frequency of approximately ω_0.

The heuristic *justification* given for this technique is that if (72) holds, then $x_0(\cdot)$, given by

73 $$x_0(t) = a_0 \sin \omega_0 t$$

is an exact solution of the equation

74 $$x(t) = -(PGNx)(t)$$

where P is the operator that associates, with each periodic function, its first harmonic. Of course, (74) is not the same equation as (63), because there is no operator P in (63). However, if the operator G is a *low-pass filter*, i.e,. if G rapidly attenuates the higher harmonics of $(Nx)(\cdot)$, then (so the reasoning goes) there is not much error in replacing (63) by (74), and of course $x_0(\cdot)$ solves (74).

In practice, the describing function technique for predicting periodic solutions works well if (as stated above) $|g(j\omega)|$ decreases rapidly as $\omega \longrightarrow \infty$. It is possible to place the describing function technique on a firm mathematical foundation (see [5] Bergen and Franks); however, this involves topics in topology that are well beyond the scope of this book.

Special Case

In the case of odd nonlinearities $n(\cdot)$ that satisfy

75 $$\sigma n(\sigma) \geq 0, \qquad \forall\, \sigma \in R$$

i.e., in the case of the so-called *first and third quadrant nonlinearities*, the harmonic balance equation (72) is particularly easy to solve. In this case, the describing function η is real nonnegative and is independent of ω. Thus (72) reduces to

76 $$1 + \hat{g}_r(\omega)\eta(a) + j\hat{g}_i(\omega) = 0$$

Separating (76) into real and imaginary parts gives

77 $$1 + \hat{g}_r(\omega)\eta(a) = 0$$

78 $$\hat{g}_i(\omega) = 0$$

Now, (78) shows that if (a_0, ω_0) solves (76), then $\hat{g}(j\omega_0)$ must be real. Further, because $\eta(a) \geq 0\ \forall\ a$, (77) shows that $\hat{g}(j\omega_0)$ must in fact be real and negative. Thus we first make a plot of $\operatorname{Re}\hat{g}(j\omega)$ vs. $\operatorname{Im}\hat{g}(j\omega)$ and determine the frequencies at which $\hat{g}(j\omega)$ is real and negative. Suppose there are m such frequencies, $\omega_1, \ldots, \omega_m$. At each such frequency ω_i, (77) can be solved to give

79 $$\eta(a) = \frac{-1}{\hat{g}(j\omega_i)}$$

Let $a_1^{(i)}, \ldots, a_{k_i}^{(i)}$ be the solutions of (79), if any. Then we predict that the system (63) has periodic solutions at frequencies $\omega_1, \ldots, \omega_m$, with amplitudes $a_1^{(i)}, \ldots, a_{k_i}^{(i)}$ corresponding to the frequency ω_i.

Two things are to be noted about the above procedure. (1) The possible frequencies of oscillation $\omega_1, \ldots, \omega_m$ do not depend on the nonlinearity $n(\cdot)$ in any way; rather, they are determined solely by the transfer function $\hat{g}(\cdot)$. (2) Once the possible frequencies of oscillation $\omega_1, \ldots, \omega_m$ are determined, the possible amplitudes corresponding to each frequency are easily found from (79). Thus, in the case of odd nonlinearities, the harmonic balance equation (72) is effectively *decoupled.*

80 **Example.** Consider the system of Fig. 4.8, with

$$\hat{g}(s) = \frac{1}{s(s+1)(s+2)}$$

$$n(\xi) = \xi^3$$

Then it is easy to verify that

$$\eta(a) = \frac{3a^2}{4}$$

and $\eta(a)$ is independent of ω. Thus the harmonic balance equation (11) becomes, in this case,

$$1 + \frac{1}{j\omega(1 + j\omega)(2 + j\omega)} \frac{3a^2}{4} = 0 \tag{81}$$

If we separate (81) into its real and imaginary parts, we get

$$1 + \hat{g}_r(\omega)\eta(a) = 0$$

$$\hat{g}_i(\omega) = 0 \tag{82}$$

First, (82) can be solved for ω in an entirely straightforward manner, which gives

$$\omega = \sqrt{2}$$

Note that we ignore the solution $\omega = -\sqrt{2}$ because, in physical terms, it is no different from the solution $\omega = \sqrt{2}$. Now, substituting $\omega = \sqrt{2}$ into (80) gives [because $g_r(\sqrt{2}) = -1/6$]

$$1 + \frac{-1}{6} \frac{3a^2}{4} = 0 \tag{83}$$

i.e., $a = \sqrt{8}$. Hence, for this system, we expect to find a periodic solution with an amplitude of approximately $\sqrt{8}$ and an angular velocity of approximately $\sqrt{2}$ radians per second.

Problem 4.1. Given two operators N_1 and N_2 of the type studied in Sec. 4.1.1, we define their sum $N_1 + N_2$ to be the operator defined by

$$[(N_1 + N_2)x](t) = (N_1x)(t) + (N_2x)(t) \tag{84}$$

Let $x_0(\cdot)$ be a reference input. Show that if $w_1(\cdot)$ is an optimal quasilinearization of N_1 with respect to the reference input $x_0(\cdot)$ and $w_2(\cdot)$ is an optimal quasilinearization of N_2 with respect to $x_0(\cdot)$, then $w_1(\cdot) + w_2(\cdot)$ is an optimal quasilinearization of $N_1 + N_2$ with respect to $x_0(\cdot)$.

Problem 4.2. Specialize the result of Problem 4.1 to describing functions: Let $n_1, n_2: R \longrightarrow R$ be two given functions, and define their *sum* $(n_1 + n_2): R \longrightarrow R$ by

$$(n_1 + n_2)(\sigma) = n_1(\sigma) + n_2(\sigma) \tag{85}$$

Show that the describing function of $n_1 + n_2$ is the sum of the describing functions of n_1 and of n_2.

Problem 4.3. Verify that the nonlinearity $n(\cdot)$ in Fig. 4.2 is the sum of the nonlinearities $n_2(\cdot)$ and $n_3(\cdot)$ (in Fig. 4.4 and 4.5, respectively). Verify that the describing function $\eta(\cdot)$ is the sum of $\eta_2(\cdot)$ and $\eta_3(\cdot)$.

Problem 4.4. Using the results of Problems 4.2 and 4.3, derive an expression for the describing function of the dead-zone limiter shown in Fig. 4.9. (*Answer:* $\eta_4(a) = m[f(\delta_2/a) - f(\delta_1/a)]$.)

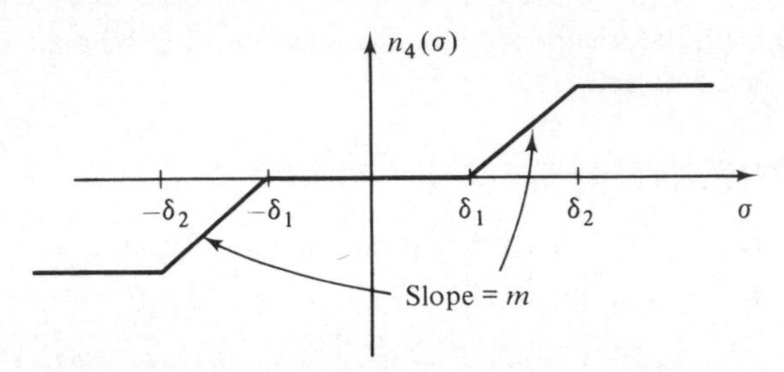

FIG. 4.9

Problem 4.5. Using the results of Problems 4.2 and 4.3, derive an expression for the describing function of the piecewise-linear element shown in Fig. 4.10. [*Answer:* $\eta_5(a) = (m_1 - m_2)f(\delta_1/a) + (m_2 - m_3)f(\delta_2/a) + m_3$.] Generalize the answers to a piecewise-linear element with l different slopes.

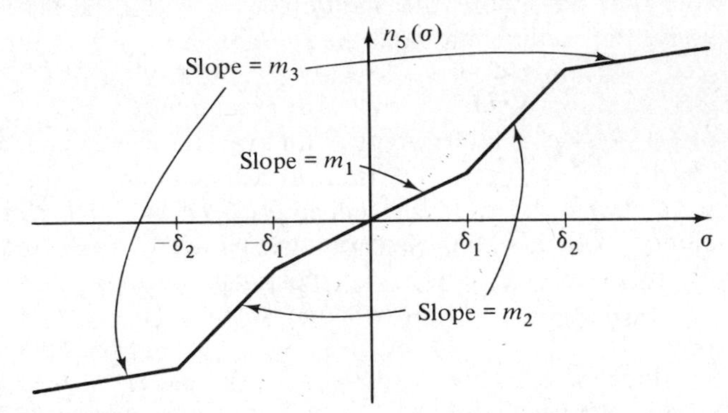

FIG. 4.10

Problem 4.6. Let $n(\cdot)$, $n_l(\cdot)$, and $n_u(\cdot)$ be odd functions such that

86 $$\sigma n_l(\sigma) \leq \sigma n(\sigma) \leq \sigma n_u(\sigma), \qquad \forall\, \sigma \in R$$

and let $\eta(\cdot)$, $\eta_l(\cdot)$, $\eta_u(\cdot)$, denote the describing functions of $n(\cdot)$, $n_l(\cdot)$, and $n_u(\cdot)$, respectively. Show that

87 $$\eta_l(a) \leq \eta(a) \leq \eta_u(a), \qquad \forall\, a \geq 0$$

Problem 4.7. Consider the system (63) with

88 $$\hat{g}(s) = \frac{5(s+1)}{s^2(s+2)^2}$$

Analyze the possible existence of periodic solutions in the cases where

(a) $n(\cdot)$ is the sign nonlinearity of Example (49),

(b) $n(\cdot)$ is the dead-zone nonlinearity of Example (51), with $\delta = 1$ and $m_2 = 2$, and

(c) $n(\cdot)$ is the five-segment nonlinearity of Fig. 4.10, with $m_1 = 2$, $m_2 = 1$, $m_3 = 0$, $\delta_1 = 1$, and $\delta_2 = 2$.

4.2 NUMERICAL SOLUTION TECHNIQUES

In this section, we shall discuss some methods for approximately solving a given set of nonlinear vector differential equations. The techniques presented here are particularly well suited for implementation on a digital computer. Due to limitations of space, we shall discuss here only the rationale of the various methods and omit the more detailed numerical analytic aspects, e.g., an analysis of the errors created by the approximations. For an in-depth discussion of these points, see [X] Chua and Lin. [8]

Throughout this section, we are concerned with finding (approximately) the solution to the nonlinear vector differential equation

1 $$\dot{\mathbf{x}}(t) = \mathbf{f}[t, \mathbf{x}(t)], \quad t \geq 0; \qquad \mathbf{x}(0) = \mathbf{x}_0^* \quad \text{(given)}$$

We assume that the function $f(t, \cdot)$ is globally Lipschitz-continuous, which assures (see Theorem [3.4(25)]) that (1) has a unique solution over $[0, \infty)$. Let $\mathbf{x}^*(t)$ denote the *actual* solution of (1) and let $(t_n)_1^\infty$ be a monotonically increasing sequence of real numbers with t_n approaching ∞ as $n \longrightarrow \infty$. Our objective is to find various expressions that approximate $\mathbf{x}^*(t_n)$, $n = 1, 2, \ldots$. That is, instead of attempting to approximate the function $\mathbf{x}^*(t)$, which is defined for all values of $t \geq 0$, we attempt to approximate the *sequence of vectors* $[\mathbf{x}^*(t_n)]_1^\infty$. This is in keeping with the modern trend toward the increased use of digital computers. In many instances, the numbers t_n are *equally spaced*, i.e.,

2 $$t_n = nh$$

for some number $h > 0$, which is referred to as the *step size*.

4.2.1 Taylor Series Methods

As the name implies, the Taylor series methods are based on expanding $\mathbf{x}(\cdot)$ in a Taylor series about some point. Suppose we know $\mathbf{x}^*(t_n)$ for some n; i.e., we know the *exact* solution at time t_n. Let us denote $\mathbf{x}^*(t_n)$ by $\mathbf{x}_n^*$, in the interests of brevity. Knowing $\mathbf{x}_n^*$, we can compute $\mathbf{x}_{n+1}^*$ by using a Taylor series, as follows:

3 $$\mathbf{x}_{n+1}^* = \mathbf{x}_n^* + h\dot{\mathbf{x}}^*(t_n) + \frac{h^2}{2}\ddot{\mathbf{x}}^*(t_n) + \ldots$$

The various derivative terms in (3) can now be calculated using the differential equation (1). For instance,

4 $$\dot{\mathbf{x}}^*(t_n) = \mathbf{f}(t_n, \mathbf{x}_n^*)$$

$$\ddot{\mathbf{x}}^*(t_n) = \frac{d}{dt}\{\mathbf{f}[t, \mathbf{x}(t)]\}_{t=t_n,\, \mathbf{x}=\mathbf{x}_n^*} \tag{5}$$
$$= \mathbf{f}_\mathbf{x}(t_n, \mathbf{x}_n^*) \cdot \mathbf{f}(t_n, \mathbf{x}_n^*) + \mathbf{f}_t(t_n, \mathbf{x}_n^*)$$

(where the subscripts denote partial differentiation), and so on.

Let us now suppose that, instead of having at hand the *actual* solution $\mathbf{x}_n^*$, we have available an *approximation* to $\mathbf{x}_n^*$, which we denote by $\mathbf{x}_n$. With the aid of this information, we can construct $\mathbf{x}_{n+1}$, which forms an approximation to $\mathbf{x}_{n+1}^*$, by a formula parallel to (3), namely,

$$\mathbf{x}_{n+1} = \mathbf{x}_n + h\mathbf{f}(t_n, \mathbf{x}_n) + \frac{h^2}{2}[\mathbf{f}_\mathbf{x}(t_n, \mathbf{x}_n) \cdot \mathbf{f}(t_n, \mathbf{x}_n) + \mathbf{f}_t(t_n, \mathbf{x}_n)] + \ldots \tag{6}$$

We can now make a further approximation by *truncating* the infinite series in (6) after a finite number of terms. The number of terms in (6) that are retained determine the *order* of the Taylor series method—if the series in (6) is truncated after the term involving h^k, where k is some integer, then the Taylor series method is said to be of kth order, or of order k. For instance, the Taylor series method of first order (also known as the forward Euler method) is given by the approximating formula

$$\mathbf{x}_{n+1} = \mathbf{x}_n + h\mathbf{f}(t_n, \mathbf{x}_n) \tag{7}$$

while the second-order Taylor series method is described by

$$\mathbf{x}_{n+1} = \mathbf{x}_n + h\mathbf{f}(t_n, \mathbf{x}_n) + \frac{h^2}{2}[\mathbf{f}_\mathbf{x}(t_n, \mathbf{x}_n) \cdot \mathbf{f}(t_n, \mathbf{x}_n) + \mathbf{f}_t(t_n, \mathbf{x}_n)] \tag{8}$$

Note that to initiate the Taylor series method(s), we use the initial condition in (1) and set $\mathbf{x}_0 = \mathbf{x}_0^*$.

While the Taylor series methods are a useful starting point for understanding more sophisticated methods, they are not of much computational use. The first-order method (8) is too inaccurate, as it corresponds to simple linear extrapolation, while higher-order methods require the calculation of a lot of partial derivatives. Also, if we are dealing with vector differential equations and we use anything higher than a second-order method, then some of the partial derivatives involved (e.g., $f_{\mathbf{xx}}$) leave the realm of matrices and become higher-order tensors. Furthermore, the computation of all these derivatives becomes unwieldy.

4.2.2 Runge–Kutta Methods

The aforementioned problems are alleviated by the Runge–Kutta methods, which duplicate the accuracy of the Taylor series methods but do not require the calculation of partial derivatives. To bring out the main idea of the Runge–Kutta methods, we begin by studying the second-order Runge–Kutta method, which employs the approximation formula

$$\mathbf{x}_{n+1} = \mathbf{x}_n + h\{a_1\mathbf{f}(t_n, \mathbf{x}_n) + a_2\mathbf{f}[t_n + \alpha h, \mathbf{x}_n + \beta\mathbf{f}(t_n, \mathbf{x}_n)]\} \tag{9}$$

In (9), the real numbers a_1, a_2, α, and β are to be chosen in such a way that the right-hand side of (9) approximates the right-hand side of (8) with an error of $\mathbf{o}(h^2)$. If a_1, a_2, α, and β can be so chosen, then because the right-hand side of (8) approximates the right-hand side of (6) with an error of $\mathbf{o}(h^2)$, we have that formula (9) accurately reproduces the first three terms of the Taylor series (6). This does *not* mean that formulas (8) and (9) produce identical results. Rather, what it means is that by employing the formula (9) instead of (6) we get an error of $\mathbf{o}(h^2)$, just as if we had employed (8) instead of (6).

We turn now to the method of selecting a_1, a_2, α, and β. Expanding $\mathbf{f}[t_n + \alpha h, \mathbf{x}_n + \beta\mathbf{f}(t_n, \mathbf{x}_n)]$ in a Taylor series about $(t_n, \mathbf{x}_n)$, we get

$$\begin{aligned}\mathbf{f}[t_n + \alpha h, \mathbf{x}_n + \beta\mathbf{f}(t_n, \mathbf{x}_n)] &= \mathbf{f}(t_n, \mathbf{x}_n) + \alpha h\mathbf{f}_t(t_n, \mathbf{x}_n) \\ &\quad + \beta\mathbf{f}_{\mathbf{x}}(t_n, \mathbf{x}_n)\cdot\mathbf{f}(t_n, \mathbf{x}_n) + \mathbf{o}(h^2)\end{aligned} \tag{10}$$

Hence

$$\begin{aligned}&h\{a_1\mathbf{f}(t_n, \mathbf{x}_n) + a_2\mathbf{f}(t_n + \alpha h, \mathbf{x}_n + \beta\mathbf{f}[t_n(t_n, \mathbf{x}_n)]]\} \\ &= (a_1 + a_2)h\mathbf{f}(t_n, \mathbf{x}_n) + a_2\alpha h^2\mathbf{f}_t(t_n, \mathbf{x}_n) \\ &\quad + a_2\beta h\mathbf{f}_{\mathbf{x}}(t_n, \mathbf{x}_n)\cdot\mathbf{f}(t_n, \mathbf{x}_n) + \mathbf{o}(h^2)\end{aligned} \tag{11}$$

Comparing the right-hand sides of (8) and (11), we see that they differ only by $\mathbf{o}(h^2)$, provided

$$a_1 + a_2 = 1 \tag{12}$$

$$a_2\alpha = \frac{1}{2} \tag{13}$$

$$a_2\beta = \frac{h}{2} \tag{14}$$

Because (12)–(14) represent three equations in four unknowns, we can assign one number arbitrarily. Thus (12)–(14) can be rearranged as

$$a_1 = 1 - a_2 \tag{15}$$

$$\alpha = \frac{1}{2a_2} \tag{16}$$

$$\beta - \frac{h}{2a_2} \tag{17}$$

provided $a_2 \neq 0$. Therefore the general second-order Runge–Kutta algorithm is described by

$$\mathbf{x}_{n+1} = \mathbf{x}_n + h\left\{(1 - a_2)\mathbf{f}(t_n, \mathbf{x}_n) + a_2\mathbf{f}\left[t_n + \frac{h}{2a_2}, \mathbf{x}_n + \frac{h\mathbf{f}(t_n, \mathbf{x}_n)}{2a_2}\right]\right\} \tag{18}$$

As a special case, by setting $a_2 = \frac{1}{2}$, we get the so-called *Heun's* or *modified trapezoidal* algorithm, which is described by

19 $$\mathbf{x}_{n+1} = \mathbf{x}_n + h\left\{\frac{\mathbf{f}(t_n, \mathbf{x}_n)}{2} + \frac{1}{2}\mathbf{f}[t_n + h, \mathbf{x}_n + h\mathbf{f}(t_n, \mathbf{x}_n)]\right\}$$

Similarly, by setting $a_2 = 1$, we get the *modified Euler–Cauchy* algorithm, which is described by

20 $$\mathbf{x}_{n+1} = \mathbf{x}_n + h\mathbf{f}\left[t_n + \frac{h}{2}, \mathbf{x}_n + \frac{h\mathbf{f}(t_n, \mathbf{x}_n)}{2}\right]$$

It should be once again emphasized that formulas (19) and (20) do not give identical results. However, if we start from the same $\mathbf{x}_n$, the $\mathbf{x}_{n+1}$ given by (19) and that given by (20) would differ only by $\mathbf{o}(h^2)$.

The most commonly used of the Runge–Kutta algorithms is one of order 4, which is described by[3]

21 $$\mathbf{x}_{n+1} = \mathbf{x}_n + h\mathbf{g}_4(t_n, \mathbf{x}_n, h)$$

where

22 $$\mathbf{g}_4(t_n, \mathbf{x}_n, h) = \frac{\mathbf{g}_0 + 2\mathbf{g}_1 + 2\mathbf{g}_2 + \mathbf{g}_3}{6}$$

23 $$\mathbf{g}_0 = \mathbf{f}(t_n, \mathbf{x}_n)$$

24 $$\mathbf{g}_1 = \mathbf{f}\left(t_n + \frac{h}{2}, \mathbf{x}_n + \frac{h\mathbf{g}_0}{2}\right)$$

25 $$\mathbf{g}_2 = \mathbf{f}\left(t_n + \frac{h}{2}, \mathbf{x}_n + \frac{h\mathbf{g}_1}{2}\right)$$

26 $$\mathbf{g}_3 = \mathbf{f}(t_n + h, \mathbf{x}_n + h\mathbf{g}_2)$$

Note that $\mathbf{g}_0, \ldots, \mathbf{g}_3$ are computed sequentially, beginning with $\mathbf{g}_0$. Because the above algorithm is of order 4, it faithfully reproduces the first five terms (including the constant term) in the Taylor series expansion for $\mathbf{x}_{n+1}$. However, it is computationally not very efficient, because for each value of n, we need to calculate four vectors $\mathbf{g}_0, \ldots, \mathbf{g}_3$, which are never used again. In contrast, the multistep algorithms, to be discussed in the next subsection, do make use of previously computed vectors.

27 **Example.** Consider the linear matrix differential equation

28 $$\dot{\mathbf{M}}(t) = \mathbf{A}\mathbf{M}(t); \qquad \mathbf{M}(0) = \mathbf{I}$$

where $\mathbf{A} \in \mathbf{R}^{n \times n}$ is a given matrix. It is well known (see Appendix II)

[3]This is only the most common of an infinite family of fourth-order Runge–Kutta algorithms.

that the solution of (28) is

$$\mathbf{M}(t) = e^{\mathbf{A}t} = \sum_{i=0}^{\infty} \mathbf{A}^i \frac{t^i}{i!} \tag{29}$$

We shall now show that when applied to the differential equation (28) the fourth-order Runge–Kutta algorithm exactly reproduces the first five terms of the Taylor series (29).

From (29), we have

$$\mathbf{M}(h) = \sum_{i=0}^{\infty} \mathbf{A}^i \frac{h^i}{i!} \tag{30}$$

Applying (21)–(26) with $\mathbf{f}(t, \mathbf{x}) = \mathbf{A}\mathbf{x}$, $t_N = 0$, and $\mathbf{x}_N = \mathbf{I}$ gives

$$\mathbf{G}_0 = \mathbf{A} \cdot \mathbf{I} = \mathbf{A} \tag{31}$$

$$\mathbf{G}_1 = \mathbf{A}\left(\mathbf{I} + \mathbf{A}\frac{h}{2}\right) = \mathbf{A} + \mathbf{A}^2\frac{h}{2} \tag{32}$$

$$\mathbf{G}_2 = \mathbf{A}\left[\mathbf{I} + \left(\mathbf{A} + \mathbf{A}^2\frac{h}{2}\right) \cdot \frac{h}{2}\right] = \mathbf{A} + \mathbf{A}^2\frac{h}{2} + \mathbf{A}^3\frac{h^2}{4} \tag{33}$$

$$\begin{aligned} \mathbf{G}_3 &= \mathbf{A}\left[\mathbf{I} + \left(\mathbf{A} + \mathbf{A}^2\frac{h}{2} + \mathbf{A}^3\frac{h^2}{4}\right)h\right] \\ &= \mathbf{A} + \mathbf{A}^2 h + \mathbf{A}^3\frac{h^2}{2} + \mathbf{A}^4\frac{h^3}{4} \end{aligned} \tag{34}$$

$$\mathbf{G}_4 = \frac{\mathbf{G}_0 + 2\mathbf{G}_1 + 2\mathbf{G}_2 + \mathbf{G}_3}{6} = \mathbf{A} + \mathbf{A}^2\frac{h}{2} + \mathbf{A}^3\frac{h^2}{6} + \mathbf{A}^4\frac{h^3}{24} \tag{35}$$

$$\mathbf{X}_{N+1} = \mathbf{I} + \mathbf{A}h + \mathbf{A}^2\frac{h^2}{2} + \mathbf{A}^3\frac{h^3}{6} + \mathbf{A}^4\frac{h^4}{24} \tag{36}$$

Thus, clearly, $\mathbf{X}_{N+1}$ contains the first five terms in the Taylor series (30).

Incidentally, this example illustrates that the Runge–Kutta algorithm is applicable not only to vector differential equations but also to matrix differential equations.

In the practical implementation of the Runge–Kutta algorithms, the following procedure sometimes works well. Given t_n and $\mathbf{x}_n = \mathbf{x}(t_n)$,

1. Pick a step size h_1, and let $t_{n+1}^{(1)} = t_n + h_1$.
2. Compute $\mathbf{x}_{n+1}^{(1)}$ using a Runge–Kutta algorithm of the desired order.
3. Halve the step size h_1 by selecting $h_2 = h_1/2$, $t_{n+1}^{(2)} = t_n + h_2$, $t_{n+2}^{(2)} = t_n + 2h_2$ $(= t_{n+1}^{(1)})$. Compute $\mathbf{x}_{n+1}^{(2)}$ and $\mathbf{x}_{n+2}^{(2)}$ using the same Runge–Kutta algorithm as in step 2.
4. Compare $\mathbf{x}_{n+2}^{(2)}$ with $\mathbf{x}_{n+1}^{(1)}$. If their difference is less than some preselected error bound, accept the value $\mathbf{x}_{n+1}^{(1)}$ and proceed to step 1; otherwise, repeat step 3.

4.2.3 Multistep Algorithms

The basic idea behind multistep algorithms is to approximate the solution of a given differential equation (1) by a polynomial in t. This is justified on the basis of the Weierstrass approximation theorem, which states that any continuous function over a finite interval can be uniformly approximated to any desired degree of accuracy by a polynomial of appropriate degree.

Specifically, given the differential equation (1) and the *uniformly spaced* time intervals t_n given by (2), we attempt to approximate $\mathbf{x}_{n+1}$ by an expression of the type

37 $$\mathbf{x}_{n+1} = \sum_{i=0}^{k} a_i \mathbf{x}_{n-i} + \sum_{i=-1}^{k} b_i h \mathbf{f}(t_{n-i}, \mathbf{x}_{n-i})$$

where k is an integer and a_i, b_i are real numbers to be selected later. Note that if $b_{-1} \neq 0$, then (37) gives only an *implicit* formula for the unknown quantity $\mathbf{x}_{n+1}$, because it appears on both sides of the equation. On the other hand, if $b_{-1} = 0$, then (37) gives an explicit formula for $\mathbf{x}_{n+1}$. An algorithm of the type (37) is called a $(k + 1)$-*step* algorithm, because the value of $\mathbf{x}$ at the $(n + 1)$st step is given in terms of the values of $\mathbf{x}$ at the previous $k + 1$ steps (namely, $\mathbf{x}_{n-k}, \ldots, \mathbf{x}_n$). In a sense, one can think of formula (37) as arising from discretizing the expression

38 $$\mathbf{x}(t) = \mathbf{x}_0^* + \int_0^t \mathbf{f}[\tau, \mathbf{x}(\tau)]\, d\tau$$

The *order* of algorithm (37) is the degree of the highest-degree polynomial for which (37) gives an exact expression for $\mathbf{x}_{n+1}$. To see how this definition of order is used, we shall consider what is meant by a second-order algorithm. Suppose[4] $x(\cdot)$ is a polynomial of degree 2, i.e., that

39 $$x(t) = c_0 + c_1 t + c_2 t^2$$

Then

40 $$x_n = c_0 + c_1 nh + c_2 n^2 h^2$$

Also, (39) implies that

41 $$\dot{x}(t) = c_1 + 2c_2 t$$

so that

42 $$f(t_n, x_n) = c_1 + 2c_2 nh$$

[4]To avoid notational clutter, we let x be a scalar. However, all arguments remain valid in the case of vector $\mathbf{x}$.

Substituting from (40) and (42) into (37) and choosing $n = 0$ for convenience, we obtain what is known as the *exactness constraint* for the algorithm (37), namely,

$$c_0 + c_1 h + c_2 h^2 = \sum_{i=0}^{k} a_i[c_0 + c_1(-ih) + c_2(-ih)^2] + \sum_{i=-1}^{k} hb_i[c_1 + 2c_2(-ih)] \tag{43}$$

Because we would like (43) to hold independently of the step size h, we equate the coefficients of like powers of h. This gives

$$c_0 = \sum_{i=0}^{k} a_i c_0 \tag{44}$$

$$c_1 = \sum_{i=0}^{k} -ia_i c_1 + \sum_{i=-1}^{k} b_i c_1 \tag{45}$$

$$c_2 = \sum_{i=0}^{k} i^2 a_i c_2 + \sum_{i=-1}^{k} -2ib_i c_2 \tag{46}$$

Once again, because we would like (43) to hold for *all* second-order polynomials, (44)–(46) must hold for *all* c_0, c_1, c_2. This implies that

$$\sum_{i=0}^{k} a_i = 1 \tag{47}$$

$$\sum_{i=0}^{k} -ia_i + \sum_{i=-1}^{k} b_i = 1 \tag{48}$$

$$\sum_{i=0}^{k} i^2 a_i + \sum_{i=-1}^{k} -2ib_i = 1 \tag{49}$$

Thus (47)–(49) give three constraints on the $2k + 3$ constants $a_0, \ldots, a_k, b_{-1}, \ldots, b_k$, and any choice of these constants such that (47)–(49) hold makes the corresponding algorithm (37) *exact* for second-degree polynomials. In other words, whenever (47)–(49) hold, the corresponding algorithm (37) is of *second order*.

It is clear that in order for (47)–(49) to hold we must have $2k + 3 \geq 3$. The second-order algorithm containing the smallest number of constants a_i, b_i is obtained by setting $2k + 3 = 3$, i.e., $k = 0$. In this case, (47)–(49) take the form

$$a_0 = 1 \tag{50}$$

$$b_{-1} + b_0 = 1 \tag{51}$$

$$2b_{-1} = 1 \tag{52}$$

which can be solved to give

$$a_0 = 1, \qquad b_{-1} = \frac{1}{2}, \qquad b_0 = \frac{1}{2} \tag{53}$$

Hence a second-order numerical integration algorithm is given by the formula

54 $$x_{n+1} = x_n + \frac{hf(t_n, x_n) + hf(t_{n+1}, x_{n+1})}{2}$$

Algorithm (54) is also known as the trapezoidal algorithm.

The derivation of the exactness constraints in the general case proceeds in a manner entirely analogous to the above. Suppose $x(\cdot)$ is a polynomial of degree m, i.e.,

55 $$x(t) = \sum_{j=0}^{m} c_j t^j$$

Then

56 $$\dot{x}(t) = \sum_{j=1}^{m} j c_j t^{j-1}$$

Therefore,

57 $$x_i = \sum_{j=0}^{m} c_j i^j h^j$$

58 $$f(t_i, x_i) = \sum_{j=1}^{m} j c_j i^{j-1} h^{j-1}$$

Substituting from (57) and (58) into (37) and taking $n = 0$ for convenience, we obtain

59 $$\sum_{j=0}^{m} c_j h^j = \sum_{i=0}^{k} \left[a_i \sum_{j=0}^{m} c_j (-ih)^j \right] + \sum_{i=-1}^{k} \left[b_i \sum_{j=1}^{m} j c_j (-ih)^{j-1} h \right]$$

Equating the coefficients of h^j on both sides and dividing the resulting equation by c_j, we obtain

60 $$1 = \sum_{i=0}^{k} a_i (-i)^j + \sum_{i=-1}^{k} b_i j (-i)^{j-1}, \qquad j = 1, \ldots, m$$

61 $$1 = \sum_{i=0}^{k} a_i \quad \text{(corresponding to } j = 0\text{)}$$

Note that in applying (60) we must take $i^j = 1$ when $i = j = 0$.

The set of equations (60)–(61) provide $m + 1$ equations constraining the $2k + 3$ constants $a_0, \ldots, a_k, b_{-1}, \ldots, b_k$. Thus, clearly, in order for algorithm (37) to be exact in the case of mth-degree polynomials, i.e., for (37) to be of *order m*, we must have

62 $$m + 1 \leq 2k + 3$$

If equality holds in (62), then we can solve for the a_i's and b_i's uniquely. Now, if m is even, one can always choose k such that

63 $$2k + 3 = m + 2$$

and then assign $b_{-1} = 0$, so that (37) becomes an *explicit* formula for x_{n+1}.

By way of illustrating the above, suppose we want to derive a third-order algorithm. Using (63), we choose $k = 1$ and $b_{-1} = 0$. The corresponding equations (60) and (61) can be solved to yield

64 $$a_0 = -4, \quad a_1 = 5, \quad b_0 = 4, \quad b_1 = 2$$

Hence a possible explicit third-order algorithm is given by

65 $$x_{n+1} = -4x_n + 5x_{n-1} + 4hf(t_n, x_n) + 2hf(t_{n-1}, x_{n-1})$$

However, this algorithm is not very widely used.

Adams–Bashforth Algorithms

The Adams–Bashforth algorithm of order m is an explicit algorithm obtained by choosing

66 $$k = m - 1$$

67 $$a_i = 0 \quad \text{for } i = 1, \ldots, k$$

68 $$b_{-1} = 0$$

Thus there are exactly $k + 2$ constants to be determined, namely a_0, $b_0, \ldots, b_k$. Clearly, (67) and (61) imply that

69 $$a_0 = 1$$

Thus, once we solve the $k + 1$ equations provided by (60), the algorithm is determined. The formulas corresponding to the Adams–Bashforth algorithms of orders 1 through 4 are listed below:

70 $$\mathbf{x}_{n+1} = \mathbf{x}_n + h\mathbf{f}(t_n, \mathbf{x}_n) \qquad (m = 1)$$

71 $$\mathbf{x}_{n+1} = \mathbf{x}_n + h[\tfrac{3}{2}\mathbf{f}(t_n, \mathbf{x}_n) - \tfrac{1}{2}\mathbf{f}(t_{n-1}, \mathbf{x}_{n-1})] \qquad (m = 2)$$

72 $$\mathbf{x}_{n+1} = \mathbf{x}_n + h[\tfrac{23}{12}\mathbf{f}(t_n, \mathbf{x}_n) - \tfrac{16}{12}\mathbf{f}(t_{n-1}, \mathbf{x}_{n-1}) + \tfrac{5}{12}\mathbf{f}(t_{n-2}, \mathbf{x}_{n-2})] \qquad (m = 3)$$

73 $$\mathbf{x}_{n+1} = \mathbf{x}_n + h[\tfrac{55}{24}\mathbf{f}(t_n, \mathbf{x}_n) - \tfrac{59}{24}\mathbf{f}(t_{n-1}, \mathbf{x}_{n-1}) + \tfrac{37}{24}\mathbf{f}(t_{n-2}, \mathbf{x}_{n-2}) - \tfrac{9}{24}\mathbf{f}(t_{n-3}, \mathbf{x}_{n-3})] \qquad (m = 4)$$

Adams–Moulton Algorithms

The Adams–Moulton algorithm of order m is an implicit algorithm obtained by choosing

74 $$k = m - 2$$

75 $$a_i = 0, \qquad i = 1, \ldots, k$$

As before, (75) and (61) together imply that

76 $$a_0 = 1$$

The remaining unknowns $b_{-1}, \ldots, b_k$ can be determined using (60). The formulas corresponding to the Adams–Moulton algorithms of orders 1 though 4 are listed below:

77 $$\mathbf{x}_{n+1} = \mathbf{x}_n + h\mathbf{f}(t_{n+1}, \mathbf{x}_{n+1}) \qquad (m = 1)$$

78 $$\mathbf{x}_{n+1} = \mathbf{x}_n + h[\tfrac{1}{2}\mathbf{f}(t_{n+1}, \mathbf{x}_{n+1}) + \tfrac{1}{2}\mathbf{f}(t_n, \mathbf{x}_n)] \qquad (m = 2)$$

79 $$\mathbf{x}_{n+1} = \mathbf{x}_n + h[\tfrac{5}{12}\mathbf{f}(t_{n+1}, \mathbf{x}_{n+1}) + \tfrac{8}{12}\mathbf{f}(t_n, \mathbf{x}_n) - \tfrac{1}{12}\mathbf{f}(t_{n-1}, \mathbf{x}_{n-1})] \qquad (m = 3)$$

80 $$\mathbf{x}_{n+1} = \mathbf{x}_n + h[\tfrac{9}{24}\mathbf{f}(t_{n+1}, \mathbf{x}_{n+1}) + \tfrac{19}{24}\mathbf{f}(t_n, \mathbf{x}_n) - \tfrac{5}{24}\mathbf{f}(t_{n-1}, \mathbf{x}_{n-1}) + \tfrac{1}{24}\mathbf{f}(t_{n-2}, \mathbf{x}_{n-2})] \qquad (m = 4)$$

Predictor–Corrector Algorithms

From the preceding discussion, we can see that a k-step Adams–Bashforth algorithm is exact for polynomials of order k, while a k-step Adams–Moulton algorithm is exact for polynomials of order $k + 1$. However, the Adams–Moulton algorithm is implicit, in the sense that knowledge of $\mathbf{x}_{n-k}, \ldots, \mathbf{x}_n$ gives an implicit formula only for $\mathbf{x}_{n+1}$. This implicit formula can be symbolized as

81 $$\mathbf{x}_{n+1} = \mathbf{y}_n + hb_{-1}\mathbf{f}(t_{n+1}, \mathbf{x}_{n+1})$$

where $\mathbf{y}_n$ is a quantity that depends only on $\mathbf{x}_{n-k}, \ldots, \mathbf{x}_n$ and is hence known. Now, (81) can be solved iteratively as follows: Choose an initial guess $\mathbf{x}_{n+1}^{(0)}$, and set

82 $$\mathbf{x}_{n+1}^{(i+1)} = \mathbf{y}_n + hb_{-1}\mathbf{f}(t_{n+1}, \mathbf{x}_{n+1}^{(i)}), \qquad i \geq 0$$

The iterations are terminated when $|\mathbf{x}_{n+1}^{(i+1)} - \mathbf{x}_{n+1}^{(i)}|$ is sufficiently small.

Two questions arise in connection with this procedure: (1) How does one choose the initial guess $\mathbf{x}_{n+1}^{(0)}$? (2) Under what conditions does the sequence of iterations generated by (82) converge? Tackling the second question first, suppose $\mathbf{f}$ satisfies a global Lipschitz condition; i.e., suppose there exists a finite constant M such that

83 $$\|\mathbf{f}(t, \mathbf{x}) - \mathbf{f}(t, \mathbf{y})\| \leq M\|\mathbf{x} - \mathbf{y}\|, \qquad \forall\ t, \mathbf{x}, \mathbf{y}$$

Then, for sufficiently small values of h (in fact, whenever $hb_{-1}M < 1$), the right-hand side of (81) defines a *contraction mapping* of $\mathbf{x}_{n+1}$. Hence, by theorem [3.3(1)], the sequence $(\mathbf{x}_{n+1}^{(i)})_0^\infty$ converges to the unique solution of (81), regardless of the starting value $\mathbf{x}_{n+1}^{(0)}$. However, it is clear that the convergence of $(\mathbf{x}_{n+1}^{(i)})_0^\infty$ is accelerated if we choose $\mathbf{x}_{n+1}^{(0)}$ to be reasonably close to the solution of (81).

Accordingly, we choose $\mathbf{x}_{n+1}^{(0)}$ to be the value generated by an *explicit* k-step algorithm and then apply the iterative formula (82). This is known as the *predictor–corrector* method. For example, a two-step predictor–corrector algorithm can be defined as follows: Given $\mathbf{x}_n$, $\mathbf{x}_{n-1}$, set

84 $$\mathbf{x}_{n+1}^{(0)} = \mathbf{x}_n + h[\tfrac{3}{2}\mathbf{f}(t_n, \mathbf{x}_n) - \tfrac{1}{2}\mathbf{f}(t_{n-1}, \mathbf{x}_{n-1})] \qquad \text{(predictor)}$$

85 $$\mathbf{x}_{n+1}^{(i+1)} = \mathbf{x}_n + h[\tfrac{1}{2}\mathbf{f}(t_{n+1}, \mathbf{x}_{n+1}^{(i)}) + \tfrac{1}{2}\mathbf{f}(t_n, \mathbf{x}_n)] \qquad \text{(corrector)}$$

In the foregoing, (84) *predicts* the value of $\mathbf{x}_{n+1}$ using an explicit formula, while (85) *corrects* this predicted value of $\mathbf{x}_{n+1}$. In practice, if the step size h is selected properly, relatively few applications of the correcting formula (85) are enough to determine $\mathbf{x}_{n+1}$ to a high degree of accuracy.

We shall conclude this section with a comparison of Runge–Kutta and multistep algorithms. The main advantage of a multistep algorithm is that there are no extraneous calculations, as in Runge–Kutta methods. However, to apply a multistep algorithm, it is essential for the time instances t_n to be *uniformly spaced*. This is not necessary in Runge–Kutta methods. Based on these considerations, it can be said that Runge–Kutta methods are better if one expects the solution $\mathbf{x}(\cdot)$ of a given equation to vary very rapidly at some times and not to at other times (as depicted in Fig. 4.11). In such cases, it is possible to keep adjusting the step size h as one goes along in order to always maintain high accuracy. Otherwise, multistep algorithms are preferable.

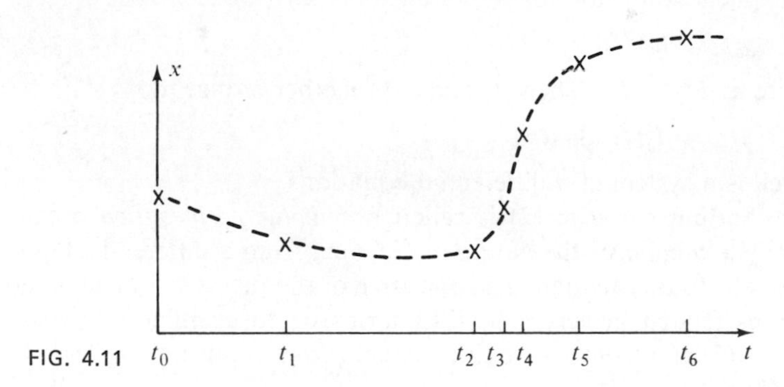

FIG. 4.11

The discussion in this section barely scratches the surface of a very well-developed subject. For a more detailed treatment, see [19] Ralston, and for an engineering viewpoint, see [8] Chua and Lin.

Problem 4.8. Show that in the case of a *linear* differential equation of the form

$$\dot{\mathbf{x}}(t) = \mathbf{A}\mathbf{x}(t) + \mathbf{v}(t)$$

the results given by a Taylor's series method of order n and a Runge–Kutta method of order n are identical.

Problem 4.9. Formulate the three-step and four-step predictor–corrector algorithms.

4.3 SINGULAR PERTURBATIONS

In this section, we shall briefly study what is known as the problem of singular perturbations. The problem can be stated as follows: Suppose we are given the system of nonlinear differential equations

1 $$\dot{\mathbf{x}}(t) = \mathbf{f}[\mathbf{x}(t), \mathbf{y}(t)]$$

2 $$\epsilon\dot{\mathbf{y}}(t) = \mathbf{g}[\mathbf{x}(t), \mathbf{y}(t)]$$

where $\mathbf{x}(t) \in R^n$, $\mathbf{y}(t) \in R^m$, $\mathbf{f}: R^n \times R^m \longrightarrow R^n$, and $\mathbf{g}: R^n \times R^m \longrightarrow R^m$. Note that for any value of ϵ other than zero the system (1)–(2) consists of $n + m$ differential equations. However, if $\epsilon = 0$, the system (1)–(2) consists of n *differential* equations and m *algebraic* equations, because with $\epsilon = 0$, (2) reduces to

3 $$\mathbf{g}[\mathbf{x}(t), \mathbf{y}(t)] = \mathbf{0}$$

Suppose it is possible to solve the m equations comprising (3) to obtain an explicit expression for $\mathbf{y}(t)$ in terms of $\mathbf{x}(t)$, of the form

4 $$\mathbf{y}(t) = \mathbf{h}[\mathbf{x}(t)]$$

where $\mathbf{h}: R^n \longrightarrow R^m$. Then (1) and (3) together reduce to

5 $$\dot{\mathbf{x}}(t) = \mathbf{f}\{\mathbf{x}(t), \mathbf{h}[\mathbf{x}(t)]\}$$

which is a system of n differential equations.

Setting $\epsilon = 0$ in (2) is called a *singular perturbation* because it completely changes the nature of (2), viz., from a differential equation to an algebraic equation. The objective of the theory of singular perturbations (stated in very simplified terms) is to examine the *simplified* system (5) and from this to draw conclusions about the *original* system (1)–(2) with $\epsilon \neq 0$.

Physically speaking, singular perturbations arise from an attempt to approximate a high-order[5] nonlinear system with another one of lower order. The following example, due to [9] Desoer and Shensa, illustrates this point.

6 **Example.** Consider the linear circuit shown in Fig. 4.12, and suppose the capacitance of ϵ represents a stray capacitance. In accor-

[5]In terms of the number of differential equations describing it.

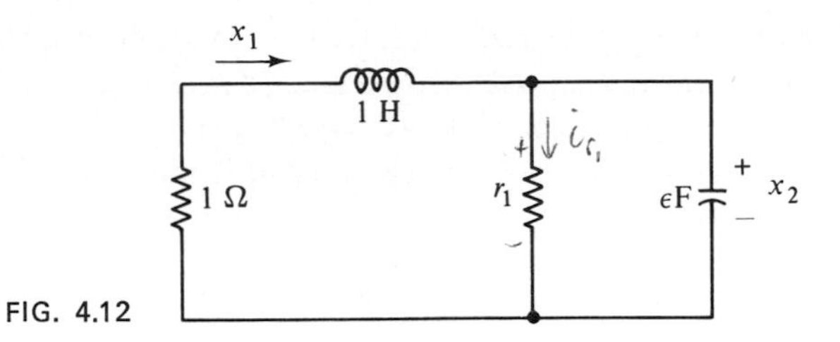

FIG. 4.12

dance with standard practice, the inductor current x_1 and the capacitor voltage x_2 are chosen as the state variables of the system. Using the current and voltage laws of Kirchhoff, the dynamic equations of the circuit in Fig. 4.12 can be written as

$$\dot{x}_1 = -x_1 - x_2 \tag{7}$$

$$\epsilon \dot{x}_2 = \frac{x_1 - x_2}{r_1} \tag{8}$$

If we wish to neglect the stray capacitance, we set $\epsilon = 0$ in (8), which gives

$$0 = \frac{x_1 - x_2}{r_1} \tag{9}$$

$$x_2 = r x_1 \tag{10}$$

Hence the new equation for x_1 is obtained from (7) as

$$\dot{x}_1 = (-1 - r)x_1 \tag{11}$$

Now, it is clear from (11) that, as long as $r > -1$, the solution x_1 of (11) decays exponentially to zero. However, returning to the original, unsimplified system (7)–(8), we have

$$\begin{bmatrix} \dot{x}_1 \\ \dot{x}_2 \end{bmatrix} = \begin{bmatrix} -1 & -1 \\ 1/\epsilon & -1/\epsilon r \end{bmatrix} \begin{bmatrix} x_1 \\ x_2 \end{bmatrix} \tag{12}$$

Letting **A** denote the 2×2 matrix in (12), one can easily verify that if $r < 0$, then for sufficiently small values of the parameter ϵ the matrix **A** has at least one eigenvalue with positive real part, so that the original system (7)–(8) is unstable in this case.[6] In other words, if $r \in (-1, 0)$, then the simplified system is stable, but the unsimplified system is unstable for sufficiently small values of ϵ.

This example not only shows that singular perturbations arise in connection with simplifying the dynamical model of a given system but also demonstrates the need to proceed with care in the simplifying process. Sometimes the simplified model can be stable, while the original system is unstable.

[6] See Sec. 5.3 for a full discussion of the stability of linear systems.

Even though the general singular perturbation problem has been posed here for *nonlinear* systems, we shall content ourselves with two results for *linear* systems, because we have not as yet developed the tools needed to rigorously analyze the stability of nonlinear systems. A discussion of the singular perturbation problem for nonlinear systems can be found in Sec. 5.4.

We shall now state concisely the two problems which we shall address next:

13 1. Given the system of linear equations

$$\begin{bmatrix} \dot{\mathbf{x}}_1 \\ \epsilon \dot{\mathbf{x}}_2 \end{bmatrix} = \begin{bmatrix} \mathbf{A}_{11} & \mathbf{A}_{12} \\ \mathbf{A}_{21} & \mathbf{A}_{22} \end{bmatrix} \begin{bmatrix} \mathbf{x}_1 \\ \mathbf{x}_2 \end{bmatrix} \tag{14}$$

find the conditions on $\mathbf{A}_{11}$, $\mathbf{A}_{12}$, $\mathbf{A}_{21}$, $\mathbf{A}_{22}$ which ensure that there exists a positive number ϵ_0 such that all eigenvalues of the matrix

$$\mathbf{A}(\epsilon) = \begin{bmatrix} \mathbf{A}_{11} & \mathbf{A}_{12} \\ \mathbf{A}_{21}/\epsilon & \mathbf{A}_{22}/\epsilon \end{bmatrix} \tag{15}$$

have negative real parts, for all ϵ in $(0, \epsilon_0)$.

16 2. Find conditions on $\mathbf{A}_{11}$, $\mathbf{A}_{12}$, $\mathbf{A}_{21}$, $\mathbf{A}_{22}$ which ensure that there exists a positive number ϵ_0 such that at least one eigenvalue of $A(\epsilon)$ has a positive real part, for all ϵ in $(0, \epsilon_0)$.

An answer to problem (13) is provided by the following theorem.

17 ***Theorem*** Suppose that all eigenvalues of A_{22} and of $\mathbf{A}_{11} - \mathbf{A}_{12}\mathbf{A}_{22}^{-1}\mathbf{A}_{21}$ have negative real parts. Then there exists a positive number ϵ_0 such that whenever ϵ belongs to $(0, \epsilon_0)$ all eigenvalues of $A(\epsilon)$ have negative real parts.

The proof depends strongly on the well-known fact (see, e.g., [22] Wilkinson) that the eigenvalues of a matrix depend continuously on the elements of the matrix. So as to avoid any ambiguities we shall state this fact precisely.

18 **fact** Let $M(\epsilon)$ be a continuous, $n \times n$ matrix-valued function of ϵ. Let $\{\lambda_{10}, \ldots, \lambda_{n0}\}$ denote the (not necessarily distinct) eigenvalues of $\mathbf{M}(0)$, and let $\{\lambda_{1\epsilon}, \ldots, \lambda_{n\epsilon}\}$ denote the (not necessarily distinct) eigenvalues of $\mathbf{M}(\epsilon)$. Then, for each $r > 0$, there exists an $\epsilon_0 > 0$ such that the inequality

$$|\lambda_{i0} - \lambda_{i\epsilon}| < r \quad \text{whenever } 0 < \epsilon < \epsilon_0 \tag{19}$$

is satisfied, by renumbering the $\lambda_{i\epsilon}$ if necessary.[7]

---

[7]Note that when we solve the equation $\det[\lambda I - M(\epsilon)] = 0$, the resulting n solutions for λ can be numbered in any order we choose.

proof of Theorem (17) In what follows, we omit the phrase "by renumbering if necessary" in the interests of brevity. For the sake of definiteness, suppose $\mathbf{A}_{11} \in R^{n\times n}$, $\mathbf{A}_{12} \in R^{n\times m}$, $\mathbf{A}_{21} \in R^{m\times n}$, $\mathbf{A}_{22} \in R^{m\times m}$. The $n+m$ eigenvalues of $A(\epsilon)$ are the $n+m$ zeroes of the polynomial

$$\Delta(\epsilon;\lambda) = \det\begin{bmatrix} \mathbf{A}_{11} - \lambda\mathbf{I} & \mathbf{A}_{12} \\ \mathbf{A}_{21}/\epsilon & \mathbf{A}_{22}/\epsilon - \lambda\mathbf{I} \end{bmatrix} \tag{20}$$

Let $\{\alpha_1, \ldots, \alpha_m\}$ denote the (not necessarily distinct) eigenvalues of $\mathbf{A}_{22}$, and let $\{\beta_1, \ldots, \beta_n\}$ denote the (not necessarily distinct) eigenvalues of $\mathbf{A}_{11} - \mathbf{A}_{12}\mathbf{A}_{22}^{-1}\mathbf{A}_{21}$.[8] We claim that (1) n of the eigenvalues of $\mathbf{A}(\epsilon)$ are "close" (in a sense to be made precise next) to $\{\beta_1, \ldots, \beta_n\}$ and (2) the remaining m eigenvalues of $\mathbf{A}(\epsilon)$ are "close" to $\{\alpha_1/\epsilon, \ldots, \alpha_m/\epsilon\}$. Specifically, we claim that, given any $r > 0$, there exists an $\epsilon_0 > 0$ such that the inequalities

$$|\lambda_i(\epsilon) - \beta_i| < r, \qquad i = 1, \ldots, n \tag{21}$$

$$\left|\lambda_i(\epsilon) - \frac{\alpha_{i-n}}{\epsilon}\right| < \frac{r}{\epsilon}, \qquad i = n+1, \ldots, n+m \qquad \text{whenever } 0 < \epsilon < \epsilon_0 \tag{22}$$

hold, where $\{\lambda_1(\epsilon), \ldots, \lambda_{n+m}(\epsilon)\}$ are the eigenvalues of $\mathbf{A}(\epsilon)$.

The proof of this claim proceeds as follows: We have

$$\epsilon\mathbf{A}(\epsilon) = \begin{bmatrix} \epsilon\mathbf{A}_{11} & \epsilon\mathbf{A}_{12} \\ \mathbf{A}_{21} & \mathbf{A}_{22} \end{bmatrix} \triangleq \mathbf{B}(\epsilon) \tag{23}$$

It is clear that the matrix $\mathbf{B}(\epsilon)$ is continuous in ϵ and that

$$\mathbf{B}(0) = \begin{bmatrix} \mathbf{0} & \mathbf{0} \\ \mathbf{A}_{21} & \mathbf{A}_{22} \end{bmatrix} \tag{24}$$

Because the $n+m$ eigenvalues of $\mathbf{B}(0)$ are $\{0, \ldots, 0, \alpha_1, \ldots, \alpha_m\}$, we have by fact (18) that, given any $r > 0$, there exists an $\epsilon_1 > 0$ such that n eigenvalues of $\mathbf{B}(\epsilon)$ are "close" to 0 and that the remaining m eigenvalues of $\mathbf{B}(\epsilon)$ are close to $\{\alpha_1, \ldots, \alpha_m\}$. Because λ is an eigenvalue of $\mathbf{A}(\epsilon)$ if and only if $\epsilon\lambda$ is an eigenvalue of $\mathbf{B}(\epsilon)$, we can conclude that m of the eigenvalues of $\mathbf{A}(\epsilon)$, say $\{\lambda_{n+1}, \ldots, \lambda_{n+m}\}$, satisfy

$$|\epsilon\lambda_i - \alpha_{i-n}| < r, \qquad i = n+1, \ldots, n+m, \quad \text{whenever } 0 < \epsilon < \epsilon_1 \tag{25}$$

or, equivalently,

$$\left|\lambda_i - \frac{\alpha_{i-n}}{\epsilon}\right| < \frac{r}{\epsilon}, \qquad i = n+1, \ldots, n+m, \quad \text{whenever } 0 < \epsilon < \epsilon_1 \tag{26}$$

The remaining n eigenvalues $\lambda_1, \ldots, \lambda_n$ satisfy

$$|\epsilon\lambda_i| < r, \qquad i = 1, \ldots, n, \quad \text{whenever } 0 < \epsilon < \epsilon_1 \tag{27}$$

even though this is not of much importance.

---

[8]Note that the hypotheses ensure that A_{22}^{-1} exists.

Next, for any $\epsilon \neq 0$, we have that λ is a zero of the characteristic polynomial $\Delta(\epsilon; \lambda)$ if and only if

$$\det \begin{bmatrix} \mathbf{A}_{11} - \lambda\mathbf{I} & \mathbf{A}_{12} \\ \mathbf{A}_{21} & \mathbf{A}_{22} - \epsilon\lambda\mathbf{I} \end{bmatrix} \triangleq \Delta_1(\epsilon; \lambda) = 0 \tag{28}$$

Note that the only change between (20) and (28) is that the second "row" of matrices has been multiplied by ϵ, which is permissible because $\epsilon \neq 0$. Now, $\Delta_1(\epsilon; \lambda)$ is continuous in ϵ at $\epsilon = 0$, and

$$\Delta_1(0; \lambda) = \det \begin{bmatrix} \mathbf{A}_{11} - \lambda\mathbf{I} & \mathbf{A}_{12} \\ \mathbf{A}_{21} & \mathbf{A}_{22} \end{bmatrix} \tag{29}$$

Noting that the determinant of the matrix

$$\begin{bmatrix} \mathbf{I} & \mathbf{0} \\ -\mathbf{A}_{22}^{-1}\mathbf{A}_{21} & \mathbf{I} \end{bmatrix} \tag{30}$$

is 1, we have

$$\begin{aligned} \Delta_1(0; \lambda) &= \det \left\{ \begin{bmatrix} \mathbf{A}_{11} - \lambda\mathbf{I} & \mathbf{A}_{12} \\ \mathbf{A}_{21} & \mathbf{A}_{22} \end{bmatrix} \begin{bmatrix} \mathbf{I} & \mathbf{0} \\ -\mathbf{A}_{22}^{-1}\mathbf{A}_{21} & \mathbf{I} \end{bmatrix} \right\} \\ &= \det \begin{bmatrix} \mathbf{A}_{11} - \mathbf{A}_{12}\mathbf{A}_{22}^{-1}\mathbf{A}_{21} - \lambda\mathbf{I} & \mathbf{A}_{12} \\ \mathbf{0} & \mathbf{A}_{22} \end{bmatrix} \\ &= \det(\mathbf{A}_{11} - \mathbf{A}_{12}\mathbf{A}_{22}^{-1}\mathbf{A}_{21} - \lambda\mathbf{I}) \quad \det \mathbf{A}_{22} \end{aligned} \tag{31}$$

Because $\det \mathbf{A}_{22}$ is a nonzero constant, we see that $\Delta_1(0; \lambda)$ is a polynomial of degree n whose zeroes are $\{\beta_1, \ldots, \beta_n\}$. Hence for $\epsilon \neq 0$, n of the zeroes of $\Delta_1(\epsilon; \lambda)$ [and hence those of $\Delta(\epsilon; \lambda)$] are "close" to $\{\beta_1, \ldots, \beta_m\}$; i.e., given any $r > 0$, there exists an $\epsilon_2 > 0$ such that

$$|\lambda_{i_j} - \beta_j| < r, \qquad j = 1, \ldots, n, \quad \text{whenever } 0 < \epsilon < \epsilon_2 \tag{32}$$

where $\{i_1, \ldots, i_n\}$ is a subset of $\{1, \ldots, n + m\}$.

Now, if we pick the numbers r, ϵ_1, and ϵ_2 carelessly, it is possible for the *same* λ_i to satisfy both (26) and (32). However, suppose we choose r as

$$r = \frac{-\min_i \{\operatorname{Re} \alpha_i\}}{2} \tag{33}$$

and choose ϵ_0 such that

$$\frac{r}{\epsilon_0} > |\beta_i|, \qquad i = 1, \ldots, n, \quad \textit{and} \tag{34}$$

$$\epsilon_0 \leq \min\{\epsilon_1, \epsilon_2\} \tag{35}$$

where ϵ_1 and ϵ_2 are as in (26) and (32), respectively. Then whenever $0 < \epsilon < \epsilon_0$, we have that m eigenvalues of $\mathbf{A}(\epsilon)$ satisfy (26) and that the *remaining* n eigenvalues of $\mathbf{A}(\epsilon)$ satisfy (32).

We can now complete the proof. Because (26) and (32) account for all $n + m$ eigenvalues of $\mathbf{A}(\epsilon)$, and because all these eigenvalues have negative real parts, we conclude that all eigenvalues of $\mathbf{A}(\epsilon)$ have negative real parts whenever $0 < \epsilon < \epsilon_0$. ∎

36 REMARKS: The proof of Theorem (17) actually gives a geometrical insight into what happens to the eigenvalues of $\mathbf{A}(\epsilon)$ as $\epsilon \longrightarrow 0$. As $\epsilon \longrightarrow 0$, n eigenvalues of $\mathbf{A}(\epsilon)$ approach the eigenvalues of $\mathbf{A}_{11} - \mathbf{A}_{12}\mathbf{A}_{22}^{-1}\mathbf{A}_{21}$, while the remaining m eigenvalues of $\mathbf{A}(\epsilon)$ approach the eigenvalues of $\mathbf{A}_{22}$, divided by ϵ. This is the import of the inequalities (21) and (22). This geometrical insight enables us to provide an answer to problem (16), which is given next.

37 ***Theorem*** If at least one of the eigenvalues of $\mathbf{A}_{22}$ or at least one of the eigenvalues of $\mathbf{A}_{11} - \mathbf{A}_{12}\mathbf{A}_{22}^{-1}\mathbf{A}_{21}$ has a positive real part, then there exists an $\epsilon_0 > 0$ such that at least one eigenvalue of $\mathbf{A}(\epsilon)$ has a positive real part whenever $0 < \epsilon < \epsilon_0$.

proof Immediate from (21), (22), and remarks (36). ∎

38 REMARKS: The behavior of the *RLC* network considered in Example (6) can now be explained in terms of Theorem (37). In that example, the eigenvalue $a_{11} - a_{12}a_{22}^{-1}a_{21}$ was negative, but a_{22} itself was positive. Accordingly, as predicted by Theorem (37), the system is unstable for sufficiently small values of ϵ.

In [9], Desoer and Shensa actually extend the results of Theorems (17) and (37) to the case of linear systems containing *both large and small* parameters, i.e., to systems described by

$$\begin{bmatrix} \dot{\mathbf{x}}_1 \\ \epsilon\dot{\mathbf{x}}_2 \\ \mu\dot{\mathbf{x}}_3 \end{bmatrix} = \begin{bmatrix} \mathbf{A}_{11} & \mathbf{A}_{12} & \mathbf{A}_{13} \\ \mathbf{A}_{21} & \mathbf{A}_{22} & \mathbf{A}_{23} \\ \mathbf{A}_{31} & \mathbf{A}_{32} & \mathbf{A}_{33} \end{bmatrix} \begin{bmatrix} \mathbf{x}_1 \\ \mathbf{x}_2 \\ \mathbf{x}_3 \end{bmatrix} \tag{39}$$

They study the behavior of system (39) as $\epsilon \longrightarrow 0$ and $\mu \longrightarrow \infty$. In this case, the second equation in (39) becomes an algebraic equation, while the third equation reduces to

$$\dot{\mathbf{x}}_3 = \mathbf{0} \tag{40}$$

The results proved by Desoer and Shensa are as follows: Define

$$\mathbf{B} = \mathbf{A}_{11} - \mathbf{A}_{12}\mathbf{A}_{22}^{-1}\mathbf{A}_{21} \tag{41}$$

and let $\mathbf{C}$ be the matrix resulting from simplifying the set of equations

$$\begin{aligned} \mathbf{0} &= \mathbf{A}_{11}\mathbf{x}_1 + \mathbf{A}_{12}\mathbf{x}_2 + \mathbf{A}_{13}\mathbf{x}_3 \\ \mathbf{0} &= \mathbf{A}_{21}\mathbf{x}_1 + \mathbf{A}_{22}\mathbf{x}_2 + \mathbf{A}_{23}\mathbf{x}_3 \\ \mu\dot{\mathbf{x}}_3 &= \mathbf{A}_{31}\mathbf{x}_1 + \mathbf{A}_{32}\mathbf{x}_2 + \mathbf{A}_{33}\mathbf{x}_3 \end{aligned} \tag{42}$$

into the form

$$\mu\dot{\mathbf{x}}_3 = \mathbf{C}\mathbf{x}_3 \tag{43}$$

Then we have the following.

44 ***Theorem*** Suppose all eigenvalues of $\mathbf{A}_{22}$, $\mathbf{B}$, and $\mathbf{C}$ have negative real parts. Then there exist $\epsilon_0 > 0$ and $\mu_0 < \infty$ such that all eigenvalues of the matrix

$$\mathbf{A}(\epsilon; \mu) = \begin{bmatrix} \mathbf{A}_{11} & \mathbf{A}_{12} & \mathbf{A}_{13} \\ \mathbf{A}_{21}/\epsilon & \mathbf{A}_{22}/\epsilon & \mathbf{A}_{23}/\epsilon \\ \mathbf{A}_{31}/\mu & \mathbf{A}_{32}/\mu & \mathbf{A}_{33}/\mu \end{bmatrix} \tag{45}$$

have negative real parts whenever $0 < \epsilon < \epsilon_0$ and $\mu > \mu_0$.

46 ***Theorem*** Suppose at least one eigenvalue of $\mathbf{A}_{22}$, or of $\mathbf{B}$, or of $\mathbf{C}$ has a positive real part. Then there exist $\epsilon_0 > 0$ and $\mu_0 < \infty$ such that at least one eigenvalue of $\mathbf{A}(\epsilon)$ has a positive real part whenever $0 < \epsilon < \epsilon_0$ and $\mu > \mu_0$.

The proof is left as an exercise.

Even though the study in this section is restricted to linear systems, the results derived here form the basis for the results given in Sec. 5.4 for singular perturbations of nonlinear systems.

Problem 4.10. Analyze the stability of the following system using Theorems (17) and (37):

$$\begin{bmatrix} \dot{x}_{11} \\ \dot{x}_{12} \\ \epsilon\dot{x}_2 \end{bmatrix} = \begin{bmatrix} -1 & 1 & 1 \\ 2 & -5 & 1 \\ 0 & -1 & -3 \end{bmatrix} \begin{bmatrix} x_{11} \\ x_{12} \\ x_2 \end{bmatrix}$$

Problem 4.11. Analyze the stability of the following system using Theorems (44) and (46):

$$\begin{bmatrix} \dot{x}_{11} \\ \dot{x}_{12} \\ \epsilon\dot{x}_2 \\ \mu\dot{x}_{31} \\ \mu\dot{x}_{32} \end{bmatrix} = \begin{bmatrix} -1 & 0 & 1 & 1 & 2 \\ 1 & -4 & 0 & -1 & 0 \\ 0 & -2 & -1 & 1 & 0 \\ 2 & 0 & 1 & -5 & 1 \\ 0 & 0 & -1 & 2 & -3 \end{bmatrix} \begin{bmatrix} x_{11} \\ x_{12} \\ x_2 \\ x_{31} \\ x_{32} \end{bmatrix}$$

Problem 4.12. Prove Theorems (44) and (46). [*Hint:* Proceed as in the proof of Theorem (17), and show that the eigenvalues of $\mathbf{A}(\epsilon)$ can be partitioned into three sets: (a) those that approach the eigenvalues of $\mathbf{B}$; (b) those that approach $1/\epsilon$ times the eigenvalues of $\mathbf{A}_{22}$; and (c) those that approach $1/\mu$ times the eigenvalues of $\mathbf{C}$.]

5 Stability in the sense of Liapunov

In this chapter, we shall study the concept of Liapunov stability, which plays an important role in modern control theory. We saw earlier that if a system is initially in an equilibrium state, it remains in it thereafter. Liapunov stability is concerned with the trajectories of a system when the initial state is *near* an equilibrium point. From an engineering point of view, this is very important because external disturbances (such as noise, component errors, etc.) are always present in a real system. The three basic concepts of Liapunov theory are stability, asymptotic stability, and global asymptotic stability. Roughly speaking, stability corresponds to the system trajectories depending continuously on the initial state; asymptotic stability corresponds to trajectories that start sufficiently close to an equilibrium point actually converging to the equilibrium state as $t \longrightarrow \infty$; and global asymptotic stability corresponds to *every* trajectory approaching a unique equilibrium point as $t \longrightarrow \infty$.

In this chapter, we shall formulate precise definitions of the various concepts of Liapunov stability, and we shall state the basic theorems that enable one to determine the stability status of a given system.

5.1 BASIC DEFINITIONS

In this section, we shall state the precise definitions of the various concepts of Liapunov stability; in addition, we shall introduce the various kinds of positive definite functions.

5.1.1 Definitions of Stability

Consider the vector differential equation

1 $$\dot{\mathbf{x}}(t) = \mathbf{f}[t, \mathbf{x}(t)], \qquad t \geq 0$$

where $\mathbf{x}(t) \in R^n$ and $\mathbf{f}: R_+ \times R^n \longrightarrow R^n$. Throughout this chapter, we shall assume that the function $\mathbf{f}$ is of such a nature that (1) has a unique solution over $[0, \infty)$ corresponding to each initial condition for $\mathbf{x}(0)$ and that this solution depends continuously on $\mathbf{x}(0)$. This is the case, for example, if $\mathbf{f}$ satisfies a global Lipschitz condition (see Sec. 3.4). Recall that an $\mathbf{x}_0 \in R^n$ is said to be an *equilibrium point* of the system (1) at time t_0 if

2 $$\mathbf{f}(t, \mathbf{x}_0) \equiv \mathbf{0}. \qquad \forall\, t \geq t_0$$

Throughout this chapter, we shall assume that the vector $\mathbf{0}$ is an equilibrium point of the system (1). This assumption does not result in any loss of generality, because if $\mathbf{x}_0$ is an equilibrium point of (1) at time t_0, then $\mathbf{0}$ is an equilibrium point at time t_0 of the system

3 $$\dot{\mathbf{z}}(t) = \mathbf{f}_1[t, \mathbf{z}(t)]$$

where

4 $$\mathbf{f}_1(t, \mathbf{z}) = \mathbf{f}(t, \mathbf{z} - \mathbf{x}_0)$$

Also, there is a one-to-one correspondence between the solutions of (1) and of (4). Thus, without loss of generality, we assume that

5 $$\mathbf{f}(t, \mathbf{0}) = \mathbf{0}, \qquad \forall\, t \geq t_0$$

As discussed in Sec. 1.2, this means that if the system (1) is started off in the initial state $\mathbf{x}(t_0) = \mathbf{0}$, the resulting trajectory is $\mathbf{x}(t) \equiv \mathbf{0}$ $\forall\, t > t_0$. However, an important practical consideration is the following: Suppose the initial state $\mathbf{x}(t_0)$ is not $\mathbf{0}$ but is "close" to it; what is the nature of the resulting trajectory? The various concepts of Liapunov stability are addressed to this question.

6 *[definition]* The equilibrium point $\mathbf{0}$ at time t_0 of (1) is said to be *stable* at time t_0 if, for each $\epsilon > \mathbf{0}$, there exists a $\delta(t_0, \epsilon) > 0$ such that

7 $$\|\mathbf{x}(t_0)\| < \delta(t_0, \epsilon) \Longrightarrow \|\mathbf{x}(t)\| < \epsilon, \qquad \forall\, t \geq t_0$$

It is said to be *uniformly stable* over $[t_0, \infty)$ if, for each $\epsilon > 0$, there exists a $\delta(\epsilon) > 0$ such that

8 $$\|\mathbf{x}(t_1)\| < \delta(\epsilon), \qquad t_1 \geq t_0 \Longrightarrow \|\mathbf{x}(t)\| < \epsilon, \qquad \forall\, t \geq t_1$$

[i.e., the same $\delta(\epsilon)$ applies for all t_1].

9 *[definition]* The equilibrium point $\mathbf{0}$ at time t_0 is *unstable* if it is not stable at time t_0.

According to definition (6), the equilibrium point $\mathbf{0}$ at time t_0 is stable at time t_0 if, given that we do not want the norm $\|\mathbf{x}(t)\|$ of the solution of (1) to exceed a prespecified positive number ϵ, we are then able to determine an a priori bound $\delta(t_0, \epsilon)$ on the norm of the initial condition $\|\mathbf{x}(t_0)\|$ such that any solution trajectory of (1) starting at time t_0 from an initial state lying in the ball of radius $\delta(t_0, \epsilon)$ always lies in the ball of radius ϵ at all times $t \geq t_0$. Another way of stating this is as follows: Arbitrarily small perturbations (about the equilibrium state $\mathbf{0}$) of the initial state $\mathbf{x}(t_0)$ result in arbitrarily small perturbations of the corresponding solution trajectories [of (1)].

To give yet another interpretation, let $C^n[t_0, \infty)$ denote the *linear space* of continuous n-vector-valued functions on $[t_0, \infty)$, and let $BC^n[t_0, \infty)$ denote the subspace of $C^n[t_0, \infty)$ consisting of *bounded* continuous functions. If we define the norm

$$\|\mathbf{x}(\cdot)\|_s = \sup_{t \in [t_0, \infty)} \|\mathbf{x}(t)\|, \qquad \forall\, \mathbf{x}(\cdot) \in BC^n[t_0, \infty) \tag{10}$$

then $BC^n[t_0, \infty)$, together with the norm $\|\cdot\|_s$, actually becomes a Banach space. Let us define a mapping T_{t_0} from R^n into $C^n[t_0, \infty)$ in the following manner: Given any $\mathbf{v} \in R^n$, let $T_{t_0}\mathbf{v}$ [which is an element of $C^n[t_0, \infty)$] be the solution of (1) corresponding to the initial condition $\mathbf{x}(t_0) = \mathbf{v}$. In other words, T_{t_0} maps initial conditions into the corresponding solution trajectories of (1). Now, (5) implies that T_{t_0} maps the element $\mathbf{0}$ of R^n into the zero function in $C^n[t_0, \infty)$. According to definition (6), the equilibrium point $\mathbf{0}$ at time t_0 is stable at time t_0 if the following two conditions hold:

1. There is some constant $c > 0$ such that the image, under T_{t_0}, of the ball $B = \{\mathbf{v}: \|\mathbf{v}\| < c\}$ is actually contained in $BC^n[t_0, \infty)$.
2. The restricted map $T_{t_0}: B \longrightarrow BC^n[t_0, \infty)$ is continuous at $\mathbf{0} \in R^n$.

Another small point needs to be clarified. In (6), $\|\cdot\|$ is *any* norm on R^n. Because all norms on R^n are topologically equivalent (see fact [3.1(52)]), one can see that the stability status of an equilibrium point does not depend on the particular norm used to verify (6).

Once the notion of stability is understood, it is easy to understand what uniform stability means. According to definition (6), the equilibrium point $\mathbf{0}$ is stable at time t_0 if, for each $\epsilon > 0$, a corresponding δ can be found such that (6) holds. In general, this δ depends on both ϵ and t_0. However, if a δ can be found that depends only on ϵ and not on the initial time t_0, then the equilibrium point $\mathbf{0}$ is uniformly stable.

Finally, let us turn to a discussion of instability. According to definition (9), instability is merely the absence of stability. It is unfortunate that the term *instability* leads some to visualize a situation where some trajectory of the system "blows up" in the sense that $\|\mathbf{x}(t)\|$

$\rightarrow \infty$ as $t \rightarrow \infty$. While this is one way in which instability can occur, it is by no means the only way. Stability of the equilibrium point $\underset{\sim}{0}$ means that, given *any* $\epsilon > 0$, one can always find a corresponding δ such that (7) holds. Therefore, the equilibrium point $\mathbf{0}$ is *unstable* if, for *some* $\epsilon > 0$, *no* $\delta > 0$ can be found such that (7) holds. Physically speaking, the equilibrium point $\mathbf{0}$ is unstable if there is some ball B_ϵ of radius ϵ centered at $\mathbf{0}$ such that for *every* $\delta > 0$, no matter how small, there is a nonzero initial state $\mathbf{x}(t_0)$ in B_δ such that the trajectory starting at $\mathbf{x}(t_0)$ eventually leaves B_ϵ. This and only this is the definition of instability. It may be that some trajectories starting within B_ϵ actually "blow up" with $\|\mathbf{x}(t)\| \rightarrow \infty$ as $t \rightarrow \infty$. However, this is not necessary for instability. This point is illustrated in the following.

11 **Example.** Consider the Van der Pol oscillator, described by

$$\dot{x}_1 = -x_2 \tag{12}$$

$$\dot{x}_2 = -x_1 + (1 - x_1^2)x_2 \tag{13}$$

which is also studied in Chap. 2. The point $x_1 = x_2 = 0$ is an equilibrium point of this system, and as such, the solution trajectory starting from $x_1(0) = x_2(0) = 0$ is given by $x_1(t) = x_2(t) \equiv 0 \ \forall \ t \geq 0$. However, solution trajectories starting from any nonzero initial state all approach the limit cycle, as shown in Fig. 5.1. Now, let us study the

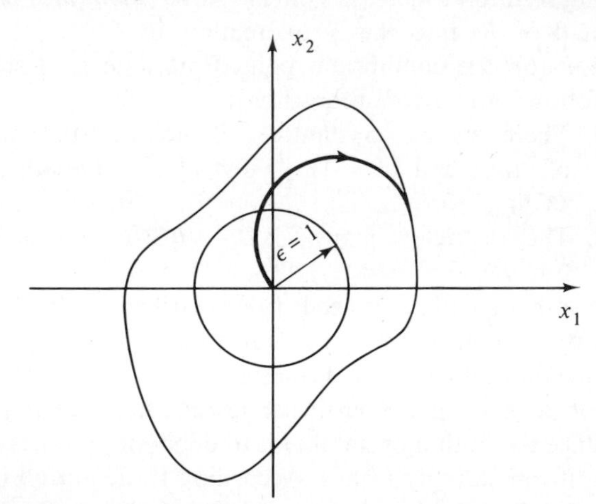

FIG. 5.1

stability status of the equilibrium point (0, 0) using definitions (6) and (9). If we choose $\epsilon = 1$, say, then $B_\epsilon = \{(x_1, x_2): (x_1^2 + x_2^2)^{1/2} \leq 1\}$ lies entirely within the region enclosed by the limit cycle. Therefore all trajectories starting from a nonzero initial state within B_ϵ eventually leave B_ϵ. Thus if we choose $\epsilon = 1$, no $\delta > 0$ can be found such that (7) is satisfied. Accordingly, the equilibrium point (0, 0) is unstable.

Note that all trajectories of the system, no matter what the initial state of the system is, are bounded, and none "blows up," so the system is well behaved in this sense.

In definition (6), stability is defined at an instant of time t_0, and uniform stability is defined over an interval $[t_0, \infty)$. However, if t_0 is clear from the context, we can just speak of stability and uniform stability. Some authors assume $t_0 = 0$, on the rationale that if $t_0 \neq 0$, one can always "translate" the time variable.

We turn now to asymptotic stability and global asymptotic stability.

14 *[definition]* The equilibrium point $\mathbf{0}$ at time t_0 is *asymptotically stable* at time t_0 if (1) it is stable at time t_0, and (2) there exists a number $\delta_1(t_0) > 0$ such that

15 $$\|\mathbf{x}(t_0)\| < \delta_1(t_0) \Longrightarrow \|\mathbf{x}(t)\| \longrightarrow 0 \qquad \text{as } t \longrightarrow \infty$$

It is *uniformly asymptotically stable* over $[t_0, \infty)$ if (1) it is uniformly stable over $[t_0, \infty)$, and (2) there exists a number $\delta_1 > 0$ such that

16 $$\|\mathbf{x}(t_1)\| < \delta_1, \qquad t_1 \geq t_0 \Longrightarrow \|\mathbf{x}(t)\| \longrightarrow 0 \qquad \text{as } t \longrightarrow \infty$$

Moreover, the convergence is uniform with respect to t_1.[1]

According to definition (14), $\mathbf{0}$ is an asymptotically stable equilibrium point at time t_0 if it is stable at time t_0 and in addition all trajectories starting from an initial state $\mathbf{x}(t_0)$ sufficiently close to $\mathbf{0}$ actually approach $\mathbf{0}$ as $t \longrightarrow \infty$. In this case, we can think of the set $B_{\delta_1(t_0)}$ defined by

17 $$B_{\delta_1(t_0)} = \{\mathbf{x} \in R^n : \|\mathbf{x}\| < \delta_1(t_0)\}$$

as a *region of attraction* for the equilibrium point $\mathbf{0}$, in the sense that all trajectories starting at time t_0 from an initial state within $B_{\delta_1(t_0)}$ eventually converge to $\mathbf{0}$. Notice, however, that (15) does *not* imply that any trajectory starting within $B_{\delta_1(t_0)}$ is confined to $B_{\delta_1(t_0)}$ for all $t \geq t_0$: It is quite possible that some trajectory starts at time t_0 from an initial state within $B_{\delta_1(t_0)}$ but leaves $B_{\delta_1(t_0)}$ at some later time. All that (15) implies is that any such trajectory will ultimately return to $B_{\delta_1(t_0)}$ after a finite amount of time,[2] will thereafter be confined to $B_{\delta_1(t_0)}$, and will approach $\mathbf{0}$ as $t \longrightarrow \infty$. The equilibrium point $\mathbf{0}$ is uniformly asymptotically stable over $[0, \infty)$ if it is uniformly stable over $[0, \infty)$

---

[1]The last statement can be made more precise, as follows: Given any $\epsilon > 0$, there exists a $T(\epsilon) < \infty$ such that

$$\|\mathbf{x}(t_1)\| < \delta_1 \Rightarrow \|\mathbf{x}(t_1 + t)\| < \epsilon \quad \text{whenever } t > T(\epsilon)$$

[2]During this finite period of time, the trajectory can leave and enter $B_{\delta_1(t_0)}$ any number of times.

and if one can find a nontrivial *ball of attraction* that is independent of initial time.

A word of caution in applying definition (14). One might be tempted to think that condition 2 of definition (14) implies the stability at time t_0 of the equilibrium point **0** and that condition 1 of definition (14) is therefore superfluous. However, a simple example shows that this is not the case. Consider a system whose trajectories take the form shown in Fig. 5.2, where all trajectories starting from nonzero initial points within the disk of radius δ_1 first touch the curve $\mathcal{C}$ before converging to the origin. This system satisfies condition 2 of definition (14). However, condition 1 is violated, because **0** is an unstable equilibrium point. This is most easily seen by applying definition (6) with $\epsilon = \delta_1$, as shown in Fig. 5.2; then no $\delta > 0$ can be found such that (7) holds.

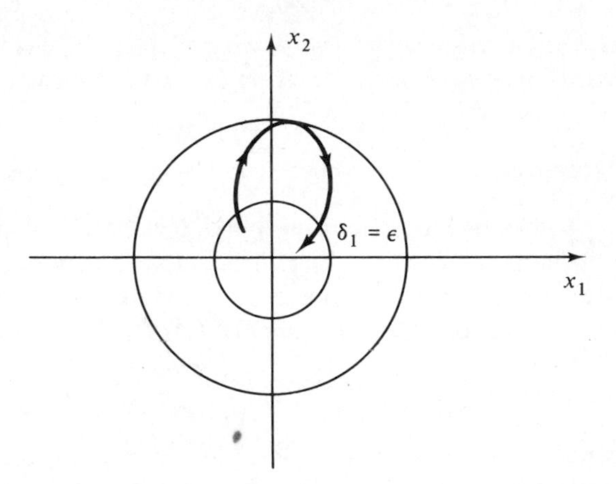

FIG. 5.2

We shall now give one last definition, which corresponds to the strongest kind of stability that we shall study in this book.

18 *[definition]* The equilibrium point **0** at time t_0 is *globally asymptotically stable* if $\mathbf{x}(t) \longrightarrow \mathbf{0}$ as $t \longrightarrow \infty$ [regardless of what $\mathbf{x}(t_0)$ is].

Thus global asymptotic stability results if *all* trajectories of the system converge to the equilibrium point **0** as $t \longrightarrow \infty$, i.e., if the sphere of attraction is the entire state space R^n.

In talking about stability, instability, and asymptotic stability, one should be very careful to associate these properties with a specific equilibrium point of the system, because in general a system can have more than one equilibrium point, each of which has its own stability status. However, if **0** is a globally asymptotically stable equilibrium point at time t_0 of a particular system, then it is clearly the *only* equilibrium point at time t_0 (see Problem 5.1). Hence, one can say, without fear of confusion, that the *system* is globally asymptotically stable as well as that the *equilibrium point* $\mathbf{0}$ is globally asymptotically stable. The distinction is that stability, asymptotic stability, and instability are basically *local* concepts, dealing with the trajectories of the system in the vicinity of an equilibrium point, whereas global asymptotic stability, as the name implies, is a *global* concept, having to do with the behavior of *all* the trajectories of the system.

19 **Example.** The traditional example for illustrating the various concepts of stability is that of a ball rolling on a curved table top under the influence of gravity. To keep matters simple, we assume a one-dimensional motion, with the table top being frictionless, or "with friction" as the example demands. It is clear that the profile of the table top, within a scale factor, corresponds exactly to the profile of the potential energy of the ball.

As our first example, we shall consider the case of the table top being absolutely flat, as depicted in Fig. 5.3. In this case, $r = 0$, $\dot{r} = 0$

Position r

FIG. 5.3 $r = 0$

is an equilibrium point, as is $r = r_0$, $\dot{r} = 0$ for any r_0, because no matter where we initially place the ball, it does not move. If we define the norm of the state vector $x = [r \ \dot{r}]'$ as

$$\|\mathbf{x}\| = (r^2 + \dot{r}^2)^{1/2} \tag{20}$$

then it is clear that $\|\mathbf{x}(t)\| = \|\mathbf{x}(0)\|$ for all $t \geq 0$. Hence the condition (7) holds for every $\epsilon > 0$ if we choose $\delta = \epsilon$. Accordingly, the equilibrium point $\mathbf{x} = \mathbf{0}$ is stable, by definition (6). Every other equilibrium point $[r_0 \ 0]'$ is also stable.

Now, suppose the table top is as shown in Fig. 5.4 and is *frictionless*. Then once again $r = 0$, $\dot{r} = 0$ is an equilibrium state. Now, suppose the ball is displaced to the dotted position and is let go (with zero initial velocity). In the absence of friction, the ball will roll down through the equilibrium position and up the other side to the same height

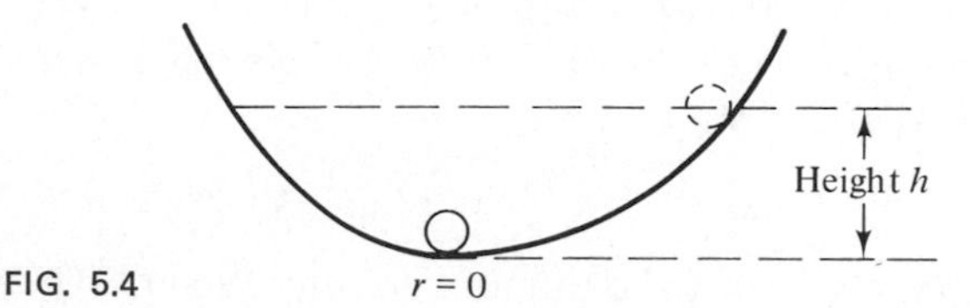

FIG. 5.4

h and back again, as there is a sustained oscillation resulting from a continuous exchange between potential energy and kinetic energy. In this case, once again, $\mathbf{x} = \mathbf{0}$ is a stable equilibrium point.

Now consider the situation in Fig. 5.4, and suppose that there is a frictional force proportional to $\dot{r}$. In this case, after the ball is displaced to the dotted position and released, it will roll back and forth, but because of the energy dissipation due to friction, the amplitude of the oscillations gradually decreases and approaches 0 as $t \longrightarrow \infty$. In this case, the equilibrium point $\mathbf{0}$ is asymptotically stable and is indeed globally asymptotically stable.

Suppose now that the table top is shaped as shown in Fig. 5.5, and suppose there is a frictional force proportional to $\dot{r}$. Then the equilibrium point at $r = r_1$, $\dot{r} = 0$ is asymptotically stable, the equilibrium point $r = r_2$, $\dot{r} = 0$ is unstable, while the equilibrium point at $r = r_3$, $\dot{r} = 0$ is stable. The details are left as an exercise.

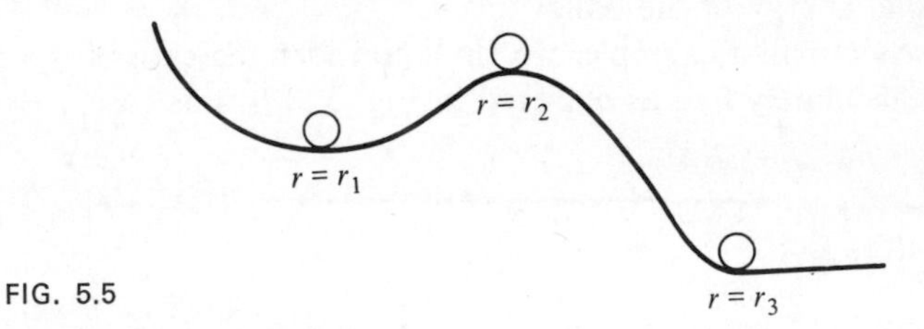

FIG. 5.5

21 **Example.** To illustrate the distinction between stability and uniform stability, consider the scalar differential equation

22 $$\dot{x}(t) = (6t \sin t - 2t)x(t)$$

The solution to (22) is given by

23 $$x(t) = x(t_0) \exp \{6 \sin t - 6t \cos t - t^2 - 6 \sin t_0 \cos t_0 + t_0^2\}$$

The system (22) has an equilibrium point at $x = 0$. We shall show that this equilibrium point is stable at any time $t_0 \geq 0$ but is not uniformly stable over $[0, \infty)$. First, let $t_0 \geq 0$ be any fixed initial time. Then

24 $$\left| \frac{x(t)}{x(t_0)} \right| = \exp \{6 \sin t - 6t \cos t - t^2 - 6 \sin t_0 + 6t_0 \cos t_0 + t_0^2\}$$

Now, if $t - t_0 > 6$, the quantity on the right-hand side of (24) is bounded above by $\exp [12 + T(6 - T)]$, where $T = t - t_0$. Because it is a continuous function of time, it is also bounded over the interval $[t_0, t_0 + 6]$. Hence if we define

25 $$c(t_0) = \sup_{t \geq t_0} \exp \{6 \sin t - 6t \cos t - t^2 - 6 \sin t_0 + 6t_0 \cos t_0 + t_0^2\}$$

we know that $c(t_0)$ is a finite number, for any fixed t_0. Thus, given any

$\epsilon > 0$, if we choose $\delta(\epsilon, t_0) = \epsilon/c(t_0)$, then (7) is satisfied, showing that $x = 0$ is a stable equilibrium point for all times t_0. On the other hand, if we choose $t_0 = 2n\pi$, then (23) yields

26 $$x[(2n + 1)\pi] = x(2n\pi) \exp\{(4n + 1)(6 - \pi)\pi\}$$

This shows that

27 $$c(2n\pi) \geq \exp\{(4n + 1)(6 - \pi)\pi\}$$

Hence $c(t_0)$ is unbounded as a function of t_0. Thus, given $\epsilon > 0$, it is *not* possible to choose a single $\delta(\epsilon)$, independent of the initial time, such that (8) holds. Therefore the equilibrium point $x = 0$ is *not* uniformly stable over $[0, \infty)$.

We shall close this subsection[3] by showing that, in the case of autonomous and periodic systems, stability is equivalent to uniform stability and that asymptotic stability is equivalent to uniform asymptotic stability. Actually, it is enough to prove these statements for the case of periodic systems, because an autonomous system can be thought of as a periodic system with arbitrary period. The following preliminary result is useful in its own right.

28 ***lemma*** Suppose the equilibrium point $\mathbf{0}$ at time t_0 of the system (1) is stable at some time $t_1 > t_0$. Then $\mathbf{0}$ is also a stable equilibrium point at all times $\tau \in [t_0, t_1]$.

proof Let $\mathbf{s}(t, \mathbf{x}_0, \tau)$ denote the solution of (1) together with the initial condition

29 $$\mathbf{x}(\tau) = \mathbf{x}_0$$

as evaluated at the time t. The fact that the equilibrium point $\mathbf{0}$ is stable at time t_1 means that, for every $\epsilon > 0$, there exists a $\delta > 0$ such that

30 $$\|\mathbf{x}_0\| < \delta \Longrightarrow \|\mathbf{s}(t, \mathbf{x}_0, t_1)\| < \epsilon, \qquad \forall\, t \geq t_1$$

Let $\tau \in [t_0, t_1]$, and define

31 $$m(r) = \sup_{\|\mathbf{x}_0\| \leq r} \|\mathbf{s}(t_1, \mathbf{x}_0, \tau)\|$$

By assumption, $m(r)$ is continuous in r, and $m(0) = 0$. Therefore one can find an $r_0 > 0$ such that

32 $$m(r) < \delta \quad \text{whenever } r < r_0$$

Finally, suppose $\|\mathbf{x}_0\| < r_0$; then (32) and (31) imply that

33 $$\|\mathbf{s}(t_1, \mathbf{x}_0, \tau)\| < \delta$$

Next, (33) and (30) together imply that

34 $$\|\mathbf{s}[t, \mathbf{s}(t_1, \mathbf{x}_0, \tau), t_1]\| < \epsilon, \qquad \forall\, t \geq t_1$$

[3] The remainder of Sec. 5.1.1 may be skipped without loss of continuity.

However, it is easy to show that

35 $$\mathbf{s}[t, \mathbf{s}(t_1, \mathbf{x}_0, \tau), t_1] = \mathbf{s}(t, \mathbf{x}_0, \tau)$$

(see Problem 5.3). Therefore we conclude that

36 $$\|\mathbf{s}(t, \mathbf{x}_0, \tau)\| < \epsilon, \qquad \forall\, t \geq \tau \text{ whenever } \|\mathbf{x}_0\| < r_0$$

Here we have made use of the obvious fact that $\delta \leq \epsilon$. Because the above argument can be repeated for any $\epsilon > 0$, it follows that $\mathbf{0}$ is a stable equilibrium point at time τ. ■

37 ***Theorem*** Consider the system (1) and suppose $\mathbf{f}$ satisfies

38 $$\mathbf{f}(t, \mathbf{x}) = \mathbf{f}(t + T, \mathbf{x}), \qquad \forall\, \mathbf{x} \in R^n, \forall\, t \geq 0$$

for some positive number T. (If the system is autonomous, T is arbitrary.) Under these conditions, the following two statements are equivalent:

(i) The equilibrium point $\mathbf{0}$ of the system (1)[4] is stable at some time $t_0 \geq 0$.

(ii) The equilibrium point $\mathbf{0}$ of the system (1) is uniformly stable over the interval $[0, \infty)$.

proof Clearly (ii) implies (i), so it remains only to show that (i) implies (ii). Accordingly, assume (i) holds. Because of the periodicity of $\mathbf{f}$, it follows readily that

39 $$\mathbf{s}(t, \mathbf{x}_0, t_0) = \mathbf{s}(t + nT, \mathbf{x}_0, t_0 + nT)$$

for all nonnegative integers n. Therefore, if (i) holds, then $\mathbf{0}$ is a stable equilibrium point at all times $t_0 + nT$, $n > 0$. Now, applying lemma (28) shows that $\mathbf{0}$ is a stable equilibrium point at all times $\tau \in [0, T]$. Define

40 $$\mu(\mathbf{x}_0, \tau) = \sup_{\gamma \geq 0} \|\mathbf{s}(\tau + \gamma, \mathbf{x}_0, \tau)\|$$

Because $\mathbf{0}$ is a stable equilibrium point all times $\tau \in [0, T]$, the function $\mu(\mathbf{x}_0, \tau)$ is continuous at $\mathbf{x}_0 = \mathbf{0}$, for each τ (see Problem 5.4). Therefore, if

41 $$\eta(\mathbf{x}_0) \triangleq \sup_{\tau \in [0, T]} \mu(\mathbf{x}_0, \tau)$$

then $\eta(\cdot)$ is continuous at $\mathbf{x}_0 = \mathbf{0}$. Thus, given any $\epsilon > 0$, we can find a $\delta > 0$ such that

42 $$\|\mathbf{x}_0\| < \delta \Longrightarrow \eta(\mathbf{x}_0) < \epsilon$$

But (42), together with the definition of $\eta(\cdot)$, means that

43 $$\|\mathbf{x}_0\| < \delta, \qquad \tau \in [0, T] \Longrightarrow \|\mathbf{s}(\tau + \gamma, \mathbf{x}_0, \tau)\| < \epsilon \qquad \forall\, \gamma \geq 0$$

[4]Note that, because of the periodicity of the function $\mathbf{f}$, we have that if $\mathbf{0}$ is an equilibrium point at some time, it is an equilibrium point it *all* times.

Finally, because of the periodicity of $\mathbf{f}$, (43) can be broadened to

44 $$\|\mathbf{x}_0\| < \delta \Longrightarrow \|\mathbf{s}(\tau + \gamma, \mathbf{x}_0, \tau)\| < \epsilon, \qquad \forall\, \tau \geq 0,\ \forall\, \gamma \geq 0$$

Because this argument can be repeated for any $\epsilon > 0$, (44) establishes that the equilibrium point $\mathbf{0}$ is uniformly stable over the interval $[0, \infty)$. ∎

A result similar to Theorem (37) can be proved for asymptotic stability as well.

45 ***Theorem*** Consider the system (1), and suppose $\mathbf{f}$ satisfies

46 $$\mathbf{f}(t, \mathbf{x}) = \mathbf{f}(t + T, \mathbf{x}), \qquad \forall\, \mathbf{x} \in R^n,\ \forall\, t \geq 0$$

for some positive number T. Under these conditions, the following two statements are equivalent:

(i) The equilibrium point $\mathbf{0}$ of the system (1) is asymptotically stable at some time $t_0 \geq 0$.

(ii) The equilibrium point $\mathbf{0}$ of the system (1) is uniformly asymptotically stable over the interval $[0, \infty)$.

The proof is omitted here in the interests of brevity and can be found in [13] Hahn (see also Problem 5.5).

5.1.2 Definite and Locally Definite Functions

In this subsection, we shall present some concepts, such as positive definite functions, that are central to the development of Liapunov stability theory, as well as the notion of the derivative along a trajectory.

47 *[definition]* A continuous function $V: R_+ \times R^n \longrightarrow R$ is said to be a *locally positive definite function* (l.p.d.f.) if there exists a continuous nondecreasing function $\alpha: R \longrightarrow R$ such that

48 (i) $\alpha(0) = 0,\ \alpha(p) > 0$ whenever $p > 0$

49 (ii) $V(t, \mathbf{0}) = 0, \quad \forall\, t \geq 0$

50 (iii) $V(t, \mathbf{x}) \geq \alpha(\|\mathbf{x}\|),\ \forall\, t \geq 0$, and for all $\mathbf{x}$ belonging to some ball

51 $$B_r = \{\mathbf{x}: \|\mathbf{x}\| \leq r\}, \qquad r > 0$$

V is said to be a *positive definite function* (p.d.f.) if (50) holds for all $\mathbf{x} \in R^n$ and in addition $\alpha(p) \longrightarrow \infty$ as $p \longrightarrow \infty$.

The kinds of functions α satisfying (48) above arise quite frequently in Liapunov theory—so much so that it is desirable to give a name to this class of functions.

52 *[definition]* A continuous function $\alpha: R \longrightarrow R$ is said to belong to class K if

(i) $\alpha(\cdot)$ is nondecreasing,
(ii) $\alpha(0) = 0$, and
(iii) $\alpha(p) > 0$ whenever $p > 0$.

Note that some authors replace (i) above by the more stringent requirement that $\alpha(\cdot)$ is strictly increasing. It turns out, however, that both definitions are equally good in proving stability theorems.

Given a continuous function $V: R_+ \times R^n \longrightarrow R$, it is rather difficult to determine, using definition (47), whether or not V is a p.d.f. or an l.p.d.f. The main source of difficulty is the need to exhibit the function $\alpha(\cdot)$. Lemmas (53) and (61) give an equivalent characterization of l.p.d.f.'s and p.d.f.'s and have the advantage that the conditions given therein are more readily verifiable than those in definition (47).

53 ***lemma*** A continuous function $W: R^n \longrightarrow R$ is an l.p.d.f. if and only if

54 (i) $W(\mathbf{0}) = 0$

55 (ii) $W(\mathbf{x}) > 0, \quad \forall\, \mathbf{x} \neq \mathbf{0}$ belonging to some ball B_r, $r > 0$ [where the ball B_r is defined in (51)].

W is a p.d.f. if and only if

56 (iii) $W(\mathbf{0}) = 0$

57 (iv) $W(\mathbf{x}) > 0, \quad \forall\, \mathbf{x} \neq \mathbf{0}$

58 (v) $W(\mathbf{x}) \longrightarrow \infty$ as $\|\mathbf{x}\| \longrightarrow \infty$, uniformly in $\mathbf{x}$.

proof We shall first consider l.p.d.f.'s. Suppose W is an l.p.d.f. in the sense of definition (47). Then clearly (i) and (ii) above hold. To prove the converse, suppose (i) and (ii) above hold, and define

59
$$\alpha(p) = \inf_{p \leq \|\mathbf{x}\| \leq r} W(\mathbf{x})$$

Then clearly $\alpha(0) = 0$, α is continuous, and $\alpha(\cdot)$ is nondecreasing, because as p increases, the infimum is taken over a smaller region. Also, suppose $p > 0$; then the infimum in (59) is also positive, because W is a continuous function.[5] Therefore $\alpha(\cdot)$ is a function of class K. Because $W(\mathbf{x}) \geq \alpha(\|\mathbf{x}\|)$ $\forall\, \mathbf{x} \in B_r$, W is an l.p.d.f. in the sense of definition (47).

In the case of p.d.f.'s, the necessity of conditions (iii), (iv), and (v) is immediate from definition (47). To prove the sufficiency, define

60
$$\alpha(p) = \inf_{p \leq \|\mathbf{x}\|} W(\mathbf{x})$$

The rest of the proof is the same as in the case of l.p.d.f.'s and is left as an exercise. ∎

[5] The proof of this statement involves the so-called compactness property of the unit ball in R^n and can be found in any standard text on real analysis; see, e.g., [20] Royden.

61 ***lemma*** A continuous function $V: R_+ \times R^n \longrightarrow R$ is an l.p.d.f. if and only if there exists an l.p.d.f. $W: R^n \longrightarrow R$ such that

62 $$V(t, \mathbf{x}) \geq W(\mathbf{x}), \qquad \forall\, t \geq 0,\ \forall\, \mathbf{x} \in B_r, \qquad r > 0$$

where B_r is a ball in R^n. A continuous function $V: R_+ \times R^n \longrightarrow R$ is a p.d.f. if and only if there exists a p.d.f. $W: R^n \longrightarrow R$ such that

63 $$V(t, \mathbf{x}) \geq W(\mathbf{x}), \qquad \forall\, t \geq 0,\ \forall\, \mathbf{x} \in R^n$$

proof We shall give the proof only for l.p.d.f.'s, because the proof for p.d.f.'s is entirely parallel. Suppose W is an l.p.d.f. and that (62) holds. Then it is easy to verify that V is an l.p.d.f. in the sense of definition (47). Conversely, suppose V is an l.p.d.f. in the sense of definition (47), and let $\alpha(\cdot)$ be the function of class K such that (50) holds. Then $W(\mathbf{x}) = \alpha(\|\mathbf{x}\|)$ is an l.p.d.f. such that (62) holds. ∎

64 REMARKS:

1. The conditions given in lemma (53) are easier to verify than those in definition (47).
2. If $W(\mathbf{x})$ is a polynomial in the components of $\mathbf{x}$, then one can systematically check, in a finite number of operations, whether or not (55) is satisfied; see [15] Jury for details.
3. Lemma (61) shows that a continuous function of t and $\mathbf{x}$ is an l.p.d.f. if and only if it dominates, at each instant of time and over some ball in R^n, an l.p.d.f. of $\mathbf{x}$ alone. Similarly, a continuous function of t and $\mathbf{x}$ is a p.d.f. if and only if it dominates, for all t and $\mathbf{x}$, a p.d.f. of $\mathbf{x}$ alone.
4. Suppose $W(\mathbf{x}) = \mathbf{x}'\mathbf{M}\mathbf{x}$, where $\mathbf{M}$ is a real symmetric $n \times n$ matrix. Then it is easy to show that W is a positive definite *function* if and only if $\mathbf{M}$ is a positive definite *matrix* (see also Problem 5.7).

65 *definition* A continuous function $V: R_+ \times R^n \longrightarrow R$ is said to be *decrescent* if there exists a function $\beta(\cdot)$ belonging to class K such that

66 $$V(t, \mathbf{x}) \leq \beta(\|\mathbf{x}\|), \qquad \forall\, t \geq 0,\ \forall\, \mathbf{x} \in B_r, \qquad r > 0$$

where B_r is a ball in R^n.

In other words, a function $V: R_+ \times R^n \longrightarrow R$ is decrescent if and only if, for each p in some interval $(0, r)$, we have

67 $$\sup_{\|\mathbf{x}\| \leq p}\ \sup_{t \geq 0} V(t, \mathbf{x}) < \infty$$

68 **Example.** The function

$$W_1(x_1, x_2) = x_1^2 + x_2^2$$

is a simple example of a p.d.f. Clearly W_1 satisfies (56) and (57); it can be seen to satisfy (58) also, because

$$W_1(\mathbf{x}) = \|\mathbf{x}\|^2$$

if we take $\|\cdot\|$ to be the Euclidean norm on R^2.

69 **Example.** The function

$$V_1(t, x_1, x_2) = (t + 1)(x_1^2 + x_2^2)$$

is a p.d.f., because it dominates the *time-invariant* p.d.f. W_1. However, it is not decrescent, because for each $(x_1, x_2) \neq (0, 0)$, $V_1(t, x_1, x_2)$ is unbounded as t increases.

70 **Example.** The function

$$V_2(t, x_1, x_2) = e^{-t}(x_1^2 + x_2^2)$$

is not a p.d.f., because no p.d.f. $W: R^n \longrightarrow R$ exists such that (63) holds. This can be seen from the fact that, for each (x_1, x_2), $V(t, x_1, x_2) \longrightarrow 0$ as $t \longrightarrow \infty$. This example shows that it is *not* possible to weaken (63) to the statement

$$V(t, \mathbf{x}) > 0, \qquad \forall\, t \geq 0, \forall\, \mathbf{x} \neq 0$$

The present function V_2 is decrescent.

71 **Example.** The function

$$W_2(x_1, x_2) = x_1^2 + \sin^2 x_2$$

is an l.p.d.f., though it is not a p.d.f. Note that (1) $W_2(0, 0) = 0$, and (2) $W_2(x_1, x_2) > 0$ whenever $(x_1, x_2) \neq (0, 0)$ and $|x_2| \leq \pi/2$. This is enough to ensure that W_2 is an l.p.d.f. However, W_2 is not a p.d.f., because it vanishes at points other than (0, 0), e.g., $(0, \pi)$.

We shall close this subsection by introducing the concept of the derivative of a function $V(t, \mathbf{x})$ along the trajectories of (1). Suppose V has continuous partial derivatives in all of its arguments, and let ∇V denote the gradient of V with respect to $\mathbf{x}$. Suppose $\mathbf{x}(\cdot)$ is a solution of (1), and consider the function $t \mapsto V[t, \mathbf{x}(t)]$. This function is differentiable with respect to t, and indeed we have

$$\frac{d}{dt} V[t, \mathbf{x}(t)] = \frac{\partial V}{\partial t}[t, \mathbf{x}(t)] + \nabla V[t, \mathbf{x}(t)]\mathbf{f}[t, \mathbf{x}(t)] \tag{72}$$

We use the symbol $\dot{V}[t, \mathbf{x}(t)]$ to denote the right-hand side of (71). This choice of symbols is justified because

$$V[t, \mathbf{x}(t)] = V[t_0, \mathbf{x}(t_0)] + \int_{t_0}^{t} \dot{V}[\tau, \mathbf{x}(\tau)]\, d\tau \tag{73}$$

whenever $\mathbf{x}(\cdot)$ is a solution of (1). This leads to the following definition.

74 *[definition]* Let $V\colon R_+ \times R^n \longrightarrow R$ be continuously differentiable with respect to all its arguments, and let ∇V denote the gradient of V with respect to $\mathbf{x}$. Then $\dot{V}\colon R_+ \times R^n \longrightarrow R$ is defined by

75
$$\dot{V}(t, \mathbf{x}) = \frac{\partial V}{\partial t}(t, \mathbf{x}) + \nabla V(t, \mathbf{x})\mathbf{f}(t, \mathbf{x})$$

and is called the *derivative of V along the trajectories of (1)*.

76 REMARKS:

1. Note that $\dot{V}$ depends not only on V but also on the system (1). If we keep the same V but change the system (1), the resulting $\dot{V}$ will in general be different.
2. The quantity $\dot{V}(t_0, \mathbf{x}_0)$ can be interpreted as follows: Suppose a solution trajectory $\mathbf{x}(\cdot)$ of (1) passess through $\mathbf{x}_0$ at time t_0. Then, at the instant t_0, the rate of change of $V[t, \mathbf{x}(t)]$ is $\dot{V}(t_0, \mathbf{x}_0)$.
3. Note that if V is independent of t and if the system (1) is autonomous, then $\dot{V}$ is also independent of t.

Problem 5.1. Show that if **0** is a globally asymptotically stable equilibrium point of (1) at time t_0, then it is the *only* equilibrium point of (1) at time t_0.

Problem 5.2. The purpose of this problem is to generalize Example (21) by developing a whole class of linear systems with equilibrium points that are stable but not uniformly stable. Consider the linear scalar system

77
$$\dot{x}(t) = a(t)x(t), \qquad t \geq 0$$

where $a(\cdot)$ is a continuous function.

(a) Verify that the general solution of (77) is

$$x(t) = x(t_0) \exp\left[\int_{t_0}^{t} a(\tau)\, d\tau\right]$$

(b) Show, using definition (6), that $x = 0$ is a stable equilibrium point at time t_0 if and only if

$$\sup_{t \geq t_0} \int_{t_0}^{t} a(\tau)\, d\tau \quad \triangleq m(t_0) < \infty$$

Show that, in this case, (7) is satisfied with

$$\delta(t, \epsilon) = \frac{\epsilon}{m(t_0)}$$

(c) Show that, if $m(t_0)$ is unbounded as a function of t_0, then 0 is *not* uniformly stable over $[0, \infty)$.

(d) Construct several functions $a(\cdot)$ such that $m(t_0)$ is finite for each t_0 but $m(\cdot)$ is an unbounded function. [*Hint:* As a start, let $a(\cdot)$ be the triangular function shown in Fig. 5.6.]

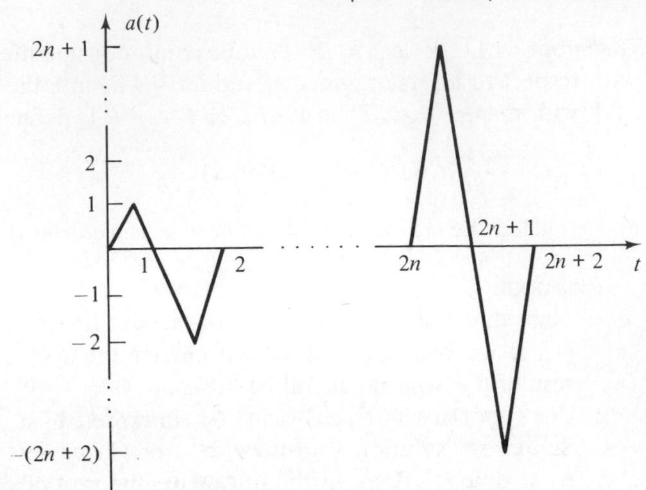

FIG. 5.6

Problem 5.3. Let $\mathbf{s}(t, \mathbf{x}_0, t_0)$ be as defined in the proof of lemma (28). Show, using the uniqueness of solutions to (1), that

78 $$\mathbf{s}[t, \mathbf{s}_0(t_1, \mathbf{x}_0, t_0), t_1] = \mathbf{s}(t, \mathbf{x}_0, t_0), \qquad \forall\, t \geq t_1 \geq t_0$$

Give a physical interpretation of (78).

Problem 5.4. Let $\mu(\mathbf{x}_0, \tau)$ be defined as in (40). Show that the stability at time τ of the equilibrium point $\mathbf{x}_0 = \mathbf{0}$ is equivalent to the continuity, at $\mathbf{x}_0 = \mathbf{0}$, of $\mu(\mathbf{x}_0, \tau)$. [*Hint:* Use (7).]

Problem 5.5. Suppose the system (1) is autonomous. Show that

79 $$\mathbf{s}(t, \mathbf{x}_0, t_0) = \mathbf{s}(t + T, \mathbf{x}_0, t_0 + T), \qquad \forall\, t \geq t_0,\ \forall\, \mathbf{x} \in R^n,\ \forall\, T \geq 0$$

Using (79), prove Theorem (45) for autonomous systems.

Problem 5.6. Complete the proofs of lemmas (53) and (61).

Problem 5.7. Suppose $V: R_+ \times R^n \longrightarrow R$ is defined by

$$V(t, \mathbf{x}) = \mathbf{x}'\mathbf{M}(t)\mathbf{x}$$

where $\mathbf{M}(t)$ is a real symmetric $n \times n$ matrix, and $t \mapsto \mathbf{M}(t)$ is continuous. Find the necessary and sufficient conditions for V to be a p.d.f. [*Answer:* $\inf_{t\geq 0} \lambda_{\min}\mathbf{M}(t) > 0$, where $\lambda_{\min}$ denotes the smallest eigenvalue.]

Problem 5.8. Determine whether or not each of the following functions is (a) an l.p.d.f., (b) a p.d.f., or (c) a decrescent function.

(i) $W(x_1, x_2) = x_1^4 + x_2^4$

(ii) $W(x_1, x_2) = x_1^2 + x_2^4$

(iii) $W(x_1, x_2) = (x_1 + x_2^2)^2$

(iv) $V(t, x_1, x_2) = t(x_1^2 + x_2^2)$

(v) $V(t, x_1, x_2) = (x_1^2 + x_2^2)/(t + 1)$

(vi) $W(x_1, x_2) = \sin^2(x_1 + x_2) + \sin^2(x_1 - x_2)$

5.2 LIAPUNOV'S DIRECT METHOD

The idea behind the various Liapunov theorems on stability, asymptotic stability, and stability is as follows: Consider a system which is "isolated" in the sense that there are no external forces acting on the system. Equation [5.1(1)] is a suitable model for such a system because no input is explicitly identified on the right-hand side of [5.1(1)]. Suppose that one can identify the equilibrium states of the system and that **0** is one of the equilibrium states (or possibly the only equilibrium state). Suppose it is also possible to define, in some suitable manner, the *total energy* of the system, which is a function having the property that it is zero at the origin and positive everywhere else. (In other words, the energy of the system has its global minimum at **0**.) Now suppose the system, which is originally at the equilibrium state **0**, is perturbed into a new nonzero initial state (where the energy level is positive, by assumption). If the system dynamics are such that the energy of the system is nonincreasing with time, then the energy level of the system never increases beyond its initial positive value. Depending on the nature of the energy function, this may be enough to imply the stability of the equilibrium point **0**. On the other hand, if the system dynamics are such that the energy of the system is monotonically decreasing with time and the energy eventually reduces to zero, it may be possible to conclude the asymptotic stability of the equilibrium state **0**, under suitable additional assumptions. The theorems of Liapunov, as well as their generalizations proved by later researchers, cast these ideas into a mathematically precise form. In addition, they offer the flexibility of allowing one to *choose* the *energy* function corresponding to a particular system.

In this section, we shall present the three basic kinds of theorems stemming from the so-called direct method of Liapunov, namely: stability theorems, asymptotic stability theorems, and instability theorems. These are illustrated by several examples, some of which are of a dynamics nature. These examples serve to show that, in some cases, the total energy stored within the system is a logical choice for the Liapunov function[6] of the system.

5.2.1 Stability Theorems

Theorem 1 is the basic stability theorem of Liapunov's direct method.

---

[6]This term is defined later.

1 ***Theorem*** The equilibrium point $\mathbf{0}$ at time t_0 of [5.1(1)] is stable if there exists a continuously differentiable l.p.d.f. V such that

2 $$\dot{V}(t, \mathbf{x}) \leq 0, \qquad \forall\, t \geq t_0,\ \forall\, \mathbf{x} \in B_r \quad \text{for some ball } B_r$$

proof Because V is an l.p.d.f., we have that

3 $$V(t, \mathbf{x}) \geq \alpha\,(\|\mathbf{x}\|), \qquad \forall\, t \geq t_0,\ \forall\, \mathbf{x} \in B_s \quad \text{for some ball } B_s$$

where $\alpha: R \longrightarrow R$ is a function of class K. To show that $\mathbf{0}$ is a stable equilibrium point at time t_0, we must show that, given any $\epsilon > 0$, we can find a $\delta(t_0, \epsilon) > 0$ such that [5.1(7)] holds. Accordingly, given $\epsilon > 0$, let $\epsilon_1 = \min\{\epsilon, r, s\}$, and pick $\delta > 0$ such that

4 $$\beta(t_0, \delta) = \sup_{\|\mathbf{x}\| \leq \delta} V(t_0, \mathbf{x}) < \alpha(\epsilon_1)$$

Such a δ can always be found, because $\alpha(\epsilon_1) > 0$ and $\beta(t_0, \delta) \longrightarrow 0$ as $\delta \longrightarrow 0$. We now show that [5.1(7)] holds if $\delta(t_0, \epsilon)$ is chosen equal to this δ. Suppose $\|\mathbf{x}(t_0)\| < \delta$. Then $V[t_0, \mathbf{x}(t_0)] \leq \beta(t_0, \delta) < \alpha(\epsilon_1)$. But because $\dot{V}(t, \mathbf{x}) \leq 0$ whenever $\|\mathbf{x}\| < \delta$,[7] we have

5 $$V[t, \mathbf{x}(t)] \leq V[t_0, \mathbf{x}(t_0)] < \alpha(\epsilon_1), \qquad \forall\, t \geq t_0$$

Now, because

6 $$\alpha(\|\mathbf{x}(t)\|) \leq V[t, \mathbf{x}(t)]$$

(5) and (6) together imply that

7 $$\alpha(\|\mathbf{x}(t)\|) < \alpha(\epsilon_1)$$

Because α is a nondecreasing function, (7) implies that

8 $$\|\mathbf{x}(t)\| < \epsilon_1 \leq \epsilon, \qquad \forall\, t \geq t_0$$

Hence [5.1(7)] holds and $\mathbf{0}$ is a stable equilibrium point at time t_0. ■

9 ***Theorem*** The equilibrium point $\mathbf{0}$ at time t_0 is uniformly stable over $[t_0, \infty)$ if there exists a continuously differentiable, decrescent l.p.d.f. V such that

10 $$\dot{V}(t, \mathbf{x}) \leq 0, \qquad \forall\, t \geq t_0,\ \forall\, \mathbf{x} \in B_r \quad \text{for some ball } B_r$$

proof Because V is decrescent, the function

11 $$\beta(\delta) = \sup_{\|\mathbf{x}\| < \delta}\ \sup_{t \geq t_0} V(t, \mathbf{x})$$

is nondecreasing and satisfies, for some $d > 0$, the condition

12 $$\beta(\delta) < \infty \quad \text{whenever } 0 \leq \delta \leq d$$

Thus, letting $\epsilon_1 = \min\{\epsilon, r, s, d\}$, we can always pick $\delta > 0$ such that $\beta(\delta) < \alpha(\epsilon_1)$. One now proceeds exactly as in the proof of Theorem (1) to

---

[7] Note that $\delta \leq \epsilon_1 \leq r$.

show that [5.1(8)] holds with this choice of δ. The details are left as an exercise. ∎

13 REMARKS:

1. Theorem (1) states that if we can find a continuously differentiable l.p.d.f. V such that its derivative along the trajectories of [5.1(1)] is always nonnegative, then the equilibrium point $\mathbf{0}$ is stable at time t_0. Theorem (9) shows that in order to conclude the *uniform* stability of the equilibrium point $\mathbf{0}$ over the interval $[t_0, \infty)$, it is enough to add the assumption that V is also decrescent. It should be noted that Theorems (1) and (9) provide only sufficient conditions for stability and uniform stability. [Actually, one can prove the converse of Theorem (9). This is discussed later on.]
2. The function V is commonly known as a *Liapunov function* or a *Liapunov function candidate*. The term *Liapunov function* is is a source of great confusion, and we shall attempt to avoid it in the following manner: Suppose, for example, that we are attempting to show that $\mathbf{0}$ is a stable equilibrium point by applying Theorem (1). Then we shall refer to a function V as a *Liapunov function candidate* (L.f.c.) if it satisfies the requirements imposed on V in the hypothesis of Theorem (1), i.e., if V is continuously differentiable and V is an l.p.d.f. If, for a particular system [5.1(1)], the conditions imposed on $\dot{V}$ are also satisfied, then V is referred to as a Liapunov function. The rationale behind this convention is as follows: Theorems (1) and (9) are sufficient conditions for certain stability properties. To apply them to a particular system, it is a fairly simple matter to find a function V satisfying the requirements on V. At this stage, V is a *Liapunov function candidate*. Now, for the particular system under study and for the particular choice of V, the conditions on $\dot{V}$ may or may not be met. If the requirements on $\dot{V}$ are met, then definite conclusions can be drawn, and V then becomes a Liapunov function. On the other hand, if the requirements on $\dot{V}$ are not met, because the theorems are only sufficient conditions, no definite conclusions can be drawn, and one has to start again with another Liapunov function candidate. The examples that follow illustrate this usage.

14 **Example.** Consider the simple pendulum, which is described by

$$\ddot{\theta} + \sin\theta = 0 \tag{15}$$

after suitable normalization. The state variable representation of this system is

16 $$\dot{x}_1 = x_2$$

17 $$\dot{x}_2 = -\sin x_1$$

Now, the total energy of the pendulum is the sum of the potential and kinetic energies, which is

18 $$V(x_1, x_2) = (1 - \cos x_1) + \frac{x_2^2}{2}$$

where the first term represents the potential energy and the second represents the kinetic energy. One can readily verify that V is an l.p.d.f. and is continuously differentiable, so that V is a Liapunov function candidate for applying Theorem (1).

Next, we have

19 $$\begin{aligned}\dot{V}(x_1, x_2) &= \sin x_1 \dot{x}_1 + x_2 \dot{x}_2 \\ &= \sin x_1(x_2) + x_2(-\sin x_1) = 0\end{aligned}$$

Therefore $\dot{V}$ also satisfies the requirements of Theorem (1). Hence V is a Liapunov function, and the equilibrium point $\mathbf{0}$ is stable by Theorem (1). Actually, because the system is autonomous, $\mathbf{0}$ is a uniformly stable equilibrium point (see Theorem [5.1(37)].)

20 **Example.** Consider a mass constrained by a nonlinear spring, with friction also present. Such a system can be represented in state variable form by

21 $$\dot{x}_1 = x_2$$

22 $$\dot{x}_2 = -f(x_2) - g(x_1)$$

where $f(x_2)$ represents the friction and $g(x_1)$ represents the restoring force of the spring. It is assumed that f and g satisfy the following conditions:

(i) f and g are continuous.

(ii) Whenever σ belongs to some interval $[-\sigma_0, \sigma_0]$, we have

23 $$\sigma f(\sigma) \geq 0, \qquad \forall\, \sigma \in [-\sigma_0, \sigma_0]$$

24 $$\sigma g(\sigma) > 0, \qquad \forall\, \sigma \in [-\sigma_0, \sigma_0], \qquad \sigma \neq 0$$

It is easy to see that Example (14) is a special case of Example (20) with $f(\sigma) = 0$ and $g(\sigma) = \sin \sigma$.

The energy stored in the system consists of the kinetic energy of the mass plus the potential energy stored in the spring. Thus the Liapunov function candidate is chosen as

$$V(x_1, x_2) = \frac{x_2^2}{2} + \int_0^{x_1} g(\sigma)\, d\sigma \tag{25}$$

Note that the energy dissipated through friction is not "stored" in the system (because it is not recoverable) and is therefore not included in V. Note also that V as defined in (25) is a natural generalization of that defined in (18). Now, by virtue of the assumptions on $g(\cdot)$, it is easy to show that V is continuously differentiable and is an l.p.d.f., and is hence a suitable Liapunov function candidate for applying Theorem (1).

Next, we have

$$\begin{aligned} \dot{V}(x_1, x_2) &= x_2\dot{x}_2 + g(x_1)\dot{x}_1 \\ &= x_2[-f(x_2) - g(x_1)] + g(x_1)x_2 \\ &= -x_2 f(x_2) \end{aligned} \tag{26}$$

By assumption (23) on $f(\cdot)$, it follows that

$$\dot{V}(x_1, x_2) \leq 0 \quad \text{whenever } |x_2| \leq \sigma \tag{27}$$

Hence $\dot{V}$ also meets the requirements of Theorem (1). Therefore V is a Liapunov function for the system (21)–(22), and $\mathbf{0}$ is a (uniformly) stable equilibrium point.

28 Example. Consider the system described by

$$\ddot{y}(t) + \dot{y}(t) + (2 + \sin t)y(t) = 0 \tag{29}$$

or, in state variable form,

$$\dot{x}_1 = x_2 \tag{30}$$

$$\dot{x}_2 = -x_2 - (2 + \sin t)x_1 \tag{31}$$

Equation (29) is an example of what is known as the damped Mathieu equation. In this case, there is no physical intuition readily available to guide us in the choice of V. Thus (after possibly a great deal of trial and error), we might be led to the Liapunov function candidate

$$V(t, x_1, x_2) = x_1^2 + \frac{x_2^2}{2 + \sin t} \tag{32}$$

Then V is continuously differentiable; moreover, V dominates the p.d.f.

$$W_1(x_1, x_2) = x_1^2 + \frac{x_2^2}{3} \tag{33}$$

and is dominated by the p.d.f.

$$W_2(x_1, x_2) = x_1^2 + x_2^2 \tag{34}$$

Hence V is a p.d.f., decrescent, and thus a suitable Liapunov function candidate for applying Theorem (9). We have

$$
\begin{aligned}
\dot{V}(t, x_1, x_2) &= -x_2^2 \frac{\cos t}{(2+\sin t)^2} + 2x_1\dot{x}_1 + \frac{2x_2}{2+\sin t}\dot{x}_2 \\
&= -x_2^2 \frac{\cos t}{(2+\sin t)^2} + 2x_1x_2 \\
&\quad + \frac{2x_2}{2+\sin t}[-x_2 - (2+\sin t)x_1] \\
&= -x_2^2 \frac{\cos t + 2(2+\sin t)}{(2+\sin t)^2} \\
&= -x_2^2 \frac{4 + 2\sin t + \cos t}{(2+\sin t)^2} \\
&\leq 0, \qquad \forall\, t \geq 0,\ \forall\, x_1, x_2
\end{aligned}
\tag{35}
$$

Thus the requirements on $\dot{V}$ in Theorem (9) are also met. Hence V is a Liapunov function for the system (30)–(31), and $\mathbf{0}$ is a uniformly stable equilibrium point over the interval $[0, \infty)$.

36 **Example.** One of the main applications of Liapunov theory is in obtaining stability conditions involving the parameters of the system under study. As an illustration, consider the system

$$\ddot{y}(t) + p(t)\dot{y}(t) + e^{-t}y(t) = 0 \tag{37}$$

which can be represented in state variable form as

$$\dot{x}_1 = x_2 \tag{38}$$

$$\dot{x}_2 = -p(t)x_2 - e^{-t}x_1 \tag{39}$$

The objective is to find some conditions on the function $p(\cdot)$ that ensure the stability at time 0 of the equilibrium point $\mathbf{0}$. For this purpose, let us choose

$$V(t, x_1, x_2) = x_1^2 + e^t x_2^2 \tag{40}$$

Because V is continuously differentiable and dominates the p.d.f.

$$W(x_1, x_2) = x_1^2 + x_2^2 \tag{41}$$

we see that V is a suitable Liapunov function candidate for applying Theorem (1). However, because V is not decrescent, it is *not* a suitable Liapunov function candidate for applying Theorem (9). Hence, using this particular V-function, we cannot hope to establish uniform stability by applying Theorem (9).

Differentiating V, we get

42
$$\dot{V}(t, x_1, x_2) = e^t x_2^2 + 2x_1(x_2) + 2e^t x_2[-p(t)x_2 - e^{-t}x_1]$$
$$= e^t x_2^2[-2p(t) + 1]$$

Hence we see that $\dot{V}$ is always nonpositive provided

43
$$p(t) \geq \frac{1}{2}, \qquad \forall\, t \geq 0$$

Thus the equilibrium point $\mathbf{0}$ is stable at time 0 provided (43) holds.

It should be emphasized that by employing a different Liapunov function candidate, we might be able to obtain entirely different stability conditions involving $p(\cdot)$.

44 REMARKS: The definition of stability in the sense of Liapunov is a *qualitative* one, in the sense that given an $\epsilon > 0$, we are only required to find *some* $\delta > 0$ such that [5.1(7)] holds—or, to put it another way, we are only required to demonstrate the *existence* of a suitable δ.

In the same way, Theorems (1) and (9) are also qualitative in the sense that they provide some conditions under which the existence of a suitable δ can be concluded. However, in principle at least, the conditions $\beta(t_0, \delta) < \alpha(\epsilon)$ [if Theorem (1) is being applied] or $\beta(\delta) < \alpha(\epsilon)$ [if Theorem (9) is being applied] can be used to actually determine a suitable value of δ corresponding to a given ϵ. But in practice this is rather messy.

5.2.2 Asymptotic Stability Theorems

In this subsection, we shall present some theorems that give sufficient conditions for asymptotic stability and global asymptotic stability.

45 ***Theorem*** The equilibrium point $\mathbf{0}$ at time t_0 of [5.1(1)] is uniformly asymptotically stable over the interval $[t_0, \infty)$ if there exists a continuously differentiable decrescent l.p.d.f. V such that $-\dot{V}$ is an l.p.d.f.

proof If $-\dot{V}$ is an l.p.d.f., then clearly $\dot{V}$ satisfies the hypothesis of Theorem (9), so that $\mathbf{0}$ is a uniformly stable equilibrium point over $[t_0, \infty)$. Thus, according to definition [5.1(14)], it is only necessary to prove the uniform convergence, to $\mathbf{0}$, of solution trajectories starting sufficiently close to $\mathbf{0}$. More precisely, we have to prove the existence of a $\delta_1 > 0$ such that, for each $\epsilon > 0$, there is a $T(\epsilon) < \infty$ such that

46
$$\|\mathbf{s}(t_1 + t, \mathbf{x}_0, t_1)\| < \epsilon \quad \text{whenever } t > T(\epsilon),\ \|\mathbf{x}_0\| < \delta_1, \text{ and } \forall\, t_1 \geq t_0$$

The hypotheses on V and $\dot{V}$ imply that there are functions $\alpha(\cdot)$, $\beta(\cdot)$, and $\gamma(\cdot)$ belonging to class K such that

$$\alpha(\|\mathbf{x}\|) \leq V(t, \mathbf{x}) \leq \beta(\|\mathbf{x}\|) \tag{47}$$

$$\dot{V}(t, \mathbf{x}) \leq -\gamma(\|\mathbf{x}\|) \qquad \forall\, t \geq t_0,\ \forall\, \mathbf{x} \in B_r \quad \text{for some ball } B_r \tag{48}$$

Now, given $\epsilon > 0$, define positive constants δ_1, δ_2, and T by

$$\beta(\delta_1) < \alpha(r) \tag{49}$$

$$\beta(\delta_2) < \min\{\alpha(\epsilon), \beta(\delta_1)\} \tag{50}$$

$$T = \frac{\alpha(r)}{\gamma(\delta_2)} \tag{51}$$

Now we claim that

$$\|\mathbf{s}(t_2, \mathbf{x}_0, t_1)\| < \delta_2 \qquad \text{for some } t_2 \in [t_1, t_1 + T] \quad \text{whenever } \|\mathbf{x}_0\| < \delta_1 \text{ and } t_1 \geq t_0 \tag{52}$$

To prove (52), assume by way of contradiction that

$$\|\mathbf{s}(t, \mathbf{x}_0, t_1)\| \geq \delta_2, \qquad \forall\, t \in [t_1, t_1 + T] \tag{53}$$

Then

$$\begin{aligned} 0 < \beta(\delta_2) &\leq V[t_1 + T, \mathbf{s}(t_1 + T, \mathbf{x}_0, t_1)] && \text{by (47)} \\ &= V(t_1, \mathbf{x}_0) + \int_{t_1}^{t_1+T} \dot{V}[\tau, \mathbf{s}(\tau, \mathbf{x}_0, t_1)]\, d\tau \\ &\leq \beta(\delta_1) - T\gamma(\delta_2) && \text{by (47) and (53)} \\ &= \beta(\delta_1) - \alpha(r) && \text{by (51)} \\ &< 0 && \text{by (49)} \end{aligned} \tag{54}$$

which is a contradiction. This shows that (52) is true.

To complete the proof, suppose $t \geq t_1 + T$. Then

$$\begin{aligned} \alpha(\|\mathbf{s}(t, \mathbf{x}_0, t_1)\|) &\leq V[t, \mathbf{s}(t, \mathbf{x}_0, t_1)] && \text{by (47)} \\ &\leq V[t_2, \mathbf{s}(t_2, \mathbf{x}_0, t_1)] \end{aligned} \tag{55}$$

by the nonpositivity of $\dot{V}$. [Note that t_2 is defined in (52).] Finally,

$$V[t_2, \mathbf{s}(t_2, \mathbf{x}_0, t_1)] \leq \beta[\|\mathbf{s}(t_2, \mathbf{x}_0, t_1)\|] \leq \beta(\delta_2) \qquad \text{by (52)} \tag{56}$$

Now, (55) and (56) together imply that

$$\alpha[\|\mathbf{s}(t, \mathbf{x}_0, t_1)\|] \leq \beta(\delta_2) < \alpha(\epsilon) \qquad \text{by (50)} \tag{57}$$

The inequality (57) establishes (46). ∎

58 ***Theorem*** The equilibrium point $\mathbf{0}$ at time t_0 of [5.1(1)] is globally asymptotically stable if there exists a continuously differentiable decrescent p.d.f. V such that

$$\dot{V}(t, \mathbf{x}) \leq -\gamma(\|\mathbf{x}\|), \qquad \forall\, t \geq t_0,\ \forall\, \mathbf{x} \in R^n \tag{59}$$

where γ is a function belonging to class K.

proof Same as that of Theorem (45), except that one can take $r = \infty$ in (47). ∎

REMARKS:

1. Note that, in Theorem (58), there is no requirement that $-\dot{V}$ should be a p.d.f., because there is no requirement that $\gamma(p) \longrightarrow \infty$ as $p \longrightarrow \infty$.
2. In Theorems (45) and (58), V is required to be decrescent, but $-\dot{V}$ need not be decrescent.
3. In view of lemma [5.1(53)], condition (59) above can be replaced by an equivalent condition, as follows: There exists a continuous function $W\colon R^n \longrightarrow R$ such that

$$\dot{V}(t, \mathbf{x}) \leq -W(\mathbf{x}), \qquad \forall\, t \geq t_0,\ \forall\, \mathbf{x} \in R^n \tag{60}$$

where

$$W(\mathbf{0}) = 0; \qquad W(\mathbf{x}) > 0, \qquad \forall\, \mathbf{x} \neq \mathbf{0} \tag{61}$$

62 Example. Consider the system of equations

$$\dot{x}_1 = x_1(x_1^2 + x_2^2 - 1) - x_2 \tag{63}$$

$$\dot{x}_2 = x_1 + x_2(x_1^2 + x_2^2 - 1) \tag{64}$$

and try the Liapunov function candidate

$$V(x_1, x_2) = x_1^2 + x_2^2 \tag{65}$$

Then V is a p.d.f. Further,

$$\dot{V}(x_1, x_2) = 2(x_1^2 + x_2^2)(x_1^2 + x_2^2 - 1) \tag{66}$$

Thus $-\dot{V}$ is an l.p.d.f. over the ball $\{(x_1, x_2)\colon x_1^2 + x_2^2 \leq 1\}$. Hence by Theorem (45), it follows that $\mathbf{0}$ is a (uniformly) asymptotically stable equilibrium point of this system. However, because $-\dot{V}$ is not an l.p.d.f. over all of R^n, Theorem (58) does not apply, and so we cannot conclude global asymptotic stability. [Of course, since Theorem (58) is only a *sufficient* condition, the fact that Theorem (58) is inapplicable does *not* mean that the system is not globally asymptotically stable. However, for this particular system, a quick sketch of the vector field associated with equations (62) and (63) reveals that not all trajectories do converge to $\mathbf{0}$.]

67 Example. Consider now the system of equations

$$\dot{x}_1(t) = -x_1(t) - e^{-2t}x_2(t) \tag{68}$$

$$\dot{x}_2(t) = x_1(t) - x_2(t) \tag{69}$$

and choose the Liapunov function candidate

$$V(t, x_1, x_2) = x_1^2 + (1 + e^{-2t})x_2^2 \tag{70}$$

It is easy to verify that V is a decrescent p.d.f. Moreover,

71 $$\dot{V}(t, x_1, x_2) = -2[x_1^2 + x_1x_2 + x_2^2(1 + 2e^{-2t})]$$

Hence $\dot{V}$ satisfies (60) with

72 $$W(x_1, x_2) = 2(x_1^2 + x_1x_2 + x_2^2)$$

Because W satisfies (61), it follows that $\mathbf{0}$ is a globally asymptotically stable equilibrium point.

The rest of this subsection is devoted to what is known as LaSalle's Theorem. LaSalle's theorem is more powerful than Theorems (45) and (58) because it enables one to conclude asymptotic stability even in cases where $-\dot{V}$ is not an l.p.d.f. However, it applies only to autonomous or periodic systems.

73 *[definition]* A set S in R^n is said to be the *positive limit set* of a trajectory $\mathbf{s}(\cdot, \mathbf{x}_0, t_0)$ of [5.1(1)] if, for every $\mathbf{y} \in S$, there exists a sequence $(t_n)_1^\infty$ such that $t_n \longrightarrow \infty$ and $\mathbf{s}(t_n, \mathbf{x}_0, t_0) \longrightarrow \mathbf{y}$.

Note that we have previously encountered the concept of the positive limit set in Chap. 2, in connection with Poincaré's theorems on limit cycles. Also, note that if $\mathbf{0}$ is a globally asymptotically stable equilibrium point of [5.1(1)], then the positive limit set of *every* trajectory of [5.1(1)] is the single element set $\{\mathbf{0}\}$.

74 *[definition]* A set M in R^n is said to be an *invariant set* of the system [5.1(1)] if, whenever $\mathbf{y} \in M$ and $t_0 \geq 0$, we have

75 $$\mathbf{s}(t, \mathbf{y}, t_0) \in M, \qquad \forall\, t \geq t_0$$

In other words, M is an invariant set of [5.1(1)] if every trajectory that starts from an initial point in M stays within M at all future times. For example, any equilibrium point of [5.1(1)] is an invariant set. If [5.1(1)] is autonomous, every trajectory of [5.1(1)] is a invariant set.

76 ***lemma*** If $\mathbf{s}(\cdot, \mathbf{x}_0, t_0)$ is a bounded trajectory of [5.1(1)], then its positive limit set S is closed and bounded. Furthermore, $\mathbf{s}(t, \mathbf{x}_0, t_0)$ approaches S as $t \longrightarrow \infty$; i.e.,

77 $$\inf_{\mathbf{y} \in S} \|\mathbf{s}(t, \mathbf{x}_0, t_0) - \mathbf{y}\| \longrightarrow 0 \qquad \text{as } t \longrightarrow \infty$$

The proof of this lemma is relatively easy but is omitted here, because it uses a few concepts from real analysis not heretofore introduced.

78 ***lemma*** Suppose the system [5.1(1)] is autonomous or periodic, and let S be the positive limit set of any trajectory of [5.1(1)]. Then S is an invariant set of [5.1(1)].

proof We shall give the proof only in the case of autonomous systems. Suppose S is the positive limit set of the trajectory $\mathbf{s}(\cdot, \mathbf{x}_0, t_0)$ of [5.1(1)] and that $\mathbf{y} \in S$. Let $t_1 \geq 0$ be arbitrary; we must show that $\mathbf{s}(t, \mathbf{y}, t_1) \in S$ for all $t \geq t_1$. Now, by the definition of S, $\mathbf{y} \in S$ implies that there exists a sequence $(t_n)_1^\infty$ such that $t_n \longrightarrow \infty$ and $\mathbf{s}(t_n, \mathbf{x}_0, t_0) \longrightarrow \mathbf{y}$ as $n \longrightarrow \infty$. Because the solutions of [5.1(1)] are (assumed to be) continuously dependent on the initial conditions, we have, for each fixed $t \geq t_1$,

$$\mathbf{s}(t, \mathbf{y}, t_1) = \lim_{n \to \infty} \mathbf{s}[t, \mathbf{s}(t_n, \mathbf{x}_0, t_0), t_1] \tag{79}$$

However, because the system is autonomous, we have

$$\mathbf{s}[t, \mathbf{s}(t_n, \mathbf{x}_0, t_0), t_1] = \mathbf{s}(t + t_n - t_1 + t_0, \mathbf{x}_0, t_0) \tag{80}$$

(see Problem 5.5). Now, as $n \longrightarrow \infty$, $t_n \longrightarrow \infty$, so that the right-hand side of (80) converges to an element of S, by lemma (76). Hence $\mathbf{s}(t, \mathbf{y}, t_1) \in S$ for all $t \geq t_1$. ∎

81 ***lemma*** Suppose the system [5.1(1)] is autonomous. Let $V: R^n \longrightarrow R$ be continuously differentiable, and suppose that for some $c > 0$, the set

$$\Omega_c = \{\mathbf{x} \in R^n : V(\mathbf{x}) \leq c\} \tag{82}$$

is bounded. Suppose that V is bounded below over the set Ω_c and that $\dot{V}(x) \leq 0 \ \forall \ \mathbf{x} \in \Omega_c$. Let S denote the subset of Ω_c defined by

$$S = \{\mathbf{x} \in \Omega_c : \dot{V}(\mathbf{x}) = 0\} \tag{83}$$

and let M be the largest invariant set of [5.1(1)] contained in S. Then, whenever $\mathbf{x}_0 \in \Omega_c$, the solution $\mathbf{s}(t, \mathbf{x}_0, 0)$[8] of [5.1(1)] approaches M as $t \longrightarrow \infty$.

proof Suppose $\mathbf{x}_0 \in \Omega_c$. Clearly, $V[\mathbf{s}(t, \mathbf{x}_0, 0)]$ is nondecreasing as a function of t, because $\dot{V}(x) \leq 0 \ \forall \ x \in \Omega_c$. This shows, by the definition of Ω_c, that $\mathbf{s}(t, \mathbf{x}_0, 0) \in \Omega_c, \ \forall \ t \geq 0$. Also, $V[\mathbf{s}(t, \mathbf{x}_0, 0)]$ has a definite limit as $t \longrightarrow \infty$, because it is a nonincreasing function that is bounded below; let c_0 denote this limit. Let L denote the limit set of the trajectory $\mathbf{s}(\cdot, \mathbf{x}_0, 0)$. Because $V[\mathbf{s}(t, \mathbf{x}_0, 0)] \longrightarrow c_0$ as $t \longrightarrow \infty$, it follows that $V(\mathbf{y}) = c_0 \ \forall \ \mathbf{y} \in L$. Because L is an invariant set of [5.1(1)], we have that $\dot{V}(\mathbf{y}) = 0 \ \forall \ \mathbf{y} \in L$, so that L is a subset of S. Because M is the largest invariant subset in S, L is also a subset of M. By lemma (76), $\mathbf{s}(t, \mathbf{x}_0, 0)$ approaches L as $t \longrightarrow \infty$, whence we conclude that $\mathbf{s}(t, \mathbf{x}_0, 0)$ approaches M as $t \longrightarrow \infty$. ∎

[8] Note that we can take the initial time to be 0 without loss of generality, because the system is autonomous.

With the aid of lemma (81), we can state and prove a sufficient condition for asymptotic stability that is weaker than Theorem (45).

84 ***Theorem*** Suppose the system [5.1(1)] is autonomous. Let $V: R^n \longrightarrow R$ be a continuously differentiable l.p.d.f., over some ball B_r, and suppose $\dot{V}(\mathbf{x}) \leq 0 \ \forall \ \mathbf{x} \in B_r$. Let

85 $$m = \sup_{\|\mathbf{x}\| \leq r} V(\mathbf{x})$$

and define

86 $$S = \{\mathbf{x}: V(\mathbf{x}) \leq m, \dot{V}(\mathbf{x}) = 0\}$$

Suppose S contains no trajectories of [5.1(1)] other than the trivial trajectory $\mathbf{x} \equiv \mathbf{0}$. Then the equilibrium point $\mathbf{0}$ of [5.1(1)] is asymptotically stable.

proof By lemma (81), whenever $\|\mathbf{x}_0\| \leq r$, the corresponding solution $\mathbf{s}(t, \mathbf{x}_0, 0)$ approaches the largest invariant set contained in S. But by the hypothesis that S contains no nontrivial trajectories of [5.1(1)], it is clear that the largest invariant set contained in S is $\{\mathbf{0}\}$. Hence $\mathbf{S}(t, \mathbf{x}_0, 0) \longrightarrow \mathbf{0}$ as $t \longrightarrow \infty$, so that $\mathbf{0}$ is an asymptotically stable equilibrium point of [5.1(1)]. ∎

Remarks: One should be careful in applying this theorem: The set S consists of *all* points in R^n where $\dot{V}(\mathbf{x}) = 0$ and $V(\mathbf{x}) \leq m$, and as such may contain points outside the ball B_r.

The results of Theorem (84) can be readily extended to encompass global asymptotic stability.

87 ***Theorem*** Suppose the system [5.1(1)] is autonomous. Suppose $V: R^n \longrightarrow R$ is a continuously differentiable p.d.f., $\dot{V}(\mathbf{x}) \leq 0 \ \forall \ \mathbf{x} \in R^n$, and the set

$$S = \{\mathbf{x} \in R^n: \dot{V}(\mathbf{x}) = 0\}$$

contains no nontrivial trajectories of [5.1(1)]. Then $\mathbf{0}$ is a globally asymptotically stable equilibrium point of [5.1(1)].

proof Let $r = \infty$ in the proof of Theorem (84). ∎

Up to now, we have only considered autonomous systems. We shall now state, without proof, La Salle's theorem for periodic systems.

88 ***Theorem*** Suppose the system [5.1(1)] is periodic; i.e., suppose

89 $$\mathbf{f}(t, \mathbf{x}) = \mathbf{f}(t + T, \mathbf{x}), \qquad \forall \ t, \forall \ \mathbf{x} \in R^n, \quad \text{for some } T > 0$$

Suppose $V: R_+ \times R^n \longrightarrow R$ is a continuously differentiable p.d.f., with

90 $$V(t, \mathbf{x}) = V(t + T, \mathbf{x}), \qquad \forall\, \mathbf{x} \in R^n, \forall\, t \geq 0$$

Define

91 $$S = \{\mathbf{x} \in R^n : \dot{V}(t, \mathbf{x}) = 0, \forall\, t \geq 0\}$$

Suppose $\dot{V}(t, \mathbf{x}) \leq 0$, $\forall\, t \geq 0$, $\forall\, \mathbf{x} \in R^n$, and that S contains no nontrivial trajectories of [5.1(1)]. Under these conditions, the equilibrium point $\mathbf{0}$ is globally asymptotically stable.

92 **Example.** Consider once again the system of Example (20), namely,

93 $$\dot{x}_1 = +x_2$$

94 $$\dot{x}_2 = -f(x_2) - g(x_1)$$

where f and g now satisfy the following conditions:

(i) f and g are continuous.

(ii) $f(0) = g(0) = 0$; $\sigma f(\sigma) > 0$, $\sigma g(\sigma) > 0$ whenever $\sigma \neq 0$.

(iii) $\int_0^\sigma g(\xi)\, d\xi \longrightarrow \infty$ as $|\sigma| \longrightarrow \infty$.

We once again choose, as the Liapunov function candidate, the total energy of the system, namely,

95 $$V(x_1, x_2) = \frac{x_2^2}{2} + \int_0^{x_1} g(\xi)\, d\xi$$

Conditions (i)–(iii) above ensure that V is a continuously differentiable p.d.f., so that V is a suitable Liapunov function candidate for applying Theorem (87). Calculating $\dot{V}$, we get

96 $$\dot{V}(x_1, x_2) = -x_2 f(x_2)$$

Thus $\dot{V}(x_1, x_2) \leq 0$, $\forall\, (x_1, x_2) \in R^2$. Moreover,

97 $$S = \{(x_1, x_2) \in R^2 : \dot{V}(x_1, x_2) = 0\} = \{(x_1, x_2) \in R^2 : x_2 = 0\}$$

because of (ii) above. If S contains any trajectories of (93)–(94), it must have $x_2(t) \equiv 0$. This implies that (i) $x_1 =$ constant $= x_{10}$, by (93), and (ii) $\dot{x}_2(t) \equiv 0$. Now, (94) gives

98 $$-f(x_2) - g(x_{10}) = 0, \qquad \text{i.e., } -g(x_{10}) = 0$$

because $x_2 = 0$. By (ii) above, this shows that $x_{10} = 0$. In other words, the only trajectory that lies entirely within S is the trivial trajectory $x_1 \equiv x_2 \equiv 0$. Thus all the conditions of Theorem (87) are satisfied, whence $\mathbf{0}$ is a globally asymptotically stable equilibrium point.

99 **Example.** Consider once again the damped Mathieu equation of Example (28), namely,

100 $$\dot{x}_1 = x_2$$

101 $$\dot{x}_2 = -x_2 + (2 + \sin t)x_1$$

This system is clearly periodic with period 2π. Let us choose V as in (32), i.e.,

102 $$V(t, x_1, x_2) = x_1^2 + \frac{x_2^2}{2 + \sin t}$$

Then V is a continuously differentiable p.d.f. and is periodic with period 2π. Therefore, V is a suitable Liapunov function candidate for applying Theorem (88). Calculating $\dot{V}$, we get

103 $$\dot{V}(t, x_1, x_2) = -x_2^2 \frac{4 + 2 \sin t + \cos t}{(2 + \sin t)^2}$$

[see (35)]. Thus $\dot{V} \leq 0$ for all t, (x_1, x_2). Moreover, the set S defined in (91) is given by

104 $$S = \{(x_1, x_2): x_2 = 0\}$$

As in the previous example, if a trajectory of (100)–(101) is entirely contained in S, it must have $x_2(t) \equiv 0$, which in turn implies that x_1 is a constant [by (100)], and that

105 $$(2 + \sin t)x_1 \equiv 0$$

Keeping in mind that x_1 is a constant, (105) implies that $x_1 = 0$. Hence the only trajectory entirely contained within S is the trivial trajectory $x_1 \equiv x_2 \equiv 0$. Because all the hypotheses of Theorem (88) are met, it follows that (0, 0) is a globally asymptotically stable equilibrium point.

5.2.3 Instability Theorems

In the two previous subsections, we presented sufficient conditions for stability and for asymptotic stability. In this final subsection, we shall give some sufficient conditions for *instability*.

106 ***Theorem*** The equilibrium point $\mathbf{0}$ at time t_0 of [5.1(1)] is unstable if there exists a continuously differentiable decrescent function $V: R_+ \times R^n \rightarrow R$ such that (i) $\dot{V}$ is an l.p.d.f., and (ii) $V(t, \mathbf{0}) = 0$, and there exist points $\mathbf{x}$ arbitrarily close to $\mathbf{0}$ such that $V(t_0, \mathbf{x}) > 0$.

proof To demonstrate that $\mathbf{0}$ is an unstable equilibrium point, we must show that, for some $\epsilon > 0$, no $\delta > 0$ exists such that [5.1(8)] holds. Suppose $\dot{V}$ is an l.p.d.f. over some ball B_r, with $r > 0$. Because V is decrescent, there is a function β belonging to class K such that

107 $$V(t, \mathbf{x}) \leq \beta(\|\mathbf{x}\|), \qquad \forall\, t \geq t_0,\ \forall\, \mathbf{x} \in B_s \text{ for some ball } B_s$$

We shall show that, if we let $\epsilon = \min\{r, s\}$, then no matter how small we choose $\delta > 0$, we can always find a corresponding $\mathbf{x}_0$ such that $||\mathbf{x}_0|| < \delta$ and $||\mathbf{s}(t, \mathbf{x}_0, t_0)||$ eventually equals or exceeds ϵ. Thus given any $\delta > 0$, pick $\mathbf{x}_0$ such that (i) $||\mathbf{x}_0|| < \delta$, and (ii) $V(t_0, \mathbf{x}_0) > 0$. In the interests of brevity, let $\mathbf{x}(t)$ denote $\mathbf{s}(t, \mathbf{x}_0, t_0)$. Now, as long as $\mathbf{x}(t)$ remains within B_ϵ, we have $\dot{V}[t, \mathbf{x}(t)] \geq 0$, which shows that

108 $$V[t, \mathbf{x}(t)] \geq V(t_0, \mathbf{x}_0) > 0$$

The inequalities (107) and (108) together imply that $||\mathbf{x}(t)||$ is bounded away from zero, which in turn implies, by the positive definiteness of $\dot{V}$, that $\dot{V}[t, \mathbf{x}(t)]$ is also bounded away from zero. Hence, within a finite amount of time, $V[t, \mathbf{x}(t)]$ will exceed the value $\beta(\epsilon)$, which means that $||\mathbf{x}(t)||$ exceeds ϵ. Thus $\mathbf{0}$ is an unstable equilibrium point. ∎

Note that, in contrast with previous theorems, the function V in Theorem (106) can assume positive or negative values. Also, the inequality (107), i.e., the requirement that V is a decrescent function places no restrictions on the behavior of $V(t, \mathbf{x})$ when it assumes negative values.

109 **Example.** Consider the system of equations

110 $$\dot{x}_1 = x_1 - x_2 + x_1 x_2$$

111 $$\dot{x}_2 = -x_2 - x_2^2$$

and choose the Liapunov function candidate

112 $$V(x_1, x_2) = (2x_1 - x_2)^2 - x_2^2$$

Even though V assumes both positive and negative values, it has the property that it assumes positive values arbitrarily close to the origin. Hence it is a suitable Liapunov function candidate for applying Theorem (106). Differentiating V, we get

113 $$\begin{aligned}\dot{V}(x_1, x_2) &= 2(2x_1 - x_2)(2\dot{x}_1 - \dot{x}_2) - 2x_2\dot{x}_2 \\ &= [(2x_1 - x_2)^2 + x_2^2](1 + x_2)\end{aligned}$$

Thus $\dot{V}$ is an l.p.d.f. over the ball $B_{1-\delta}$ for any $\delta \in (0, 1)$. Hence V is a Liapunov function, and $\mathbf{0}$ is an unstable equilibrium point, by Theorem (106).

114 REMARKS: Some authors prove a less efficient version of Theorem (106) by showing that $\mathbf{0}$ is a unstable equilibrium point if one can find a function V such that both V and $\dot{V}$ are l.p.d.f.'s. Actually, it can be shown that if one can find such a function V, then $\mathbf{0}$ is what is known as a *completely unstable* equilibrium point; i.e., there exists an $\epsilon > 0$ such that *every* trajectory $\mathbf{x}(\cdot)$ other than the trivial trajectory $\mathbf{x}(t) \equiv \mathbf{0}$ satisfies $||\mathbf{x}(t)|| \geq \epsilon$ for some t. Such a hopeless situation rarely occurs in practice, and as a result, a theorem of this type is of very little use.

Theorems (115) and (126) give alternative sufficient conditions for instability. They also have certain other advantages, which are discussed later.

115 ***Theorem*** The equilibrium point $\mathbf{0}$ at time t_0 of [5.1(1)] is unstable if there exists a continuously differentiable decrescent function $V: R_+ \times R^n \longrightarrow R$ such that (i) $V(t_0, \mathbf{0}) = 0$, and $V(t_0, \mathbf{x})$ assumes positive values arbitrarily close to the origin, and (ii) $\dot{V}(t, \mathbf{x})$ is of the form

116
$$\dot{V}(t, \mathbf{x}) = \lambda V(t, \mathbf{x}) + V_1(t, \mathbf{x})$$

where $\lambda > 0$ is a constant, and $V_1: R_+ \times R^n \longrightarrow R$ satisfies the condition

117
$$V_1(t, \mathbf{x}) \geq 0, \qquad \forall\, t \geq t_0, \forall\, \mathbf{x} \in B_r \quad \text{for some ball } B_r$$

proof We show that if we choose $\epsilon = r$, then [5.1(7)] cannot be satisfied for any choice of $\delta > 0$. Thus, given $\delta > 0$, choose $\mathbf{x}_0$ such that (i) $\|\mathbf{x}_0\| < \delta$, and (ii) $V(t_0, \mathbf{x}_0) > 0$. Let $\mathbf{x}(\cdot)$ denote the resulting solution trajectory $\mathbf{s}(\cdot, \mathbf{x}_0, t_0)$ of [5.1(1)]. Because, whenever $\|\mathbf{x}(t)\| \leq r$,

118
$$\frac{d}{dt}\{V[t, \mathbf{x}(t)]\} = \lambda V[t, \mathbf{x}(t)] + V_1[t, \mathbf{x}(t)] \geq \lambda V[t, \mathbf{x}(t)]$$

it follows that

119
$$\frac{d}{dt}\{e^{-\lambda t} V[t, \mathbf{x}(t)]\} \geq 0$$

Hence

120
$$V[t, \mathbf{x}(t)] \geq e^{\lambda(t-t_0)} V(t_0, \mathbf{x}_0)$$

Hence $V[t, \mathbf{x}(t)]$ increases without bound, and therefore $\|\mathbf{x}(t)\|$ must eventually equal ϵ. This shows that $\mathbf{0}$ is an unstable equilibrium point. ■

121 **Example.** Consider the system of equations

122
$$\dot{x}_1 = x_1 + 2x_2 + x_1 x_2^2$$

123
$$\dot{x}_2 = 2x_1 + x_2 - x_1^2 x_2$$

and let

124
$$V(x_1, x_2) = x_1^2 - x_2^2$$

Then V is a suitable Liapunov function candidate for applying Theorem (115). Calculating $\dot{V}$, we get

125
$$\begin{aligned} \dot{V}(x_1, x_2) &= 2x_1\dot{x}_1 - 2x_2\dot{x}_2 = 2x_1^2 - 2x_2^2 + 4x_1^2x_2^2 \\ &= 2V(x_1, x_2) + 4x_1^2x_2^2 \end{aligned}$$

Because $4x_1^2x_2^2 \geq 0\ \forall\, (x_1, x_2) \in R^2$, it follows by Theorem (115)

that $\mathbf{0}$ is an unstable equilibrium point. Note that neither V nor $\dot{V}$ is an l.p.d.f. in this example.

In Theorems (106) and (115), the function $\dot{V}$ is required to satisfy certain conditions at all points belonging to some neighborhood of the origin in R^n. In Theorem (126), the condition on $\dot{V}$ is only required to hold in some region for which the origin is a boundary point (and not an interior point). Theoretically, however, Theorems (115) and (126) are equivalent, as shown later. Theorem (126) is generally known as Cetaev's theorem.

126 ***Theorem*** The equilibrium point $\mathbf{0}$ at time t_0 of [5.1(1)] is unstable if the following conditions hold: There exist a continuously differentiable function $V: R_+ \times R^n \rightarrow R$ and a closed set Ω containing $\mathbf{0}$ in its interior such that

(i) There is an open subset Ω_1 of Ω containing $\mathbf{0}$ on its boundary,

127 (ii) $V(t, \mathbf{x}) > 0, \quad \forall\, t \geq t_0, \forall\, \mathbf{x} \in \Omega_1$

128 $V(t, \mathbf{x}) = 0, \quad \forall\, t \geq t_0, \forall\, \mathbf{x} \in \partial\Omega_1$

(the boundary of Ω_1 in Ω)

(iii) $V(t, \mathbf{x})$ is bounded above in Ω, uniformly in t, and

129 (iv) $\dot{V}(t, \mathbf{x}) \geq \gamma(\|\mathbf{x}\|), \quad \forall\, t \geq t_0, \forall\, \mathbf{x} \in \Omega_1$

where γ is a function belonging to class K.

proof Pictorially, the situation can be represented as in Fig. 5.7. In view of the assumptions on V and $\dot{V}$, it is clear that along any trajectory starting in Ω_1, V increases indefinitely and cannot cross the set $\partial\Omega_1$. Because V is bounded above on Ω, the trajectory must eventually reach the boundary of Ω, regardless of its starting point in Ω_1. This means that $\mathbf{0}$ is an unstable equilibrium point. ∎

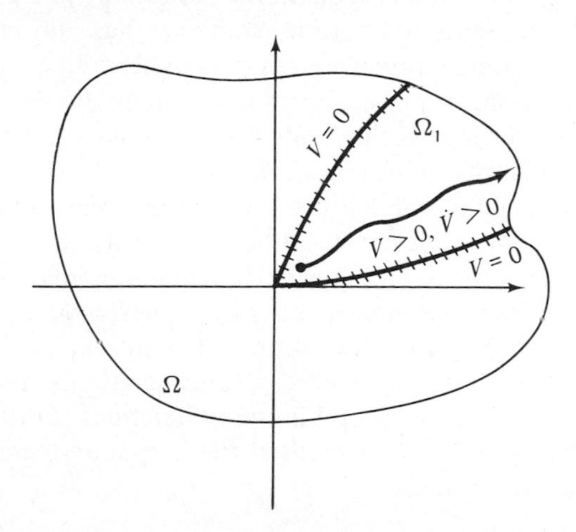

FIG. 5.7

130 **Example.** Consider the system of equations

131 $$\dot{x}_1 = 2x_1 + x_2 + x_1^2$$

132 $$\dot{x}_2 = x_1 - x_2 + x_1x_2$$

and choose the Liapunov function candidate

133 $$V(x_1, x_2) = x_1x_2$$

Let Ω be the set

134 $$\Omega = \{(x_1, x_2): x_1^2 + x_2^2 \leq 1\}$$

and let Ω_1 be the subset

135 $$\Omega_1 = \{(x_1, x_2) \in \Omega: x_1 \geq 0, x_2 \geq 0\}$$

Then conditions (i)–(iii) of Theorem (126) are satisfied. To verify condition (iv), we differentiate V to get

136 $$\begin{aligned}\dot{V}(x_1, x_2) &= x_1(x_1 - x_2 + x_1x_2) + x_2(2x_1 + x_2 + x_1^2) \\ &= x_1^2 + x_2^2 + 2x_1x_2 + 2x_1^2x_2 > 0, \quad \forall\, (x_1, x_2) \in \Omega_1/\{\mathbf{0}\}\end{aligned}$$

Hence by Theorem (126), (0, 0) is an unstable equilibrium point.

5.2.4 Concluding Remarks

In this section, we have presented various theorems in Liapunov stability theory. The key feature of all these theorems is that they enable one to draw conclusions about the stability status of an equilibrium point, *without solving the system equations*, by constructing a suitable Liapunov function. The favorable aspects of these theorems are

1. There is no need to solve the system equations (as mentioned above), and
2. The Liapunov function V, especially in the theorems on stability and asymptotic stability, has an intuitively appealing geometric interpretation as the *energy* of the system.

In fact, in many physical problems, the total energy stored within the system is the logical candidate for a Liapunov function. This is brought out in Examples (14) and (20).

The unfavorable aspects of the theorems given here are (1) they present only sufficient conditions for various forms of stability, so that if a particular Liapunov function candidate V fails to meet the requirements on $\dot{V}$, no conclusions can be drawn, and the testing process must be begun once again, with a different Liapunov function candidate. (2) In a general system of nonlinear equations, there is no systematic procedure for generating Liapunov functions. Both of these are serious drawbacks and have prevented the Liapunov theory from being more widely used than it is.

In this context, it is worthwhile to note two points: First, Liapunov presented his original theorems only as sufficient conditions. However, later researchers have shown that the *existence* of a Liapunov function is also a *necessary* condition for various forms of stability. In particular, the existence of a function V satisfying the hypotheses of Theorems (9), (45), (115), and (126) is a necessary as well as sufficient condition for the equilibrium point $\mathbf{0}$ to have the stability status mentioned in those theorems. [This means, in particular, that Theorems (115) and (126) are equivalent.] However, knowing that a Liapunov function exists still does not help one to find it. As a result, the so-called *converse* theorems, which establish the existence of a Liapunov function, are of theoretical interest only.

The second point is that, even though in general it is a very difficult task to find a Liapunov function for a given system, there are two special cases where the choice of a Liapunov function is relatively easy, namely: (1) linear systems, and (2) "weakly" nonlinear systems. Linear systems are thoroughly explored in Sec. 5.3, while weakly nonlinear systems are studied in Sec. 5.4. In his original work, Liapunov also derived a method whereby the stability of a nonlinear system can be ascertained by linearizing the nonlinear system about an equilibrium point and studying the resulting linear system. This method is known as Liapunov's first or indirect method and is covered in Sec. 5.4. Many authors present the indirect method before the direct method, but we prefer to derive the indirect method as a consequence of the direct method and of the Liapunov theory for linear systems.

Problem 5.9. A phase-locked loop in communication networks can be described by

$$\ddot{y}(t) + [a + b\cos y(t)]\dot{y}(t) + c\sin y(t) = 0$$

(a) Transform this equation into state variable form by choosing $x_1 = y, x_2 = \dot{y}$.

(b) Suppose $c > 0$. Using the Liapunov function candidate

$$V(x_1, x_2) = c(1 - \cos x_1) + x_2^2/2$$

show that (0, 0) is a stable equilibrium point if $a \geq b \geq 0$, and that (0, 0) is an asymptotically stable equilibrium point if $a > b \geq 0$. (Hint: In the latter case, use LaSalle's theorem.)

Problem 5.10. Consider the differential equation

$$\ddot{y}(t) + f[y(t)]\dot{y}(t) + g[y(t)] = 0$$

(a) Transform this equation into state variable form by choosing $x_1 = y, x_2 = \dot{y}$.

(b) Suppose that, in some interval $[-\sigma_0, \sigma_0]$, we have

$$\sigma g(\sigma) > 0 \quad \text{whenver } \sigma \in [-\sigma_0, \sigma_0], \sigma \neq 0$$

Generalize the Liapunov function candidate of Problem 5.9 to the present system. Show that (0, 0) is a stable equilibrium point if $f(\sigma) \geq 0$ $\forall\, \sigma \in [-\sigma_0, \sigma_0]$, and that (0, 0) is an asymptotically stable equilibrium point if $f(\sigma) > 0 \quad \forall\, \sigma \in [-\sigma_0, \sigma_0]$.

Problem 5.11. Consider the system

$$\dot{x}_1 = +x_1 + 2x_2^2$$

$$\dot{x}_2 = 2x_1x_2 + x_2^3$$

Using the Liapunov function candidate

$$V(x_1, x_2) = x_1^2 - x_2^2$$

show that (0, 0) is an unstable equilibrium point.

Problem 5.12. Consider the system

$$\dot{x}_1 = x_1^2 - x_1x_2$$

$$\dot{x}_2 = -x_1^2 - 2x_1x_2$$

using the Liapunov function candidate

$$V(x_1, x_2) = x_1(x_1 - x_2)$$

and Theorem (115), show that (0, 0) is an unstable equilibrium point.

5.3 STABILITY OF LINEAR SYSTEMS

In this section, we shall study the Liapunov stability of systems described by linear vector differential equations. The results presented here not only enable us to obtain necessary and sufficient conditions for the stability of linear systems but also pave the way to deriving Liapunov's indirect method (which consists of determining the stability of a nonlinear system by linearization).

5.3.1 Stability and the State Transition Matrix

Consider a system described by the linear vector differential equation

$$\dot{\mathbf{x}}(t) = \mathbf{A}(t)\mathbf{x}(t), \qquad t \geq 0 \tag{1}$$

The system (1) is autonomous if $\mathbf{A}(t)$ is constant as a function of time; otherwise it is nonautonomous. It is clear that $\mathbf{0}$ is always an equilibrium point of the system (1), at any time $t_0 \geq 0$. Further, $\mathbf{0}$ is an *isolated* equilibrium point at time t_0 if $\mathbf{A}(t)$ is nonsingular for some $t \geq t_0$.

The general solution of (1) is given by (see Appendix II)

2 $$\mathbf{x}(t) = \mathbf{\Phi}(t, t_0)\mathbf{x}(t_0)$$

where $\mathbf{\Phi}(\cdot, \cdot)$ is the state transition matrix associated with $\mathbf{A}(\cdot)$ and is the unique solution of the equation

3 $$\frac{d}{dt}\mathbf{\Phi}(t, t_0) = \mathbf{A}(t)\mathbf{\Phi}(t, t_0)$$

4 $$\mathbf{\Phi}(t_0, t_0) = \mathbf{I}$$

With the aid of this explicit characterization for the solutions of (1), it is possible to derive some useful conditions for the stability of the equilibrium point $\mathbf{0}$. Because these conditions involve the state transition matrix $\mathbf{\Phi}$, they are not of much *computational* value, because it is in general impossible to derive an analytical expression for $\mathbf{\Phi}$. Nevertheless, they are of *conceptual* value, enabling one to understand the mechanisms of stability in linear systems.

5 ***Theorem*** The equilibrium point $\mathbf{0}$ at time t_0 is stable at time t_0 if and only if[9]

6 $$\sup_{t \geq t_0} \|\mathbf{\Phi}(t, t_0)\|_i \triangleq m(t_0) < \infty$$

proof

(i) "*If*" Suppose (6) is true, and let $\epsilon > 0$ be specified. If we define $\delta(\epsilon, t_0)$ as $\epsilon/m(t_0)$, then, whenever $\|\mathbf{x}(t_0)\| < \delta(\epsilon, t_0)$, we have

7 $$\|\mathbf{x}(t)\| = \|\mathbf{\Phi}(t, t_0)\mathbf{x}(t_0)\| \leq \|\mathbf{\Phi}(t, t_0)\|_i \|\mathbf{x}(t_0)\| < m(t_0)\cdot\delta(\epsilon, t_0) = \epsilon$$

so that [5.1(7)] is satisfied. This shows that the equilibrium point $\mathbf{0}$ is stable at time t_0.

(ii) "*Only If*" Suppose (6) is false, so that $\|\Phi(\cdot, t_0)\|_i$ is an unbounded function. We shall show that $\mathbf{0}$ is an unstable equilibrium point. Accordingly, let $\epsilon > 0$ be any given number; we show that, no matter how small we choose $\delta > 0$, it is always possible to find an $\mathbf{x}(t_0)$ with $\|\mathbf{x}(t_0)\| < \delta$ such that $\|\mathbf{x}(t)\| \geq \epsilon$ for some $t \geq t_0$. Thus suppose $\delta > 0$ is fixed. Because $\|\Phi(\cdot, t_0)\|_i$ is an unbounded function, there exists a $t \geq t_0$ such that

8 $$\|\Phi(t, t_0)\|_i > \frac{\epsilon}{\delta_1}$$

where $\delta_1 < \delta$ is some positive number. Next, select $\mathbf{v}$ to be a vector of norm 1 such that

9 $$\|\Phi(t, t_0)\mathbf{v}\| = \|\Phi(t, t_0)\|_i$$

[9]Recall that $\|\cdot\|_i$ denotes the induced norm of a matrix.

This is possible in view of the definition of the induced matrix norm (definition [3.2(5)]). Finally, let $\mathbf{x}(t_0) = \delta_1 \mathbf{v}$. Then $\|\mathbf{x}(t_0)\| < \delta_1$; moreover,

10
$$\begin{aligned}\|\mathbf{x}(t)\| &= \|\mathbf{\Phi}(t, t_0)\mathbf{x}(t_0)\| = \|\delta_1 \mathbf{\Phi}(t, t_0)\mathbf{v}\| \\ &= \delta_1 \|\mathbf{\Phi}(t, t_0)\|_i > \epsilon\end{aligned}$$

Hence the equilibrium point $\mathbf{0}$ is unstable. ∎

11 ***Theorem*** With regard to the system (1), the following three conditions are equivalent:

(i) The equilibrium point $\mathbf{0}$ is asymptotically stable at time t_0.
(ii) The equilibrium point $\mathbf{0}$ is globally asymptotically stable at time t_0.
12 (iii) $\lim_{t \to \infty} \|\mathbf{\Phi}(t, t_0)\|_i = 0$.

proof **(ii) ⇒ (i)** Obvious.

(iii) ⇒ (ii) Suppose (12) holds. Then, because $\|\Phi(\cdot, t_0)\|_i$ is continuous and approaches 0 as $t \longrightarrow \infty$, it follows that $\|\Phi(\cdot, t_0)\|_i$ is bounded over the interval $[t_0, \infty)$. Hence, by Theorem (5), the equilibrium point $\mathbf{0}$ is stable at time t_0. Furthermore, for arbitrary $\mathbf{x}(t_0) \in R^n$, we have

$$\|\mathbf{x}(t)\| \leq \|\mathbf{\Phi}(t, t_0)\|_i \|\mathbf{x}(t_0)\| \longrightarrow 0 \qquad \text{as } t \longrightarrow \infty$$

so that $\mathbf{x}(t) \longrightarrow \mathbf{0}$ as $t \longrightarrow \infty$. This shows that the equilibrium point $\mathbf{0}$ is globally asymptotically stable at time t_0.

(i) ⇒ (iii) Suppose (iii) is false. Then at least one component of $\mathbf{\Phi}(t, t_0)$, say $\phi_{ij}(t, t_0)$, does not approach zero as $t \longrightarrow \infty$. Let $\mathbf{x}(t_0) = \delta \mathbf{e}_j$, where $\mathbf{e}$ is the jth elementary unit vector (i.e., $\mathbf{e}_j$ has all zero components, except for the jth component, which is 1). Then, regardless of how small we make δ, it is clear that the ith component of the corresponding solution $\mathbf{x}(\cdot)$ does *not* approach 0 as $t \longrightarrow \infty$. Thus, if (iii) is false, then there are trajectories starting from initial conditions arbitrarily close to $\mathbf{0}$ that do *not* converge to $\mathbf{0}$. Thus (i) is also false. ∎

13 ***Theorem*** With regard to the system (1), the following two statements are equivalent.

(i) The equilibrium point $\mathbf{0}$ is stable at time t_0.
(ii) The equilibrium point $\mathbf{0}$ is stable at all times $t_1 \geq t_0$.

proof **(ii) ⇒ (i)** Obvious.

(i) ⇒ (ii) Suppose (i) is true. By Theorem (5), this implies that

14
$$\sup_{t \geq t_0} \|\mathbf{\Phi}(t, t_0)\|_i < \infty$$

Now, suppose $t_0 \geq t_1$. By the semigroup property of the state transition matrix (see Appendix II), we have

15
$$\mathbf{\Phi}(t, t_1) = \mathbf{\Phi}(t, t_0) \cdot [\mathbf{\Phi}(t_1, t_0)]^{-1}$$

Hence

16 $$\|\mathbf{\Phi}(t, t_1)\|_i \leq \|\mathbf{\Phi}(t, t_0)\|_i \cdot \|[\mathbf{\Phi}(t_1, t_0)]^{-1}\|_i$$

However, because t_1 and t_0 are *finite* numbers, $\|[\mathbf{\Phi}(t_1, t_0)]^{-1}\|_i$ is also a finite number. This fact, together with (14), shows that

17 $$\sup_{t \geq t_1} \|\mathbf{\Phi}(t, t_1)\|_i < \infty$$

Hence the equilibrium point $\mathbf{0}$ is stable at time t_1. Because this argument can be repeated for any t_1, (ii) is established. ∎

18 REMARKS: Theorem (13) does *not* say that stability of time t_0 is equivalent to *uniform stability* over $[t_0, \infty)$. Rather, Theorem (13) is to be contrasted with lemma [5.1(28)]. The latter shows that, for *any* system, stability of the equilibrium point $\mathbf{0}$ at some time t_0 implies stability at all times *prior* to t_0. In contrast, Theorem (13) shows that, in the special case of *linear* systems, stability at time t_0 implies stability at all times *after* t_0. The key point in the proof of Theorem (13) is that $\|[\Phi(t, t_0)]^{-1}\|_i$ is finite. In a general nonlinear system, the transition map $\mathbf{x}_0 \mapsto \mathbf{s}(t_1, \mathbf{x}_0, t_0)$ is bounded, but (1) it may not have an inverse, and (2) even if it has an inverse, the inverse may not be bounded. Thus Theorem (13) is a special result for linear systems.[10]

The necessary and sufficient conditions for uniform stability are now given.

19 ***Theorem*** The equilibrium point $\mathbf{0}$ is uniformly stable over the interval $[0, \infty)$ if and only if

20 $$\sup_{t_0 \geq 0} m(t_0) = \sup_{t_0 \geq 0} \sup_{t \geq t_0} \|\Phi(t, t_0)\|_i \triangleq m_0 < \infty$$

21 REMARKS: Recall, from Theorem (5), that a necessary and sufficient condition for the equilibrium point $\mathbf{0}$ to be stable at time t_0 is that $m(t_0)$ is a finite number. From Theorem (13), if $m(t_0)$ is finite for *some* value of t_0, it is finite for *all* values of t_0. However, it is still possible for $m(t_0)$ to be an unbounded function of t_0. See Problem 5.2.

proof

(i) "*If*" Suppose m_0 is finite; then, for any $\epsilon > 0$, [5.1(8)] is satisfied with $\delta = \epsilon/m_0$.

---

[10]Theorem (13) also holds in the case of nonlinear systems of the form $\dot{\mathbf{x}}(t) = \mathbf{f}[t, \mathbf{x}(t)]$, where $\mathbf{f}$ is continuous in t and Lipschitz-continuous in $\mathbf{x}$; this is because in this case also the map $\mathbf{x}_0 \mapsto \mathbf{s}(t_1, \mathbf{x}_0, t_0)$ has a bounded inverse; see Theorem [3.4(50)].

(ii) "*Only If*" Suppose $m(t_0)$ is unbounded as a function of t_0. Then at least one component of $\Phi(\cdot, \cdot)$, say $\phi_{ij}(\cdot, \cdot)$, has the property that

22 $$\sup_{t \geq t_0} |\phi_{ij}(t, t_0)| \text{ is unbounded as a function of } t_0$$

Let $\mathbf{x}_0 = \mathbf{e}_j$, the jth elementary vector. Then, because of (22), the ratio $\|\Phi(t, t_0)\mathbf{x}_0\|/\|\mathbf{x}_0\|$ cannot be bounded independently of t_0. Thus the equilibrium point $\mathbf{0}$ is not uniformly stable over $[0, \infty)$. ∎

23 ***Theorem*** The equilibrium point $\mathbf{0}$ of (1) is uniformly asymptotically stable over $[0, \infty)$ if and only if

24 $$\sup_{t_0 \geq 0} \sup_{t \geq t_0} \|\Phi(t, t_0)\|_i < \infty$$

25 $$\|\Phi(t, t_0)\|_i \longrightarrow 0 \quad \text{as } t \longrightarrow \infty, \qquad \text{uniformly in } t_0$$

REMARKS: The condition (25) can be equivalently expressed as follows: For each $\epsilon > 0$, there exists a $T(\epsilon) < \infty$ such that

26 $$\|\Phi(t, t_0)\|_i < \epsilon, \qquad \forall\, t \geq t_0 + T(\epsilon),\ \forall\, t_0 \geq 0$$

proof

(i) "*If*" By Theorem (19), if (24) holds, then the equilibrium point $\mathbf{0}$ is uniformly stable over $[0, \infty)$. Similarly, if (25) holds, the ratio $\|\mathbf{\Phi}(t, t_0)\mathbf{x}_0\|/\|\mathbf{x}_0\|$ approaches zero as $t \longrightarrow \infty$, uniformly in t_0 and $\mathbf{x}_0$. Thus both conditions of definition [5.1(14)] are satisfied, whence $\mathbf{0}$ is uniformly asymptotically stable over the interval $[0, \infty)$.

(ii) "*Only If*" This part of the proof is left as an exercise. ∎

Theorem (27) below gives an alternative equivalent condition for uniform asymptotic stability.

27 ***Theorem*** The equilibrium point $\mathbf{0}$ of (1) is uniformly asymptotically stable over the interval $[0, \infty)$ if and only if there exist positive constants m and λ such that

28 $$\|\mathbf{\Phi}(t, t_0)\|_i \leq m e^{-\lambda(t-t_0)}, \qquad \forall\, t \geq t_0,\ \forall\, t_0$$

REMARKS: Some authors refer to a system satisfying (28) as being *exponentially stable*, because, if (28) holds, then the solutions of (1) decay exponentially. The import of Theorem (27) is thus to show that exponential stability is equivalent to uniform asymptotic stability over $[0, \infty)$.

proof

(i) "*If*" Suppose (28) holds. Then clearly (24) and (25) also hold, so that the equilibrium point $\mathbf{0}$ is uniformly asymptotically stable.

(ii) "*Only If*" Suppose (24) and (25) hold. Then there exist m and T such that

$$\|\Phi(t, t_1)\|_i \leq m_0, \qquad \forall\, t \geq t_1, \forall\, t_1 \tag{29}$$

$$\|\Phi(t, t_1)\|_i \leq \frac{1}{2}, \qquad \forall\, t \geq t_1 + T, \forall\, t_1 \tag{30}$$

In particular, (30) implies that

$$\|\Phi(t_1 + T, t_1)\|_i \leq \frac{1}{2}, \qquad \forall\, t_1 \tag{31}$$

Now, given any t_0 and any $t \geq t_0$, pick k such that $t_0 + kT \leq t \leq t_0 + kT + T$. Then we have

$$\Phi(t, t_0) = \Phi(t, t_0 + kT)\cdot\Phi(t_0 + kT, t_0 + kT - T) \ldots \Phi(t_0 + T, t_0) \tag{32}$$

Hence

$$\|\Phi(t, t_0)\|_i \leq \|\Phi(t, t_0 + kT)\|_i \cdot \prod_{j=1}^{k} \|\Phi(t_0 + jT, t_0 + jT - T)\|_i \tag{33}$$

where the empty product is taken as 1. Substituting from (29) and (31) in (33), we get

$$\|\Phi(t, t_0)\|_i \leq m_0 \cdot 2^{-k} = (2m_0)2^{-(k+1)} \leq 2m_0 \cdot 2^{-(t-t_0)/T} \tag{34}$$

Hence (28) is satisfied, if we choose

$$m = 2m_0 \tag{35}$$

$$\lambda = \frac{\log 2}{T} \tag{36}$$

∎

In conclusion, in this subsection we have presented several results that relate the state transition matrix of a linear system to its stability properties. These results are not of much use for testing purposes, because finding the state transition matrix of a linear system is in general a formidable task. However, some conceptual insight is provided by these results. For example, Theorem (27), which shows that exponential stability is equivalent to uniform asymptotic stability, is not very obvious on the surface.

5.3.2 Autonomous Systems

Throughout this subsection, we shall restrict ourselves to linear autonomous systems, i.e., systems described by

$$\dot{\mathbf{x}}(t) = \mathbf{A}\mathbf{x}(t) \tag{37}$$

In this special case, Liapunov theory is very complete, as demonstrated by the following series of theorems.

38 ***Theorem*** The equilibrium point **0** of (37) is (globally) asymptotically stable if and only if all eigenvalues of **A** have negative real parts.

The equilibrium point **0** of (37) is stable if and only if all eigenvalues of **A** have zero real parts and every eigenvalue of **A** having a zero real part is a simple zero of the minimal polynomial of **A**.

proof From Appendix II, the state transition matrix $\mathbf{\Phi}(t, t_0)$ of the system (37) is given by

39 $$\mathbf{\Phi}(t, t_0) = \exp\,[\mathbf{A}(t - t_0)]$$

where $\exp\,[\mathbf{A}(\cdot)]$ is the matrix exponential of **A**. Furthermore, $\exp\,(\mathbf{A}t)$ can be expressed as

40 $$\exp\,(\mathbf{A}t) = \sum_{i=1}^{r} \sum_{j=1}^{m_i} t^{j-1} \exp\,(\lambda_i t) p_{ij}(\mathbf{A})$$

where r is the number of distinct eigenvalues of **A**, $\lambda_1, \ldots, \lambda_r$ are the (distinct) eigenvalues of **A**, and m_i is the multiplicity of λ_i as a zero of the minimal polynomial of **A**, and p_{ij} are the interpolating polynomials. The stated conditions for stability and asymptotic stability now follow readily by applying Theorems (5) and (11). ∎

Thus, in the case of linear time-invariant systems of the form (37), the stability status of the equilibrium point **0** can be determined by studying the eigenvalues of **A**. However, it is possible to formulate an entirely different approach to the problem, namely by the use of quadratic Liapunov functions. This theory is of interest in itself and is moreover useful in the study of the first method of Liapunov (the indirect method).

Given the system (37), the idea is to choose a Liapunov function candidate of the form

41 $$V(\mathbf{x}) = \mathbf{x}'\mathbf{P}\mathbf{x}$$

where **P** is a real symmetric matrix. Then $\dot{V}$ is given by

42 $$\begin{aligned}\dot{V}(\mathbf{x}) &= \dot{\mathbf{x}}'\mathbf{P}\mathbf{x} + \mathbf{x}'\mathbf{P}\dot{\mathbf{x}} = \mathbf{x}'(\mathbf{A}'\mathbf{P} + \mathbf{P}\mathbf{A})\mathbf{x} \\ &= -\mathbf{x}'\mathbf{Q}\mathbf{x}\end{aligned}$$

where

43 $$\mathbf{A}'\mathbf{P} + \mathbf{P}\mathbf{A} = -\mathbf{Q}$$

Equation (43) is commonly known as the Liapunov matrix equation. By means of this equation, it is possible to study the stability properties of the equilibrium point **0** of (37). For example, if a pair of matrices (**P**, **Q**) satisfying (43) can be found such that both **P** and **Q** are positive definite, then both V and $-\dot{V}$ are positive definite functions. Hence, by Theorem [5.2(58)], the equilibrium point **0** of (37) is globally asymptotically stable. Alternatively, if a pair (**P**, **Q**) can be found such that **Q** is positive definite and **P** has at least one negative eigenvalue, then $-\dot{V}$ is a positive definite function, and $-V$ assumes positive values arbitrarily close to the origin. Hence, by Theorem [5.2(106)], **0** is an unstable equilibrium point of (37).

This, then, is the rationale behind studying equation (43). There are clearly two possible ways in which (43) can be tackled: (1) Given a particular matrix **A**, one can pick a particular matrix **P** and study the properties of the resulting matrix **Q**, or (2) given **A**, one can pick **Q** and study the resulting **P**. We shall pursue the latter approach, for two reasons—one pragmatic and the other philosophical. The pragmatic reason is that the second approach is the one for which the theory is better developed. On a more philosophical level, we can reason as follows: Given a matrix **A**, we presumably do not know ahead of time whether **0** is a stable or unstable equilibrium point. If we pick **P** and study the resulting **Q**, we would be obliged (because of our available stability theorems) to make an a priori guess as to the stability status of **0** (we would pick **P** to be positive definite if we thought **0** were asymptotically stable and **P** to have some negative eigenvalues if we thought **0** were unstable). On the other hand, if we were to pick **Q**, we could noncommittally choose **Q** to be positive definite and study the resulting **P**: If **P** were to be positive definite, **0** would be asymptotically stable (by Theorem [5.2(58)]), while if **P** were to have some negative eigenvalues, **0** would be unstable (by Theorem [5.2(106)]).

One difficulty with selecting **Q** and trying to find the corresponding **P** is that, depending on **A**, (43) may not have a unique solution for **P**. Theorem (44) gives necessary and sufficient conditions for (43) to have a unique solution for **P** corresponding to every $\mathbf{Q} \in R^{n \times n}$.

44 ***Theorem*** Let $\mathbf{A} \in R^{n \times n}$ and let $\{\lambda_1, \ldots, \lambda_n\}$ be the (not necessarily distinct) eigenvalues of **A**. Then (43) has a unique solution for **P** corresponding to every $\mathbf{Q} \in R^{n \times n}$ if and only if

45 $$\lambda_i + \lambda_j^* \neq 0, \qquad \forall\, i, j$$

(where * denotes complex conjugation).

The proof of this theorem is not difficult but requires some concepts from linear algebra not heretofore covered, and the reader is therefore referred to [7] Chen for the proof.

Using Theorem (44), we can state the following corollary.

46 ***corollary*** A necessary condition for all eigenvalues of **A** to have negative real parts is that (43) has a unique solution for **P** corresponding to each $\mathbf{Q} \in \mathbf{R}^{n \times n}$.

proof If all eigenvalues of **A** have negative real parts, then (45) is satisfied. ∎

The following lemma provides an alternative characterization of the solutions of (43).

47 ***lemma*** Let $\mathbf{A}$ be a matrix whose eigenvalues all have negative real parts. Then the unique solution for $\mathbf{R}$ of the equation

48 $$\mathbf{A}'\mathbf{R} + \mathbf{R}\mathbf{A} = -\mathbf{M}$$

where $\mathbf{M}$ is a given real symmetric matrix, is

49 $$\mathbf{R} = \int_0^\infty e^{\mathbf{A}'t}\mathbf{M}e^{\mathbf{A}t}\,dt$$

proof Consider the linear matrix differential equation

50 $$\dot{\mathbf{S}}(t) = \mathbf{A}'\mathbf{S}(t) + \mathbf{S}(t)\mathbf{A} + \mathbf{M}; \qquad \mathbf{S}(0) = \mathbf{0}$$

Because the right-hand side of (50) is affine in $\mathbf{S}$, and is thus globally Lipschitz-continuous in $\mathbf{S}$, it follows from Theorem [3.4(25)] that (50) has exactly one solution. We shall now show that this solution is in fact given by

51 $$\mathbf{S}(t) = \int_0^t e^{\mathbf{A}'\tau}\mathbf{M}e^{\mathbf{A}\tau}\,d\tau$$

Differentiating both sides of (51), we get

52 $$\dot{\mathbf{S}}(t) = e^{\mathbf{A}'t}\mathbf{M}e^{\mathbf{A}t}$$

On the other hand,

53 $$\begin{aligned}\mathbf{A}'\mathbf{S}(t) + \mathbf{S}(t)\mathbf{A} &= \int_0^t \mathbf{A}'e^{\mathbf{A}'\tau}\mathbf{M}e^{\mathbf{A}\tau}\,d\tau + \int_0^t e^{\mathbf{A}'\tau}\mathbf{M}e^{\mathbf{A}\tau}\mathbf{A}\,d\tau \\ &= \int_0^t (\mathbf{A}'e^{\mathbf{A}'\tau}\mathbf{M}e^{\mathbf{A}\tau} + e^{\mathbf{A}'\tau}\mathbf{M}\mathbf{A}e^{\mathbf{A}\tau})\,d\tau \\ &= \int_0^t \frac{d}{d\tau}(e^{\mathbf{A}'\tau}\mathbf{M}e^{\mathbf{A}\tau})\,d\tau = e^{\mathbf{A}'t}\mathbf{M}e^{\mathbf{A}t} - \mathbf{M} \\ &= \dot{\mathbf{S}}(t) - \mathbf{M}\end{aligned}$$

so that (50) is satisfied.

Now, because all eigenvalues of $\mathbf{A}$ have negative real parts, we see from (52) that $\dot{\mathbf{S}}(t) \longrightarrow \mathbf{0}$ as $t \longrightarrow \infty$. Also, as $t \longrightarrow \infty$, the integral in (51) remains well defined and in fact approaches $\mathbf{R}$ as given in (49). Thus, as $t \longrightarrow \infty$, in the steady state, (50) reduces to

54 $$\mathbf{0} = \mathbf{A}'\mathbf{R} + \mathbf{R}\mathbf{A} + \mathbf{M}$$

which is equivalent to (48). Hence (49) provides a solution to (48). By Theorem (44), it is in fact the unique solution. ∎

We can now state one of the main results for the Liapunov matrix equation.

55 ***Theorem*** Given a matrix $\mathbf{A} \in R^{n\times n}$, the following three statements are equivalent:

(i) All eigenvalues of $\mathbf{A}$ have negative real parts.

(ii) There exists *some* positive definite matrix $\mathbf{Q} \in R^{n\times n}$ such that (43) has a unique solution for $\mathbf{P}$, and this solution is positive definite.

(iii) For *every* positive definite matrix $\mathbf{Q} \in R^{n\times n}$, (43) has a unique solution for $\mathbf{P}$, and this solution is positive definite.

proof **(iii) ⇒ (ii)** Obvious.

(ii) ⇒ (i) Suppose (ii) is true for some particular matrix $\mathbf{Q}$. Then we can apply Theorem [5.2(58)] with the Liapunov function $V(\mathbf{x}) = \mathbf{x}'\mathbf{P}\mathbf{x}$ to conclude that $\mathbf{0}$ is an asymptotically stable equilibrium of (37). Thus, by Theorem (38), it follows that all eigenvalues of $\mathbf{A}$ have negative real parts.

(i) ⇒ (iii) Suppose all eigenvalues of $\mathbf{A}$ have negative real parts. Then by lemma (47), corresponding to each $\mathbf{Q} \in R^{n\times n}$, (43) has the unique solution for $\mathbf{P}$ given by

56 $$\mathbf{P} = \int_0^\infty e^{\mathbf{A}'t}\mathbf{Q}e^{\mathbf{A}t}\,dt$$

It only remains to show that $\mathbf{P}$ is positive definite whenever $\mathbf{Q}$ is positive definite. Thus suppose $\mathbf{Q}$ is positive definite, and express $\mathbf{Q}$ in the form $\mathbf{Q} = \mathbf{M}'\mathbf{M}$, where $\mathbf{M} \in R^{n\times n}$ is nonsingular. We show that $\mathbf{P}$ is positive definite by showing that

57 $$\mathbf{x}'\mathbf{P}\mathbf{x} \geq 0, \qquad \forall\, \mathbf{x} \in R^n$$

58 $$\mathbf{x}'\mathbf{P}\mathbf{x} = 0 \Longrightarrow \mathbf{x} = \mathbf{0}$$

With $\mathbf{Q} = \mathbf{M}'\mathbf{M}$, $\mathbf{P}$ becomes

59 $$\mathbf{P} = \int_0^\infty e^{\mathbf{A}'\tau}\mathbf{M}'\mathbf{M}e^{\mathbf{A}\tau}\,d\tau$$

Thus, for any $\mathbf{x} \in R^n$, we have

60 $$\mathbf{x}'\mathbf{P}\mathbf{x} = \int_0^\infty \mathbf{x}'e^{\mathbf{A}'\tau}\mathbf{M}'\mathbf{M}e^{\mathbf{A}\tau}\mathbf{x}\,d\tau = \int_0^\infty \|\mathbf{M}e^{\mathbf{A}\tau}\mathbf{x}\|_2^2\,d\tau \geq 0$$

where $\|\cdot\|_2$ denotes the Euclidean norm. Also, (60) shows that if $\mathbf{x}'\mathbf{P}\mathbf{x} = 0$, then

61 $$\mathbf{M}e^{\mathbf{A}t}\mathbf{x} = 0, \qquad \forall\, t \geq 0$$

If we take $t = 0$ in (61), we get $\mathbf{M}\mathbf{x} = \mathbf{0}$, which implies that $\mathbf{x} = \mathbf{0}$ because $\mathbf{M}$ is nonsingular. Thus $\mathbf{P}$ is positive definite and (i) → (iii). ∎

62 REMARKS:

1. Theorem (55) is very important in that it enables one to unambiguously determine whether or not $\mathbf{0}$ is an asymptotically stable equilibrium point, in the following manner: Given

$\mathbf{A} \in R^{n \times n}$, pick $\mathbf{Q} \in R^{n \times n}$ to be *any* positive definite matrix (a logical choice is $\mathbf{Q} = \mathbf{I}$ or some other diagonal matrix), and solve (43) for $\mathbf{P}$. Suppose that (43) has no solution or that it has more than one solution (i.e., it does *not* have a unique solution); then $\mathbf{0}$ is not asymptotically stable. Suppose that (43) does have a unique solution for $\mathbf{P}$ and that this solution is not positive definite; then once again $\mathbf{0}$ is not asymptotically stable. On the other hand, if this unique solution for $\mathbf{P}$ is positive definite, then $\mathbf{0}$ is an asymptotically stable equilibrium point of (37).

2. Theorem (55) states that whenever $\mathbf{Q}$ is positive definite (and all eigenvalues of $\mathbf{A}$ have negative real parts) the corresponding $\mathbf{P}$ is also positive definite. It does *not* say that whenever $\mathbf{P}$ is positive definite the corresponding $\mathbf{Q}$ is positive definite. The latter statement is false in general.
3. As a computational device, Theorem (55) can be compared to Routh's criterion, which enables one to determine whether or not the zeroes of a given polynomial all have negative real parts without actually computing the zeroes. In the same way, Theorem (55) enables one to determine whether or not the eigenvalues of a given matrix all have negative real parts without actually computing the eigenvalues. If $\mathbf{A}$ is an $n \times n$ matrix, then (43) represents a system of $n(n+1)/2$ equations in the same number of unknowns (note that $\mathbf{P}$ and $\mathbf{Q}$ are symmetric). Thus, for small values of n, there seems to be no advantage to using the Liapunov matrix equations as a testing device. However, for intermediate values of n (say $6 \leq n \leq 15$), Theorem (55) provides a useful tool. See [15] Jury for alternative computational techniques for testing whether or not all the eigenvalues of a given matrix have negative real parts.

63 **Example.**[11] Let

$$\mathbf{A} = \begin{bmatrix} -2 & 1 \\ -1 & 1 \end{bmatrix}; \qquad \mathbf{Q} = \begin{bmatrix} 1 & 0 \\ 0 & 1 \end{bmatrix}$$

and let

$$\mathbf{P} = \begin{bmatrix} p_1 & p_2 \\ p_2 & p_3 \end{bmatrix}$$

---

[11] To avoid getting lost in an avalanche of numbers, we shall illustrate the use of the Liapunov matrix equation only for second-order systems, even though for $n = 2$ this is the hard way to do it.

be the matrix of unknowns. Then (43) becomes

$$\begin{bmatrix} -2p_1 - p_2 & -2p_2 - p_3 \\ p_1 + p_2 & p_2 + p_3 \end{bmatrix} + \begin{bmatrix} -2p_1 + p_2 & p_1 + p_2 \\ -2p_2 - p_3 & p_2 + p_3 \end{bmatrix}$$
$$= \begin{bmatrix} -1 & 0 \\ 0 & -1 \end{bmatrix}$$

which translates into the three equations

$$\begin{aligned} -4p_1 - 2p_2 \qquad &= -1 \\ p_1 - p_2 - p_3 &= 0 \\ 2p_2 + 2p_3 &= -1 \end{aligned}$$

The solutions of these equations are

$$p_1 = -\frac{1}{2}, \qquad p_2 = \frac{3}{2}, \qquad p_3 = -2$$

Thus

$$\mathbf{P} = \begin{bmatrix} -1/2 & 3/2 \\ 3/2 & -2 \end{bmatrix}$$

Even though (43) has a unique solution in this case, the resulting matrix **P** is clearly not positive definite. Hence not all eigenvalues of **A** have negative real parts.

64 **Example.** Let

$$\mathbf{A} = \begin{bmatrix} -2 & -3 \\ 1 & 2 \end{bmatrix}, \qquad \mathbf{Q} = \begin{bmatrix} 2 & 0 \\ 0 & 2 \end{bmatrix}, \qquad \mathbf{P} = \begin{bmatrix} p_1 & p_2 \\ p_2 & p_3 \end{bmatrix}$$

Then (43) reduces to the three equations

$$\begin{aligned} -4p_1 + 2p_2 \qquad &= -2 \\ -3p_1 \qquad + p_3 &= 0 \\ -6p_2 + 4p_3 &= -2 \end{aligned}$$

Adding 4 times the second equation and -3 times the first equation, we get

$$-6p_2 + 4p_3 = 6$$

which is clearly inconsistent with the third equation. Thus (43) has no solution in this case. Thus we conclude that not all eigenvalues of **A** have negative real parts.

65 **Example.** Let

$$\mathbf{A} = \begin{bmatrix} 1 & -1 \\ 4 & -3 \end{bmatrix}, \qquad \mathbf{Q} = \begin{bmatrix} 1 & 0 \\ 0 & 2 \end{bmatrix}$$

Then (43) has a unique solution for **P**, given by

$$\mathbf{P} = \frac{1}{4}\begin{bmatrix} 102 & -29 \\ -29 & 7 \end{bmatrix}$$

42 -11
11 5

Because **P** is positive definite, the conclusion is that all eigenvalues of **A** have negative real parts.

Theorem (55) shows that if the equilibrium point **0** of (37) is asymptotically stable, then this fact can be ascertained by choosing a quadratic Liapunov function and applying Theorem [5.2(58)]. The following result, which is stated without proof, enables us to conclude that if the equilibrium point **0** of (37) is unstable because some eigenvalue of **A** has a positive real part,[12] then this fact can also be ascertained by choosing a quadratic Liapunov function and applying Theorem [5.2(106)].

66 ***Theorem*** Consider (43), and suppose (45) is satisfied, so that (43) has a unique solution for **P** corresponding to each **Q**. Under these conditions, if **Q** is positive definite, then **P** has as many negative eigenvalues as there are eigenvalues of **A** with positive real part.

proof See [21] Taussky. ∎

REMARKS:

1. Note that the hypotheses of Theorem (66) rule out **A** having any eigenvalues with zero real part.
2. The significance of Theorem (66) lies in that it enables one to prove the instability of (37) using a quadratic Liapunov function. Specifically, suppose the matrix **A** satisfies the hypotheses of Theorem (66) and that V is defined by

$$V(\mathbf{x}) = -\mathbf{x}'\mathbf{P}\mathbf{x} \tag{67}$$

Then

$$\dot{V}(\mathbf{x}) = \mathbf{x}'\mathbf{Q}\mathbf{x} \tag{68}$$

Moreover, $\dot{V}$ is a p.d.f., and V assumes positive values arbitrarily close to **0**, because, by Theorem (66), **P** has at least one negative eigenvalue. Hence, by Theorem [5.2(106)], the equilibrium point **0** of (37) is unstable. Moreover, it turns out that a similar argument can be used in Liapunov's indirect method.

? Top of next

[12]The equilibrium point **0** of (37) can be unstable in another way, namely, all eigenvalues of **A** have nonpositive real parts, but some eigenvalue of **A** having a zero real part is a multiple zero of the minimal polynomial of **A**.

5.3.3 Nonautonomous Systems

In the case of linear time-varying systems characterized by (1), the stability status of the equilibrium point **0** can be ascertained, in theory at least, by studying the state transition matrix. This is detailed in Sec. 5.3.1. The purpose of the present subsection is twofold:

1. To present some simple sufficient conditions for stability, asymptotic stability, and instability based on the concept of the matrix measure, and
2. To prove the existence of a quadratic Liapunov function for uniformly asymptotically stable linear systems.

We shall begin by studying some stability conditions based on the matrix measure. The following result proves useful for this purpose.

69 ***lemma*** With regard to (1), the following inequalities hold:

70
$$\exp\left\{\int_{t_0}^{t} -\mu_i[-\mathbf{A}(\tau)]\,d\tau\right\} \leq \|\mathbf{\Phi}(t, t_0)\|_i \leq \exp\left\{\int_{t_0}^{t} \mu_i[\mathbf{A}(\tau)]\,d\tau\right\}$$

where $\|\cdot\|$ is any norm on R^n and $\|\cdot\|_i$, $\mu_i(\cdot)$, respectively, denote the corresponding induced matrix norm and matrix measure on $R^{n\times n}$.

proof Immediate from theorem [3.5(1)]. ∎

Many simple stability conditions can be derived from lemma (69). The proofs are left as exercises and are straightforward applications of results in Sec. 5.3.1.

71 ***lemma*** The equilibrium point **0** of (37) is stable at time t_0 if there exists a finite constant $m(t_0)$ such that

72
$$\int_{t_0}^{t} \mu_i[\mathbf{A}(\tau)]\,d\tau \leq m(t_0), \qquad \forall\, t \geq t_0$$

0 is a uniformly stable equilibrium point over $[t_0, \infty)$ if there exists a finite constant m_0 such that

73
$$\int_{t_1}^{t} \mu_i[\mathbf{A}(\tau)]\,d\tau \leq m_0, \qquad \forall\, t \geq t_1,\ \forall\, t_1 \geq t_0$$

74 ***lemma*** The equilibrium point **0** of (37) is asymptotically stable at time t_0 if

75
$$\int_{t_0}^{t} \mu_i[\mathbf{A}(\tau)]\,d\tau \longrightarrow \infty \qquad \text{as } t \longrightarrow \infty$$

0 is uniformly asymptotically stable over $[t_0, \infty)$ if, for every $M > 0$, there exists a $T < \infty$ such that

76
$$\int_{t_1}^{t_1+t} \mu_i[\mathbf{A}(\tau)]\,d\tau < -M, \qquad \forall\, t \geq T,\ \forall\, t_1 \geq t_0$$

77 ***lemma*** The equilibrium point $\mathbf{0}$ of (37) is unstable at time t_0 if

78 $$\int_{t_0}^{t} -\mu_i[-A(\tau)]\, d\tau \longrightarrow \infty \qquad \text{as } t \longrightarrow \infty$$

79 REMARKS:

1. As we saw in Secs. 3.2 and 3.5, the measure of a matrix is strongly dependent on the vector norm on R^n that is used to define the matrix measure. In lemmas (71), (74), and (77), the indicated conclusions follow if the appropriate conditions hold for *some* matrix measure $\mu_i(\cdot)$. Thus there is a great deal of flexibility in applying lemmas (71), (74), and (77), because one is free to choose the matrix measure $\mu_i(\cdot)$, and a wise user will exploit this flexibility to the fullest. The examples that follow illustrate this point.
2. The three lemmas above provide only sufficient conditions that are by no means necessary. However, they do have the advantage that they are easy to verify and do not require the determination of the state transition matrix.
3. Lemma (77) is of rather dubious value because its hypothesis actually ensures the *complete* instability of the equilibrium point $\mathbf{0}$; i.e., *every* nontrivial trajectory "blows up."

80 **Example.** Consider the system (37) with

$$\mathbf{A}(t) = \begin{bmatrix} \cos t & -\sin t \\ \sin t & \cos t \end{bmatrix}$$

and apply the measure μ_{i2}. In this particular case

$$\mathbf{A}(t) + \mathbf{A}'(t) = \begin{bmatrix} 2\cos t & 0 \\ 0 & 2\cos t \end{bmatrix}$$

Because $\mu_{i2}(\mathbf{A}) = \lambda_{\max}(\mathbf{A} + \mathbf{A}')/2$ (see Sec. 3.2), we have

$$\mu_{i2}[\mathbf{A}(t)] = \cos t$$

$$\int_{t_0}^{t} \mu_{i2}[\mathbf{A}(\tau)]\, d\tau = \sin t - \sin t_0 \leq 2, \qquad \forall\, t \geq t_0,\ \forall\, t_0 \geq 0$$

Hence, by lemma (71), the equilibrium point $\mathbf{0}$ is uniformly stable over the interval $[0, \infty)$.

81 **Example.** Consider the system (1), with

$$\mathbf{A}(t) = \begin{bmatrix} -2 & -2 + 3e^{-t} \\ -1 - 2e^{-2t} & -3 \end{bmatrix}$$

and choose the matrix measure $\mu_{i1}(\cdot)$. It is easy to see that, as $t \longrightarrow \infty$, $\mathbf{A}(t)$ approaches a constant matrix $\mathbf{A}(\infty)$ and that $\mu_{i1}[\mathbf{A}(\infty)] = -1$.

Because the matrix measure is a continuous function, it follows that $\mu_{i1}[\mathbf{A}(t)] \longrightarrow -1$ as $t \longrightarrow \infty$. Hence (one can show that) (76) is satisfied, so that, by lemma (74), the equilibrium point $\mathbf{0}$ is uniformly asymptotically stable over $[0, \infty)$.

Note that if we had employed the matrix measure $\mu_{i\infty}(\cdot)$ in this example, lemma (74) would have been inapplicable. Instead, we would have had to settle for lemma (71) and a conclusion of uniform stability over $[0, \infty)$ (verify this statement).

The Existence of Quadratic Liapunov Functions

In Sec. 5.3.2 it is shown that if the system (37) is asymptotically stable, then a quadratic Liapunov function exists for this system. We shall now prove a similar result for uniformly asymptotically stable systems of the form (1). This result is derived from using the following lemmas.

82 ***lemma*** Suppose $\mathbf{Q}(\cdot)\colon R_+ \longrightarrow R^{n \times n}$ is continuous and bounded and that the equilibrium point $\mathbf{0}$ of (1) is uniformly asymptotically stable over $[0, \infty)$. Then, for each $t \geq 0$, the matrix

83 $$\mathbf{P}(t) = \int_t^\infty \mathbf{\Phi}'(\tau, t)\mathbf{Q}(\tau)\mathbf{\Phi}(\tau, t)\, d\tau$$

is well defined; further, $\mathbf{P}(t)$ is bounded as a function of t.

proof The hypothesis of uniform asymptotic stability implies [because of Theorem (27)] that there exist positive constants m and λ such that

84 $$\|\mathbf{\Phi}(\tau, t)\|_i \leq m e^{-\lambda(\tau - t)}, \qquad \forall\, \tau \geq t$$

The bound (84), together with the boundedness of $\mathbf{Q}(\cdot)$, proves the lemma. ∎

85 ***lemma*** Suppose that, in addition to the hypotheses of lemma (82), we also have the following:

(i) $\mathbf{Q}(t)$ is symmetric and positive definite for each $t \geq 0$; moreover, there exists a positive constant α such that

86 $$\alpha \mathbf{x}'\mathbf{x} \leq \mathbf{x}'\mathbf{Q}(t)\mathbf{x}, \qquad \forall\, \mathbf{x} \in R^n, \forall\, t \geq 0$$

(ii) The matrix $\mathbf{A}(t)$ in (1) is bounded with respect to t; i.e., there exists a finite constant m_0 such that

87 $$\|\mathbf{A}(t)\|_{i2} \leq m_0, \qquad \forall\, t \geq 0$$

(where $\|\cdot\|_{i2}$ denotes the l_2-induced matrix norm). Under these conditions, the matrix $\mathbf{P}(t)$ given by (83) is positive definite for each $t \geq 0$; moreover, there exists a positive constant β such that

88 $$\beta \mathbf{x}'\mathbf{x} \leq \mathbf{x}'\mathbf{P}(t)\mathbf{x}, \qquad \forall\, \mathbf{x} \in R^n, \forall\, t \geq 0$$

proof Let $\mathbf{x} \in R^n$, and consider the triple product $\mathbf{x}'\mathbf{P}(t)\mathbf{x}$. From (83), we get

89
$$\mathbf{x}'\mathbf{P}(t)\mathbf{x} = \int_t^\infty \mathbf{x}'\mathbf{\Phi}'(\tau, t)\mathbf{Q}(t)\mathbf{\Phi}(\tau, t)\mathbf{x}\, d\tau$$
$$= \int_t^\infty \mathbf{s}'(\tau, \underset{\sim}{x}, t)\mathbf{Q}(t)\mathbf{s}(\tau, \underset{\sim}{x}, t)\, d\tau$$

where $\mathbf{s}(\tau, \mathbf{x}, t)$ denotes (as before) the solution of (1), as evaluated at time τ, corresponding to the initial condition $\mathbf{x}$ at time t. Now, (86) and (89) together imply that

90
$$\mathbf{x}'\mathbf{P}(t)\mathbf{x} \geq \alpha \int_t^\infty \|\mathbf{s}(\tau, \mathbf{x}, t)\|_2^2\, d\tau$$

where $\|\cdot\|_2$ is the Euclidean norm on R^n. Now, by Theorem [3.5(1)], we have

91
$$\|\mathbf{s}(\tau, \mathbf{x}, t)\|_2 \geq \|\mathbf{x}\|_2 \exp\left\{-\int_t^\tau \mu_{i2}[-\mathbf{A}(\theta)]\, d\theta\right\}$$
$$\geq \|\mathbf{x}\|_2 \exp\left\{-\int_t^\tau \|\mathbf{A}(\theta)\|_{i2}\, d\theta\right\} \qquad \text{by (ii) of [3.2 (19)]}$$
$$\geq \|\mathbf{x}\|_2 \exp\left[-m_0(\tau - t)\right] \qquad \text{by (87)}$$

Substituting from (91) into (90) gives

92
$$\mathbf{x}'\mathbf{P}(t)\mathbf{x} \geq \alpha \int_t^\infty \mathbf{x}'\mathbf{x}e^{-2m_0(\tau - t)}\, d\tau$$
$$= \mathbf{x}'\mathbf{x}\frac{\alpha}{2m_0}$$

The inequality (88) now follows by taking $\beta = \alpha/2m_0$. ∎

93 ***Theorem*** Suppose $\mathbf{Q}(\cdot)$ and $\mathbf{A}(\cdot)$ satisfy the hypotheses of lemmas (82) and (85). The the uniform asymptotic stability of the equilibrium point $\mathbf{0}$ of (1) can be established by selecting the Liapunov function

94
$$V(t, \mathbf{x}) = \mathbf{x}'\mathbf{P}(t)\mathbf{x}$$

where $\mathbf{P}(\cdot)$ is given by (83), and applying Theorem [5.2(45)].

proof With $V(t, \mathbf{x})$ defined by (94), we have

95
$$\dot{V}(t, \mathbf{x}) = \mathbf{x}'[\dot{\mathbf{P}}(t) + \mathbf{A}'(t)\mathbf{P}(t) + \mathbf{P}(t)\mathbf{A}(t)]\mathbf{x}$$

It is easily verified by simple differentiation of (83) that

96
$$\dot{\mathbf{P}}(t) = -\mathbf{A}'(t)\mathbf{P}(t) - \mathbf{P}(t)\mathbf{A}(t) - \mathbf{Q}(t)$$

Hence

97
$$\dot{V}(t, \mathbf{x}) = -\mathbf{x}'\mathbf{Q}(t)\mathbf{x}$$

Thus the functions V and $\dot{V}$ satisfy the various hypotheses of Theorem [5.2(45)]. ∎

REMARKS: Theorem (93) can be thought of as the extension, to the nonautonomous case, of Theorem (55). In short, Theorem (93) states that if the equilibrium point **0** of (1) is uniformly asymptotically stable over $[0, \infty)$, then this fact can be deduced by applying Theorem [5.2(45)] with a *quadratic* Liapunov function. However, whereas Theorem (55) also provides a useful testing procedure for asymptotic stability, Theorem (93) does not have any value as a testing procedure. The utility of Theorem (93) lies in the fact that it provides the basis for Liapunov's indirect method (linearization), as applied to nonautonomous systems.

Some Specialized Results

We shall close this subsection by presenting a few specialized results for nonautonomous systems.

First, suppose that the matrix $\mathbf{A}(t)$ in (1) is periodic in t. In this case, we know (by Theorems [5.1(57)] and [5.1(45)]) that the stability at time 0 of the equilibrium point **0** is equivalent to the uniform stability of **0** over $[0, \infty)$ and that the asymptotic stability at time 0 of **0** is equivalent to the uniform asymptotic stability of **0** over $[0, \infty)$. Further, it is possible to test for the stability and/or asymptotic stability of **0** by examining a *constant* matrix. This is brought out in the following.

98 ***lemma*** Suppose $\mathbf{A}(\cdot)$ in (1) satisfies

99 $$\mathbf{A}(t+T) = \mathbf{A}(t), \qquad \forall\, t \geq 0$$

for some $T > 0$. Then the corresponding state transition matrix $\mathbf{\Phi}(t, t_0)$ can be expressed as

100 $$\mathbf{\Phi}(t, t_0) = \mathbf{\Psi}(t, t_0) \exp[\mathbf{M}(t - t_0)]$$

where

101 $$\mathbf{\Psi}(t+T, t_0) = \mathbf{\Psi}(t, t_0), \qquad \forall\, t \geq 0$$

and $\mathbf{M}$ is a constant $n \times n$ matrix.

proof Define

102 $$\mathbf{R} = \mathbf{\Phi}(t_0 + T, t_0)$$

and choose $\mathbf{M}$ such that

103 $$\mathbf{R} = \exp(\mathbf{M}T)$$

We claim that $\mathbf{\Psi}(t, t_0)$ defined by

104 $$\mathbf{\Psi}(t, t_0) = \mathbf{\Phi}(t, t_0) \exp[-\mathbf{M}(t - t_0)]$$

satisfies (101). To see this, we proceed as follows:

105
$$\begin{aligned}\mathbf{\Psi}(t+T, t_0) &= \mathbf{\Phi}(t+T, t_0)\cdot\exp[-\mathbf{M}(t+T-t_0)] \\ &= \mathbf{\Phi}(t+T, t_0+T)\cdot\mathbf{\Phi}(t_0+T, t_0)\cdot\exp(-\mathbf{M}T) \\ &\quad\cdot\exp[-\mathbf{M}(t-t_0)]\end{aligned}$$

However, because of the periodicity of $\mathbf{A}(\cdot)$, we have

106
$$\mathbf{\Phi}(t+T, t_0+T) = \mathbf{\Phi}(t, t_0)$$

Hence (105) simplifies to

107
$$\mathbf{\Psi}(t+T, t_0) = \mathbf{\Phi}(t, t_0)\cdot\exp[-\mathbf{M}(t-t_0)] = \mathbf{\Psi}(t, t_0)$$

This establishes (101). ∎

Once this representation (100) for the state transition matrix is obtained, the results of Sec. 5.3.1 can be used to obtain necessary and sufficient conditions for stability. These are given next.

108 ***Theorem*** Consider the system (1), and suppose $\mathbf{A}(\cdot)$ satisfies (99). Then the equilibrium point $\mathbf{0}$ of (1) is uniformly stable over $[0, \infty)$ if and only if all eigenvalues of $\mathbf{M}$ have nonpositive real parts, and any eigenvalue of $\mathbf{M}$ having a zero real part is a simple zero of the minimal polynomial of $\mathbf{M}$. The equilibrium point $\mathbf{0}$ of (1) is uniformly asymptotically stable over $[0, \infty)$ if and only if all eigenvalues of $\mathbf{M}$ have negative real parts.

We turn now to so-called *slowly varying* systems. Consider the differential equation (1), and suppose that for each fixed t the eigenvalues of the matrix $\mathbf{A}(t)$ all have negative real parts. One might ask whether this is enough to ensure the asymptotic stability of the equilibrium point $\mathbf{0}$. The answer, unfortunately, is no, as brought out in the following.

109 **Example.** Consider the system (1), with

$$\mathbf{A}(t) = \begin{bmatrix} -1 + a\cos^2 t & 1 - a\sin t\cos t \\ -1 - a\sin t\cos t & -1 + a\sin^2 t \end{bmatrix}$$

Then it is easy to verify that

$$\mathbf{\Phi}(t, 0) = \begin{bmatrix} e^{(a-1)t}\cos t & e^{-t}\sin t \\ -e^{(a-1)t}\sin t & e^{-t}\cos t \end{bmatrix}$$

Now, the eigenvalues of $\mathbf{A}(t)$ are in fact independent of t and satisfy the characteristic equation

$$\lambda^2 + (2-a)\lambda + (2-a) = 0$$

Thus, if $1 < a < 2$, the eigenvalues of $\mathbf{A}(t)$ have negative real parts for all t, and in fact the eigenvalues of $\mathbf{A}(t)$ are bounded away from the imaginary axis; yet, the equilibrium point $\mathbf{0}$ is unstable.

The above example shows that even if the eigenvalues of $\mathbf{A}(t)$ all have negative real parts for each t, the equilibrium point $\mathbf{0}$ may still be unstable. On the other hand, it is possible to prove the following result.

110 ***Theorem*** Consider the system (1); suppose that $\mathbf{A}(\cdot)$ is continuously differentiable and that the eigenvalues of $\mathbf{A}(t)$ all have negative real parts for all $t \geq 0$. Under these conditions, the equilibrium point $\mathbf{0}$ of (1) is uniformly asymptotically stable over $[0, \infty)$, provided the quantity

111 $$\sup_{t \geq 0} \|\dot{\mathbf{A}}(t)\|_i \triangleq v$$

is sufficiently small.

Because the quantity v can be thought to provide a measure of the *rate of variation* of $\mathbf{A}(\cdot)$, the above result states that the equilibrium point $\mathbf{0}$ of (1) is uniformly asymptotically stable over $[0, \infty)$, provided (1) the eigenvalues of $\mathbf{A}(t)$ all have negative real parts for each $t \geq 0$ and (2) the rate of variation of $\mathbf{A}(\cdot)$ is sufficiently small.

Because Theorem (110) is a special case of a general result for nonlinear systems, we shall omit the proof. For the general result, see Sec. 5.6.

Problem 5.13. Using the Liapunov matrix equation, determine whether each of the matrices below is Hurwitz (i.e., whether all of its eigenvalues have negative real parts).

$$\text{(a) } \mathbf{A} = \begin{bmatrix} 2 & 1 \\ -8 & -3 \end{bmatrix} \quad \text{(b) } \mathbf{A} = \begin{bmatrix} 2 & 1 \\ -3 & -2 \end{bmatrix} \quad \text{(c) } \mathbf{A} = \begin{bmatrix} 1 & 2 \\ -2 & -5 \end{bmatrix}$$

Problem 5.14. Consider an RLC network that does not contain any capacitor loops or inductor cutsets. Such a network can be described in state variable form by choosing the capacitor voltages and inductor currents as the state variables. Specifically, if $\mathbf{x}_C$ denotes the vector of capacitor voltages and $\mathbf{x}_L$ denotes the vector of inductor currents, then the state equations are of the form

$$\begin{bmatrix} \dot{\mathbf{x}}_C \\ \dot{\mathbf{x}}_L \end{bmatrix} = \begin{bmatrix} \mathbf{C} & \mathbf{0} \\ \mathbf{0} & -\mathbf{L} \end{bmatrix}^{-1} \begin{bmatrix} \mathbf{R}_{11} & \mathbf{R}_{12} \\ -\mathbf{R}'_{12} & \mathbf{R}_{22} \end{bmatrix} \begin{bmatrix} \mathbf{x}_C \\ \mathbf{x}_L \end{bmatrix}$$

where $\mathbf{C}$ is the (diagonal positive definite) matrix of capacitance values; $\mathbf{L}$ is the (positive definite) matrix of inductance values; $\mathbf{R}_{11}, \mathbf{R}_{12}, \mathbf{R}_{22}$ are matrices arising from the resistive subnetwork; and $\mathbf{R}_{11}, \mathbf{R}_{22}$ are symmetric.

(a) Show that the total energy stored in the capacitors and inductors is given by

$$V = \frac{1}{2}\mathbf{x}_C'\mathbf{C}\mathbf{x}_C + \frac{1}{2}\mathbf{x}_L'\mathbf{L}\mathbf{x}_L$$

(b) Using this V as a Liapunov function candidate, show that $(\mathbf{0}, \mathbf{0})$ is a stable equilibrium point if $\mathbf{R}_{11}$ and $\mathbf{R}_{22}$ are both positive semidefinite, and that $(\mathbf{0}, \mathbf{0})$ is an asymptotically stable equilibrium point if $\mathbf{R}_{11}$ and $\mathbf{R}_{22}$ are both positive definite.

Problem 5.15. Using the results of Sec. 5.3.3, determine whether $\mathbf{0}$ is a (uniformly) stable or (uniformly) asymptotically stable equilibrium point, for each of the systems below.

(a) $$\begin{bmatrix} \dot{x}_1(t) \\ \dot{x}_2(t) \end{bmatrix} = \begin{bmatrix} -2 + \sin t^2 & 1 \\ \cos t & -1 \end{bmatrix} \begin{bmatrix} x_1(t) \\ x_2(t) \end{bmatrix}$$

(b) $$\begin{bmatrix} \dot{x}_1(t) \\ \dot{x}_2(t) \\ \dot{x}_3(t) \end{bmatrix} = \begin{bmatrix} 2t - t^2 & 1 - t & 3 + t \\ t^2 & -t^3 & 0 \\ 2 & 5 & 4 - t \end{bmatrix} \begin{bmatrix} x_1(t) \\ x_2(t) \\ x_3(t) \end{bmatrix}$$

Problem 5.16. A system of the form [5.1(1)] is said to be *bounded* at time t_0 if for every $\delta > 0$ there exists an $\epsilon(\delta, t_0)$ such that

$$\|\mathbf{x}(t_0)\| < \delta \Longrightarrow \|\mathbf{x}(t)\| < \epsilon(t_0, \delta) \qquad \forall\, t \geq t_0$$

It is said to be *uniformly bounded* over $[t_0, \infty)$ if for every $\delta > 0$ there exists an $\epsilon(\delta) > 0$ such that

$$\|\mathbf{x}(t_1)\| < \delta,\; t_1 \geq t_0 \Longrightarrow \|\mathbf{x}(t)\| < \epsilon(\delta) \qquad \forall\, t \geq t_1$$

(a) Show that for a *linear* system of the form (1), the equilibrium point $\mathbf{0}$ is bounded at line t_0 if and only if it is stable at time t_0. [Hint: Use Theorem (5).]

(b) Show that for a linear system of the form (1), the equilibrium point $\mathbf{0}$ is uniformly bounded over $[t_0, \infty)$ if and only if it is uniformly stable over $[t_0, \infty)$. [Hint: Use Theorem (19).]

5.4 LIAPUNOV'S INDIRECT METHOD

In this section, we shall combine the results of the earlier sections to obtain one of the most useful results in stability theory, namely Liapunov's indirect method (also known as Liapunov's first method). The great value of this method lies in the fact that, under certain conditions, it enables one to draw conclusions about a *nonlinear* system by studying the behavior of a *linear* system.

We shall begin by studying the concept of linearizing a nonlinear system around an equilibrium point. Consider first the autonomous system

$$\dot{\mathbf{x}}(t) = \mathbf{f}[\mathbf{x}(t)] \tag{1}$$

Suppose $\mathbf{f}(\mathbf{0}) = \mathbf{0}$, so that $\mathbf{0}$ is an equilibrium point of the system (1), and suppose also that $\mathbf{f}$ is continuously differentiable. Define

$$\mathbf{A} = \left[\frac{\partial \mathbf{f}}{\partial \mathbf{x}}\right]_{\mathbf{x}=\mathbf{0}} \tag{2}$$

i.e., $\mathbf{A}$ denotes the Jacobian matrix of $\mathbf{f}$ evaluated at $\mathbf{x} = \mathbf{0}$. By the definition of the Jacobian, we have that if we write

$$\mathbf{f}_1(\mathbf{x}) = \mathbf{f}(\mathbf{x}) - \mathbf{A}\mathbf{x} \tag{3}$$

then

$$\lim_{\|\mathbf{x}\| \to 0} \frac{\|\mathbf{f}_1(\mathbf{x})\|}{\|\mathbf{x}\|} = 0 \tag{4}$$

where all norms are l_2-norms. Alternatively, we can say that

$$\mathbf{f}(\mathbf{x}) = \mathbf{A}\mathbf{x} + \mathbf{f}_1(\mathbf{x}) \tag{5}$$

is the Taylor's series expansion of $\mathbf{f}(\cdot)$ around the point $\mathbf{x} = \mathbf{0}$ [recall that $\mathbf{f}(\mathbf{0}) = \mathbf{0}$]. With this notation, we refer to the system

$$\dot{\mathbf{z}}(t) = \mathbf{A}\mathbf{z}(t) \tag{6}$$

as the *linearization* or the *linearized* system, around the equilibrium point $\mathbf{0}$, of the system (1).

The development for nonautonomous systems is similar. Given the nonautonomous system

$$\dot{\mathbf{x}}(t) = \mathbf{f}[t, \mathbf{x}(t)] \tag{7}$$

suppose that

$$\mathbf{f}(t, \mathbf{0}) = \mathbf{0}, \qquad \forall\, t \geq \mathbf{0} \tag{8}$$

and that $\mathbf{f}(t, \cdot)$ is continuously differentiable. Define

$$\mathbf{A}(t) = \left[\frac{\partial \mathbf{f}(t, \mathbf{x})}{\partial \mathbf{x}}\right]_{\mathbf{x}=\mathbf{0}} \tag{9}$$

$$\mathbf{f}_1(t, \mathbf{x}) = \mathbf{f}(t, \mathbf{x}) - \mathbf{A}(t)\mathbf{x} \tag{10}$$

Now, by the definition of the Jacobian, for *each fixed $t \geq 0$*, we have

$$\lim_{\|\mathbf{x}\| \to 0} \frac{\|\mathbf{f}_1(t, \mathbf{x})\|}{\|\mathbf{x}\|} = 0 \tag{11}$$

However, it may or may not be true that

$$\lim_{\|\mathbf{x}\| \to 0} \sup_{t \geq 0} \frac{\|\mathbf{f}_1(t, \mathbf{x})\|}{\|\mathbf{x}\|} = 0 \tag{12}$$

In other words, the convergence in (11) may or may not be uniform in t. *Provided* (12) *holds*, we refer to the system

$$\dot{\mathbf{z}}(t) = \mathbf{A}(t)\mathbf{z}(t) \tag{13}$$

as the linearized system (or linearization), around the equilibrium point $\mathbf{0}$, of the system (7).

14 **Example.** Consider the system

15 $$\dot{x}_1(t) = -x_1(t) + tx_2^2$$

16 $$\dot{x}_2(t) = x_1(t) - x_2(t)$$

In this case, $\mathbf{f}(t, \cdot)$ is continuously differentiable, and

17 $$\mathbf{A}(t) = \begin{bmatrix} -1 & 0 \\ 1 & -1 \end{bmatrix}, \qquad \forall\, t \geq 0$$

However, the remainder term $\mathbf{f}_1(t, \mathbf{x})$ is given by

18 $$\mathbf{f}_1(t, \mathbf{x}) = [tx_2^2 \quad 0]'$$

so that the uniformity condition (12) does not hold. Accordingly, we do *not* refer to the system

19 $$\dot{z}_1(t) = -z_1(t)$$

20 $$\dot{z}_2(t) = z_1(t) - z_2(t)$$

as a linearization of the system (15)–(16).

Theorem (21) is the main *stability* result of Liapunov's indirect method. Because there is nothing to be gained by assuming autonomy, we shall state the result for nonautonomous systems.

21 ***Theorem*** Consider the (nonautonomous) system

22 $$\dot{\mathbf{x}}(t) = \mathbf{f}[t, \mathbf{x}(t)]$$

Suppose that

23 $$\mathbf{f}(t, \mathbf{0}) = \mathbf{0}$$

and that $\mathbf{f}(t, \cdot)$ is continuously differentiable. Define

24 $$\mathbf{A}(t) = \left[\frac{\partial \mathbf{f}(t, \mathbf{x})}{\partial \mathbf{x}}\right]_{\mathbf{x}=\mathbf{0}}$$

25 $$\mathbf{f}_1(t, \mathbf{x}) = \mathbf{f}(t, \mathbf{x}) - \mathbf{A}(t)\mathbf{x}$$

and assume that

26 $$\lim_{\|\mathbf{x}\| \to 0} \sup_{t \geq 0} \frac{\|\mathbf{f}_1(t, \mathbf{x})\|}{\|\mathbf{x}\|} = 0$$

27 $$\mathbf{A}(\cdot) \text{ is bounded}$$

Under these conditions, if the equilibrium point $\mathbf{0}$ of the system

28 $$\dot{\mathbf{z}}(t) = \mathbf{A}(t)\mathbf{z}(t)$$

is uniformly asymptotically stable over $[0, \infty)$, then the equilibrium point $\mathbf{0}$ of the system (22) is also uniformly asymptotically stable over $[0, \infty)$.

proof Because $\mathbf{A}(\cdot)$ is bounded and the equilibrium point $\mathbf{0}$ of (28) is uniformly asymptotically stable over $[0, \infty)$, it follows from lemma [5.3(85)] that the matrix

29 $$\mathbf{P}(t) = \int_t^\infty \mathbf{\Phi}'(\tau, t)\mathbf{\Phi}(\tau, t)\, d\tau$$

is well defined for all $t \geq 0$, is positive definite for all $t \geq 0$, and satisfies, for some positive α and β, the inequalities,

30 $$\alpha\mathbf{x}'\mathbf{x} \leq \mathbf{x}'\mathbf{P}(t)\mathbf{x} \leq \beta\mathbf{x}'\mathbf{x}, \qquad \forall\ \mathbf{x} \in R^n,\ \forall\ t \geq 0$$

Hence the function

31 $$V(t, \mathbf{x}) = \mathbf{x}'\mathbf{P}(t)\mathbf{x}$$

is a decrescent p.d.f. Calculating $\dot{V}$ for the system (22), we get

32 $$\begin{aligned}\dot{V}(t, \mathbf{x}) &= \mathbf{x}'\dot{\mathbf{P}}(t)\mathbf{x} + \mathbf{f}'(t, \mathbf{x})\mathbf{P}(t)\mathbf{x} + \mathbf{x}'\mathbf{P}(t)\mathbf{f}(t, \mathbf{x}) \\ &= \mathbf{x}'[\dot{\mathbf{P}}(t) + \mathbf{A}'(t)\mathbf{P}(t) + \mathbf{P}(t)\mathbf{A}(t)]\mathbf{x} + 2\mathbf{x}'\mathbf{P}(t)\mathbf{f}_1(t, \mathbf{x})\end{aligned}$$

However, from (29), it can easily be shown that

33 $$\dot{\mathbf{P}}(t) + \mathbf{A}'(t)\mathbf{P}(t) + \mathbf{P}(t)\mathbf{A}(t) = -\mathbf{I}$$

Hence

34 $$\dot{V}(t, \mathbf{x}) = -\mathbf{x}'\mathbf{x} + 2\mathbf{x}'\mathbf{P}(t)\mathbf{f}_1(t, \mathbf{x})$$

Because (26) holds, pick a number $r > 0$ such that

35 $$\|\mathbf{f}_1(t, \mathbf{x})\| \leq \frac{1}{3\beta}\|\mathbf{x}\|, \qquad \forall\ \mathbf{x} \text{ with } \|\mathbf{x}\| < r, \quad \forall\ t \geq 0$$

Then (35) and (30) together imply that

36 $$|2\mathbf{x}'\mathbf{P}(t)\mathbf{f}_1(t, \mathbf{x})| \leq \frac{2\mathbf{x}'\mathbf{x}}{3}, \qquad \forall\ t \geq 0 \ \text{ whenever } \|\mathbf{x}\| < r$$

Therefore

37 $$\dot{V}(t, \mathbf{x}) \leq \frac{-\mathbf{x}'\mathbf{x}}{3} \qquad \forall\ t \geq 0 \ \text{ whenever } \|\mathbf{x}\| < r$$

Thus all the hypotheses for Theorem [5.2(45)] are satisfied, and we conclude that the equilibrium point $\mathbf{0}$ of (22) is uniformly asymptotically stable over $[0, \infty)$. ■

38 ***corollary*** Consider the autonomous system

39 $$\dot{\mathbf{x}}(t) = \mathbf{f}[\mathbf{x}(t)]$$

Suppose that $\mathbf{f}(\mathbf{0}) = \mathbf{0}$ and that $\mathbf{f}$ is continuously differentiable. Define

40 $$\mathbf{A} = \left[\frac{\partial \mathbf{f}}{\partial \mathbf{x}}\right]_{\mathbf{x}=\mathbf{0}}$$

Under these conditions, the equilibrium point $\mathbf{0}$ of (39) is (uniformly) asymptotically stable if all eigenvalues of $\mathbf{A}$ have negative real parts.

To prove the instability counterpart to Theorem (21), we have to assume that the linearized system is autonomous, even if the original nonlinear system is not.

41 ***Theorem*** Consider the system

42 $$\dot{\mathbf{x}}(t) = \mathbf{f}[t, \mathbf{x}(t)]$$

Suppose that (23) holds and that $\mathbf{f}(t, \cdot)$ is continuously differentiable. Suppose in addition that

43 $$\mathbf{A}(t) = \left[\frac{\partial \mathbf{f}(t, \mathbf{x})}{\partial \mathbf{x}}\right]_{\mathbf{x}=\mathbf{0}} \equiv \mathbf{A}_0 \quad \text{(a constant matrix)}, \qquad \forall\, t \geq 0$$

and that (26) holds. Under these conditions, the equilibrium point $\mathbf{0}$ of (42) is unstable if at least one eigenvalue of $\mathbf{A}_0$ has a positive real part.

proof We shall give the proof only for the case where the matrix $\mathbf{A}_0$ satisfies the condition [5.3(45)]. The proof for the general case can be obtained from the one given below by using continuity arguments.

Because $\mathbf{A}_0$ is assumed to satisfy [5.3(45)], and has at least one eigenvalue with a positive real part, we have from Theorem [5.3(66)] that the equation

44 $$\mathbf{A}_0'\mathbf{P} + \mathbf{P}\mathbf{A}_0 = \mathbf{I}$$

has a unique solution for $\mathbf{P}$ and that this matrix $\mathbf{P}$ has at least one positive eigenvalue. Now, by arguments entirely analogous to those used in the proof of Theorem (21), it can be shown that if we choose

45 $$V(\mathbf{x}) = \mathbf{x}'\mathbf{P}\mathbf{x}$$

then $\dot{V}(t, \mathbf{x})$ is an l.p.d.f., so that, by Theorem [5.2(106)], the equilibrium point $\mathbf{0}$ of (42) is unstable. The details are left as an exercise. ∎

REMARKS: Theorems (21) and (41) are very useful because they enable one to draw conclusions about the stability status of the equilibrium point $\mathbf{0}$ of a given *nonlinear* system by examining a *linear* system. The advantages of these results are self-evident. Some of the limitations of these results are the following: (1) The conclusions based on linearization are purely local in nature; to study *global* asymptotic stability, it is still necessary to resort to Liapunov's direct method. (2) In the case where the linearized system is autonomous, if some eigenvalues of $\mathbf{A}$ have zero real parts and the remainder have negative real parts, linearization techniques are inconclusive, because this case falls outside the scope of both Theorems (21) and (41). (This can be rationalized as follows: In the case described above, one can say that the linearized system is on the verge of stability—or, if one is a pessimist, that it is on the verge of instability. Thus it stands to reason that the stability status of the equilibrium point **0** is actually determined by the higher-order terms, which are being neglected in the linearization.) (3) In the case where the linearized system is nonautonomous, if the equilibrium point **0** is asymptotically stable but not uniformly asymptotically stable over $[0, \infty)$, the linearization technique is once again inconclusive. It can be shown by means of examples that the assumption of *uniform* asymptotic stability in Theorem (21) is indispensable. Note that corollary (38) is the basis of the linearization technique discussed in Sec. 2.3.

46 **Example.** Consider the Van der Pol oscillator

47 $$\dot{x}_1 = x_2$$

48 $$\dot{x}_2 = \mu(1 - x_1^2)x_2 - x_1$$

The linearization of this system around (0, 0) is

49 $$\begin{bmatrix} \dot{z}_1 \\ \dot{z}_2 \end{bmatrix} = \begin{bmatrix} 0 & 1 \\ -1 & \mu \end{bmatrix} \begin{bmatrix} z_1 \\ z_2 \end{bmatrix}$$

If $\mu > 0$, the eigenvalues of this matrix **A** both have positive real parts. Hence, by Theorem (41), the equilibrium point (0, 0) of the system (47)–(48) is unstable.

50 **Example.** The purpose of this example is to show that inequality (35) can sometimes be used to obtain an estimate for the *domain of attraction* of the asymptotically stable equilibrium point **0**.

Consider the system

51 $$\dot{x}_1 = -x_1 + x_2 + x_1 x_2$$

52 $$\dot{x}_2 = -x_1 + x_2^2$$

The linearization of this system around (0, 0) is

53 $$\begin{bmatrix} \dot{z}_1 \\ \dot{z}_2 \end{bmatrix} = \begin{bmatrix} -1 & 1 \\ -1 & 0 \end{bmatrix} \begin{bmatrix} z_1 \\ z_2 \end{bmatrix}$$

If we let **A** denote the matrix in (53), we find that the eigenvalues of **A** are $(-1 \pm j\sqrt{3})/2$; hence, by Theorem (21), we know that (0, 0) is an asymptotically stable equilibrium point of (51)–(52).

To obtain an estimate for the domain of attraction r[cf. (35)], we first solve the equation

54 $$\mathbf{A'P} + \mathbf{PA} = -\mathbf{I}$$

for **P**. It can easily be verified that

$$\mathbf{P} = \begin{bmatrix} 1 & -1/2 \\ -1/2 & 3/2 \end{bmatrix} \tag{55}$$

Hence β in (30) can be chosen as the largest eigenvalue of **P**, which is $(5 + \sqrt{5})/4$, or approximately 1.81. To satisfy (35), we must choose r in such a way that

$$\|\mathbf{f}_1(\mathbf{x})\| \leq \frac{\rho}{2\beta}\|\mathbf{x}\| \quad \text{whenever } \|\mathbf{x}\| < r \tag{56}$$

where $\rho < 1$ is some number. Note that we chose $\rho = \frac{2}{3}$ in (35) for the purposes of proving the theorem, but here it is desirable to choose ρ as close to 1 as possible in order to get the least conservative value for r. In the present case, we have

$$\frac{\|\mathbf{f}_1(\mathbf{x})\|}{\|\mathbf{x}\|} = \frac{(x_1^2 x_2^2 + x_2^4)^{1/2}}{(x_1^2 + x_2^2)^{1/2}} = |x_2| \leq \|\mathbf{x}\| \tag{57}$$

Hence, to satisfy (56), we can choose r as close as possible to $1/2\beta$ or approximately 0.27. In other words, what we have shown is that every trajectory starting from an initial condition whose norm is less than 0.27 approaches **0** as $t \longrightarrow \infty$.

An Application: Feedback Stabilization of Nonlinear Control Systems[13]

As an application of Liapunov's indirect method in the design of nonlinear control systems, we shall consider the following problem: Given an autonomous system described by

$$\dot{\mathbf{x}}(t) = \mathbf{f}[\mathbf{x}(t), \mathbf{u}(t)] \tag{58}$$

where $\mathbf{f}\colon R^n \times R^m \longrightarrow R^n$, the objective is to find a *feedback control law*, i.e., a relationship of the form

$$\mathbf{u}(t) = \mathbf{g}[\mathbf{x}(t)] \tag{59}$$

in such a way that the equilibrium point **0** of the *closed-loop* system

$$\dot{\mathbf{x}}(t) = \mathbf{f}[\mathbf{x}(t), \mathbf{g}(\mathbf{x}(t))] \tag{60}$$

is asymptotically stable.

To give a solution to this problem based on Liapunov's indirect method, we make the following assumptions:

---

[13]This application requires some familiarity with linear systems and the linear regulator problem in optimal control theory.

61
1. $\mathbf{f}(\mathbf{0}, \mathbf{0}) = \mathbf{0}$.
2. $\mathbf{f}$ is continuously differentiable.
3. If we define

$$\mathbf{A} = \left[\frac{\partial \mathbf{f}(\mathbf{x}, \mathbf{u})}{\partial \mathbf{x}}\right]_{\mathbf{x}=\mathbf{0}, \mathbf{u}=\mathbf{0}} \tag{62}$$

$$\mathbf{B} = \left[\frac{\partial \mathbf{f}(\mathbf{x}, \mathbf{u})}{\partial \mathbf{u}}\right]_{\mathbf{x}=\mathbf{0}, \mathbf{u}=\mathbf{0}} \tag{63}$$

then we have

$$\text{rank}\,[\mathbf{B} \,|\, \mathbf{A}\mathbf{B} \,|\, \ldots \,|\, \mathbf{A}^{n-1}\mathbf{B}] = n \tag{64}$$

Assumption (3) above has a very simple interpretation: In much the same way that (6) is called the linearization of (1) around the point $\mathbf{x} = \mathbf{0}$, one can think of the system

$$\dot{\mathbf{z}}(t) = \mathbf{A}\mathbf{z}(t) + \mathbf{B}\mathbf{v}(t) \tag{65}$$

as the linearization of the system (58) around the point $\mathbf{x} = \mathbf{0}$, $\mathbf{u} = \mathbf{0}$. Thus, assumption (3) states that the linearized system (65) is controllable. For such systems, the following result is well known (see, e.g., [7] Chen).

66 **fact** If (64) holds, then there exists an $n \times n$ matrix $\mathbf{K}$ such that all eigenvalues of $\mathbf{A} - \mathbf{B}\mathbf{K}$ have negative real parts.

Thus fact (66) tells us that if we choose the feedback control law

$$\mathbf{v}(t) = -\mathbf{K}\mathbf{z}(t) \tag{67}$$

in system (65), it is possible to choose the matrix $\mathbf{K}$ in such a way that the equilibrium point $\mathbf{z} = \mathbf{0}$ of the closed-loop system

$$\dot{\mathbf{z}}(t) = (\mathbf{A} - \mathbf{B}\mathbf{K})\mathbf{z}(t) \tag{68}$$

is asymptotically stable. It is worth noting that fact (66) is not just of theoretical interest and that there are actually systematic procedures for finding a suitable $\mathbf{K}$, given $\mathbf{A}$ and $\mathbf{B}$.

One such procedure comes from the linear regulator problem in optimal control theory (see, e.g., [2] Athans and Falb) and can be summarized as follows: Given $\mathbf{A}$, $\mathbf{B}$ that satisfy (64), pick positive definite matrices $\mathbf{P} \in R^{n \times n}$ and $\mathbf{Q} \in R^{m \times m}$, and solve the so-called *Riccati equation*

$$-\mathbf{P} - \mathbf{A}'\mathbf{M} - \mathbf{M}\mathbf{A} + \mathbf{M}\mathbf{B}\mathbf{Q}^{-1}\mathbf{B}'\mathbf{M} = \mathbf{0} \tag{69}$$

for the unknown matrix $\mathbf{M} \in R^{n \times n}$. It can be shown that, under the stated assumptions, (69) has a unique solution for $\mathbf{M}$, which is symmetric and positive definite. Moreover, if we choose

70 $$\mathbf{K} = \mathbf{Q}^{-1}\mathbf{B}'\mathbf{M}$$

then all eigenvalues of $\mathbf{A} - \mathbf{BK}$ have negative real parts. The last statement can be proved using Liapunov theory, and we shall now do so.

71 **fact** Suppose that $\mathbf{A}$, $\mathbf{B}$ satisfy (64), that $\mathbf{P} \in R^{n \times n}$ and $\mathbf{Q} \in R^{m \times m}$ are positive definite, that $\mathbf{M}$ satisfies (69), and that $\mathbf{K}$ is defined by (70). Then all eigenvalues of $\mathbf{A} - \mathbf{BK}$ have negative real parts.

proof We make use, without proof, of the fact that the solution $\mathbf{M}$ of (69) is positive definite. We have

72 $$\mathbf{A} - \mathbf{BK} = \mathbf{A} - \mathbf{BQ}^{-1}\mathbf{B}'\mathbf{M}$$

Therefore

73 $$\begin{aligned}(\mathbf{A} - \mathbf{BK})'\mathbf{M} + \mathbf{M}(\mathbf{A} - \mathbf{BK}) &= [\mathbf{A} - \mathbf{BQ}^{-1}\mathbf{B}'\mathbf{M}]'\mathbf{M} \\ &\quad + \mathbf{M}[\mathbf{A} - \mathbf{BQ}^{-1}\mathbf{B}'\mathbf{M}] \\ &= -\mathbf{P} - \mathbf{MBQ}^{-1}\mathbf{B}'\mathbf{M}\end{aligned}$$

where we use (69) in the last step. Now, by assumption, $\mathbf{P}$ is positive definite, and $\mathbf{MBQ}^{-1}\mathbf{B}'\mathbf{M}$ is at least positive semidefinite, so that the right-hand side of (73) is negative definite. Therefore, the matrix $\mathbf{A} - \mathbf{BK}$ satisfies the Liapunov matrix equation [5.3(43)], with $\mathbf{A} - \mathbf{BK}$ playing the role of $\mathbf{A}$ (in [5.3(43)]), $\mathbf{M}$ playing the role of $\mathbf{P}$, and $\mathbf{P} + \mathbf{MBQ}^{-1}\mathbf{B}'\mathbf{M}$ playing the role of $\mathbf{Q}$. Hence, by Theorem [5.3(55)], all eigenvalues of $\mathbf{A} - \mathbf{BK}$ have negative real parts. ∎

Up to now, we have studied only the background material dealing with the stabilization of the *linear* system (65). We shall now present the application of Liapunov's indirect method, which is contained in Theorem (74). Basically, Theorem (74) shows that once a stabilizing feedback gain has been found for the *linearized* system (65), we have already found a way to stabilize the original *nonlinear* system (58).

74 ***Theorem*** Consider the autonomous system

75 $$\dot{\mathbf{x}}(t) = \mathbf{f}[\mathbf{x}(t), \mathbf{u}(t)]$$

where $\mathbf{f}: R^n \times R^m \rightarrow R^n$. Suppose that $\mathbf{f}$ is continuously differentiable and that $\mathbf{f}(\mathbf{0}, \mathbf{0}) = \mathbf{0}$. Define

76 $$\mathbf{A} = \left[\frac{\partial \mathbf{f}}{\partial \mathbf{x}}\right]_{\mathbf{x}=\mathbf{0}, \mathbf{u}=\mathbf{0}}$$

77 $$\mathbf{B} = \left[\frac{\partial \mathbf{f}}{\partial \mathbf{x}}\right]_{\mathbf{x}=\mathbf{0}, \mathbf{u}=\mathbf{0}}$$

Suppose there exists a matrix $\mathbf{K} \in R^{m \times n}$ such that all eigenvalues of $\mathbf{A} - \mathbf{BK}$ have negative real parts. Under these conditions, if we choose

78 $$\mathbf{u}(t) = -\mathbf{K}\mathbf{x}(t)$$

in (75), then $\mathbf{0}$ is an asymptotically stable equilibrium point of the resulting system

79 $$\dot{\mathbf{x}}(t) = \mathbf{f}[\mathbf{x}(t), -\mathbf{K}\mathbf{x}(t)]$$

proof Let $\mathbf{g}: R^n \longrightarrow R^n$ be the function defined by

80 $$\mathbf{g}(\mathbf{x}) = \mathbf{f}(\mathbf{x}, -\mathbf{K}\mathbf{x})$$

Then the system (79) can be represented as

81 $$\dot{\mathbf{x}}(t) = \mathbf{g}[\mathbf{x}(t)]$$

Next, we observe that

82 $$\left[\frac{\partial \mathbf{g}}{\partial \mathbf{x}}\right]_{\mathbf{x}=\mathbf{0}} = \mathbf{A} - \mathbf{B}\mathbf{K}$$

Because, by assumption, all eigenvalues of $\mathbf{A} - \mathbf{B}\mathbf{K}$ have negative real parts, it follows from corollary (38) that $\mathbf{0}$ is an asymptotically stable equilibrium point of (79). ■

Thus Theorem (74) provides a powerful design tool for nonlinear control systems. The technique based on Theorem (74) can be split into three parts:

1. Linearize the given nonlinear system.
2. Find a stabilizing control law for the linearized system.
3. Implement the *same* control law for the original nonlinear system.

83 **Example.** Consider a second-order system described by

$$\dot{x}_1 = 3x_1 + x_2^2 - \text{sat}\,(2x_2 + u)$$

$$\dot{x}_2 = \sin x_1 - x_2 + u$$

where the *sat* function is defined by

$$\text{sat}\,\sigma = \begin{cases} \sigma, & |\sigma| \leq 1 \\ \text{sign}\,\sigma, & |\sigma| \geq 1 \end{cases}$$

The linearization of this system around $\mathbf{x} = \mathbf{0}$, $u = 0$ is

$$\begin{bmatrix} \dot{z}_1 \\ \dot{z}_2 \end{bmatrix} = \begin{bmatrix} 3 & -2 \\ 1 & -1 \end{bmatrix} \begin{bmatrix} z_1 \\ z_2 \end{bmatrix} + \begin{bmatrix} -1 \\ 1 \end{bmatrix} v$$

The $\mathbf{A}$ matrix above has one eigenvalue with a positive real part. However, the linearized system can be made stable by applying the control law

$$v = [4 \quad -1] \begin{bmatrix} z_1 \\ z_2 \end{bmatrix}$$

Hence the original nonlinear system can be stabilized by choosing

$$u = 4x_1 - x_2$$

An Application: Singular Perturbations

In Sec. 4.3, we derived some results for singularly perturbed linear systems. In what follows, we shall use Liapunov's indirect method in conjunction with the results of Sec. 4.3 to derive some results for singularly perturbed nonlinear systems. The proof of Theorem (84) is omitted, because it follows readily from earlier results.

84 ***Theorem*** Consider the system

$$\dot{\mathbf{x}}(t) = \mathbf{f}[\mathbf{x}(t), \mathbf{y}(t)] \tag{85}$$

$$\epsilon\dot{\mathbf{y}}(t) = \mathbf{g}[\mathbf{x}(t), \mathbf{y}(t)] \tag{86}$$

where $\mathbf{f}: R^n \times R^m \longrightarrow R^n$, $\mathbf{g}: R^n \times R^m \longrightarrow R^m$ are continuously differentiable and satisfy

$$\mathbf{f}(\mathbf{0}, \mathbf{0}) = \mathbf{0} \tag{87}$$

$$\mathbf{g}(\mathbf{0}, \mathbf{0}) = \mathbf{0} \tag{88}$$

Define

$$\mathbf{A}_{11} = \frac{\partial \mathbf{f}}{\partial \mathbf{x}}\bigg|_{(\mathbf{0}, \mathbf{0})} \tag{89}$$

$$\mathbf{A}_{12} = \frac{\partial \mathbf{f}}{\partial \mathbf{y}}\bigg|_{(\mathbf{0}, \mathbf{0})} \tag{90}$$

$$\mathbf{A}_{21} = \frac{\partial \mathbf{g}}{\partial \mathbf{x}}\bigg|_{(\mathbf{0}, \mathbf{0})} \tag{91}$$

$$\mathbf{A}_{22} = \frac{\partial \mathbf{g}}{\partial \mathbf{y}}\bigg|_{(\mathbf{0}, \mathbf{0})} \tag{92}$$

and suppose $\mathbf{A}_{22}$ is nonsingular. Under these conditions, there exists a continuously differentiable function $\mathbf{h}: R^n \longrightarrow R^m$ such that, in some neighborhood of $(\mathbf{0}, \mathbf{0})$,

$$\mathbf{v} = \mathbf{h}(\mathbf{x}) \tag{93}$$

is a solution of

$$\mathbf{g}(\mathbf{x}, \mathbf{y}) = \mathbf{0} \tag{94}$$

Moreover, we have that

(i) If all eigenvalues of $\mathbf{A}_{22}$ and of $\mathbf{A}_{11} - \mathbf{A}_{12}\mathbf{A}_{22}^{-1}\mathbf{A}_{21}$ have negative real parts, then there is an ϵ_0 such that $(\mathbf{0}, \mathbf{0})$ is an asymptotically stable equilibrium point of (85)–(86) whenever $0 < \epsilon \leq \epsilon_0$.

(ii) If at least one eigenvalue of $\mathbf{A}_{22}$ or of $\mathbf{A}_{11} - \mathbf{A}_{12}\mathbf{A}_{22}^{-1}\mathbf{A}_{21}$ has a positive real part, then there is an ϵ_0 such that $(\mathbf{0}, \mathbf{0})$ is an unstable equilibrium point of (85)–(86) whenever $0 < \epsilon \leq \epsilon_0$.

REMARKS: The existence of the function **h** follows from the implicit function theorem.

95 **Example.** Consider the system of equations

$$\dot{x}_1 = x_1(x_1 + y - 1) + x_2^2 + y \tag{96}$$

$$\dot{x}_2 = x_1(x_1^2 - 5) + x_2(x_1 - 3) + y(x_2 + 1) \tag{97}$$

$$\epsilon \dot{y} = x_1(x_1 - 2) - x_2(x_1 - 1) - y(y + 1) \tag{98}$$

For this system, one can easily verify that

$$\mathbf{A}_{11} = \begin{bmatrix} -1 & 0 \\ -5 & -3 \end{bmatrix}$$

$$\mathbf{A}_{12} = \begin{bmatrix} 1 \\ 1 \end{bmatrix}$$

$$\mathbf{A}_{21} = [2 \quad 1]$$

$$\mathbf{A}_{22} = [-1]$$

$$\mathbf{A}_{11} - \mathbf{A}_{12}\mathbf{A}_{22}^{-1}\mathbf{A}_{21} = \begin{bmatrix} 1 & 1 \\ -3 & -2 \end{bmatrix} \triangleq \mathbf{M}$$

Since all the eigenvalues of $\mathbf{A}_{22}$ and of $\mathbf{M}$ have negative real parts, we conclude that there exists an $\epsilon_0 > 0$ such that (0, 0, 0) is an asymptotically stable equilibrium point of the system (96)–(98) whenever $0 < \epsilon \leq \epsilon_0$.

Problem 5.17. For the system of Problems 5.9–5.12, apply Liapunov's indirect method, and determine whether or not it yields conclusive results.

Problem 5.18. Consider a feedback system described by

$$\dot{\mathbf{x}}(t) = \mathbf{A}\mathbf{x}(t) + \mathbf{b}u(t)$$

$$y(t) = \mathbf{c}'\mathbf{x}(t)$$

$$u(t) = -\phi(y(t))$$

Suppose that the matrix $\mathbf{A} - \mathbf{b}k\mathbf{c}'$ is Hurwitz (i.e., has all eigenvalues with negative real parts) whenever k belongs to the interval $[0, k_0]$. Show that $\mathbf{0}$ is an asymptotically stable equilibrium point whenever $\phi(\cdot)$ is a continuously differentiable function such that

$$0 \leq [d\phi(\sigma)/d\sigma]_{\sigma=0} \leq k_0$$

5.5 THE LUR'E PROBLEM

In this section, we shall study the stability of an important class of control systems, namely, feedback systems whose forward path contains a linear time-invariant subsystem and whose feedback path contains a memoryless (possibly time-varying) nonlinearity. Such a system is shown in Fig. 5.8 and is described by the equations

$$\dot{\mathbf{x}}(t) = \mathbf{A}\mathbf{x}(t) + \mathbf{b}\xi(t), \qquad t \geq 0 \tag{1}$$

$$\sigma(t) = \mathbf{c}'\mathbf{x}(t) + d\xi(t) \tag{2}$$

$$\xi(t) = -\phi[t, \sigma(t)] \tag{3}$$

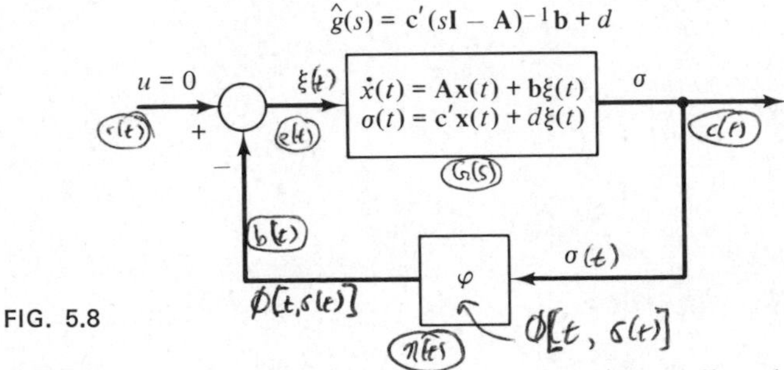

FIG. 5.8

In the above, $\mathbf{A} \in R^{n \times n}$; $\mathbf{b}, \mathbf{c}, \mathbf{x} \in R^n$; d, σ, ξ are all scalars; and the function $\phi(\cdot, \cdot)$ is continuous in both its arguments.[14] A study of systems of the form (1)–(3) is important for at least two reasons:

1. Many physical systems can be naturally interpreted as consisting of a *linear part* and a *nonlinear part*, so that the description (1)–(3) is reasonably general.
2. Many comprehensive results are available concerning the stability of such systems (of which only a few are given in this book).

All the results given in this section depend in an essential way on a theorem known as the Kalman–Yacubovitch lemma. Actually, the result used here is a generalization of this lemma due to Lefschetz. The proof of this particular theorem is not given in this book, because it utilizes several concepts not heretofore introduced. For an early treatment of this result, see [16] Lefschetz. A thorough development of the results of this section can be found in [17] Narendra and Taylor.

[14] We also assume, as in the rest of this chapter, that $\phi(\cdot, \cdot)$ is of such a nature that the system (1)–(3) has a unique solution corresponding to each initial condition $\mathbf{x}(0)$ and that this solution depends continuously on $\mathbf{x}(0)$. This is the case, for example, if $\phi(\cdot, \cdot)$ satisfies a global Lipschitz condition.

5.5.1 Problem Statement

In this subsection, we shall state the problem under study more precisely and discuss two conjectures, due to Aizerman and Kalman, respectively.

The problem dealt with in this section is as follows: Given a system described by (1)–(3), suppose

(A1) All eigenvalues of $\mathbf{A}$ have negative real parts (in this case $\mathbf{A}$ is known as a *Hurwitz* matrix), *or*

(A2) $\mathbf{A}$ has a simple eigenvalue of zero, and the rest of its eigenvalues have negative real parts.

(A3) The pair $(\mathbf{A}, \mathbf{b})$ is *controllable*; i.e.,

$$\text{rank } [\mathbf{b} \,|\, \mathbf{A}\mathbf{b} \,|\, \ldots \,|\, \mathbf{A}^{n-1}\mathbf{b}] = n \tag{4}$$

(A4) The pair $(\mathbf{A}, \mathbf{c})$ is observable; i.e.,

$$\text{rank } [\mathbf{c} \,|\, \mathbf{A}'\mathbf{c} \,|\, \ldots \,|\, (\mathbf{A}')^{n-1}\mathbf{c}] = n \tag{5}$$

(A5) The nonlinearity $\phi(\cdot, \cdot)$ satisfies

$$\phi(t, 0) = 0, \qquad \forall\, t \geq 0 \tag{6}$$

$$\sigma\phi(t, \sigma) \geq 0, \qquad \forall\, \sigma \in R, \forall\, t \geq 0 \tag{7}$$

[The condition (7) is sometimes stated as "$\phi(t, \cdot)$ is a *first and third-quadrant* nonlinearity," because (7) ensures that the graph of $\sigma \mapsto \phi(t, \sigma)$ lies in the first and third quadrants, for all $t \geq 0$.]

Subject to the above assumptions, the problem is to find conditions on $\mathbf{A}$, $\mathbf{b}$, $\mathbf{c}$, $\mathbf{d}$ which ensure that the equilibrium point $\mathbf{x} = \mathbf{0}$ of the system (1)–(3) is globally asymptotically stable whenever ϕ is *any* nonlinearity satisfying (A5).

In contrast with the systems studied in Sec. 5.2, we are concerned at present not with a *particular* system but with an entire *class* of systems, because we have not made any assumptions on ϕ other than that it satisfies (A5). For this reason, the problem under study is sometimes referred to as an *absolute stability* problem. It is also known as the *Lur'e problem*, after the Soviet scientist A. I. Lur'e.

It should be mentioned here that assumptions (A3) and (A4) do not result in any loss of generality, because given any proper rational function $\hat{h}(s)$, one can always find $\mathbf{A}$, $\mathbf{b}$, $\mathbf{c}$, d such that

$$\hat{h}(s) = \mathbf{c}'(s\mathbf{I} - \mathbf{A})^{-1}\mathbf{b} + d \tag{8}$$

and such that (A3), (A4) hold. See [7] Chen for full details.

In some cases, (7) is replaced by the following more stringent condition: There exist constants $k_2 \geq k_1 \geq 0$ such that

$$k_1\sigma^2 \leq \sigma\phi(t, \sigma) \leq k_2\sigma^2, \qquad \forall\, \sigma \in R, \forall\, t \geq 0 \tag{9}$$

Because of its importance, (9) is given a name.

10 *[definition]* The function $\phi(\cdot, \cdot)$ is said to *belong to the sector* $[k_1, k_2]$, or to *lie in the sector* $[k_1, k_2]$, if (9) holds.

Aizerman's Conjecture

In 1949, the Soviet scientist M.A. Aizerman made the following tempting conjecture: Suppose that $d = 0$, that ϕ belongs to the sector $[k_1, k_2]$, and that for each $k \in [k_1, k_2]$, the matrix $\mathbf{A} - \mathbf{bc}'k$ is Hurwitz. Then the equlibrium point $\mathbf{0}$ of the *nonlinear* system (1)–(3) is globally asymptotically stable.

The basis for this conjecture is the following reasoning: Suppose that $d = 0$ and that ϕ belongs to the sector $[k_1, k_2]$. If we replace ϕ by a *linear* element of gain k, the system (1)–(3) reduces to the linear time-invariant system

11 $$\dot{\mathbf{x}}(t) = [\mathbf{A} - \mathbf{bc}'k]\mathbf{x}(t)$$

The hypothesis is that the system (11) is globally asymptotically stable for all $k \in [k_1, k_2]$, and we would like to conclude that the original *nonlinear* system is also globally asymptotically stable.[15] If it were true, this conjecture would have been extremely useful, because it would have allowed one to deduce the stability of a nonlinear system by studying only linear systems. Unfortunately, several counterexamples have shown that this conjecture is false in general. (See, e.g., [23] Willems: also see Theorem [6.7(42)] for a related result.)

Kalman's Conjecture

In 1957, R. E. Kalman made the following conjecture, which strengthens the hypothesis of Aizerman's conjecture. Suppose the nonlinear function ϕ satisfies

12 $$k_3 \leq \frac{\partial \phi(t, \sigma)}{\partial \sigma} \leq k_4, \qquad \forall\, \sigma \in R,\ \forall\, t \geq 0$$

and suppose the matrix $\mathbf{A} - \mathbf{bc}'k$ is Hurwitz for all $k \in [k_3, k_4]$; then the nonlinear system (1)–(3) is globally asymptotically stable.

The hypothesis of Kalman's conjecture is stronger than that of Aizerman's conjecture, because in general $k_3 \leq k_1$ and $k_2 \leq k_4$. Therefore the matrix $\mathbf{A} - \mathbf{bc}'k$ is assumed to be Hurwitz for a larger range of values of k. Also, by Problem 5.18, the hypotheses of Kalman's conjecture insure that $\mathbf{0}$ is an asymptotically stable equilibrium point

[15] As pointed out in Sec. 5.1, one can unambiguously say that the *system* is globally asymptotically stable, rather than that the *equilibrium point* $\mathbf{0}$ is globally asymptotically stable.

whenever $\phi(\cdot, \cdot)$ satisfies (12). In spite of this, however, Kalman's conjecture is also false.

Even though both Aizerman's and Kalman's conjectures were ultimately shown to be false, they nevertheless served a very useful purpose, because they created a great deal of interest in the Lur'e problem and were thus indirectly responsible for the many results that are available today.

In the remainder of this section, we shall present three kinds of stability criteria for the Lur'e problem. These criteria are arranged according to the complexity of the Liapunov function employed in deriving them. That is, the circle criterion (Sec. 5.5.2) is based on the simplest Liapunov function, the Lur'e–Postnikov criterion (Sec. 5.5.3) is based on one that is rather more complicated, while Popov's criterion (Sec. 5.5.4) employs the most complicated Liapunov function of the three.

The proofs of these criteria make use of the following generalization of the Kalman–Yacubovitch lemma, which is due to Lefschetz.

13 ***Theorem*** Given a Hurwitz matrix $\mathbf{A}$, a vector $\mathbf{b}$ such that the pair $(\mathbf{A}, \mathbf{b})$ is controllable, a real vector $\mathbf{v}$, scalars $\gamma \geq 0$ and $\epsilon > 0$, and a positive definite matrix $\mathbf{Q}$, there exist a positive definite matrix $\mathbf{P}$ and a vector $\mathbf{q}$ that satisfy the equations

14 $$\mathbf{A}'\mathbf{P} + \mathbf{P}\mathbf{A} = -\mathbf{q}\mathbf{q}' - \epsilon\mathbf{Q}$$

15 $$\mathbf{P}\mathbf{b} - \mathbf{v} = \gamma^{1/2}\mathbf{q}$$

if and only if ϵ is small enough, and the scalar function

16 $$\hat{h}(s) \triangleq \gamma + h\mathbf{v}'(s\mathbf{I} - \mathbf{A})^{-1}\mathbf{b}$$

satisfies[16]

17 $$\text{Re}\, \hat{h}(j\omega) > 0, \qquad \forall\, \omega \in R$$

proof See [16] Lefschetz [pp. 115–18]. ∎

The use of this theorem in constructing Liapunov functions becomes clearer in Secs. 5.5.2–5.5.4.

5.5.2 Circle Criterion

Given the nonlinear system (1)–(3), suppose that the matrix $\mathbf{A}$ is Hurwitz [so that (A1) is satisfied] and that (A3)–(A5) hold. The circle

[16] Although we have not introduced the concept here, we mention that, with the assumptions on $\mathbf{A}$, (17) is equivalent to the requirement that $\hat{h}(s)$ is a strictly positive real function.

criterion arises out of attempting to deduce the global asymptotic stability of the system by employing a Liapunov function of the form

$$V(\mathbf{x}) = \mathbf{x}'\mathbf{P}\mathbf{x} \tag{18}$$

Note that the nonlinear characteristic ϕ does not appear explicitly in the Liapunov function (13). Accordingly, this V-function is referred to as a *common* Liapunov function, because the same Liapunov function is used with all nonlinearities ϕ. This is in contrast with the Liapunov functions used in Secs. 5.5.3 and 5.5.4.

Throughout this subsection we shall make the following *basic assumptions*:

1. ϕ belongs to the sector $[0, k]$ for some finite positive k, and
2. $1 + kd > 0$.

If, in a given problem, ϕ belongs to some general sector $[\alpha, \beta]$, one can always employ what are known as *loop transformations* to ensure that the transformed ϕ belongs to the sector $[0, \beta - \alpha]$. This is discussed later. Assumption 2 requires a little more explanation. Suppose $1 + kd \leq 0$. Then one can always find a number $k_0 \in [0, k]$ such that $1 + k_0 d = 0$. Suppose we now choose $\phi_0(\sigma) = k_0\sigma$. Then ϕ_0 belongs to the class of nonlinearities under consideration (namely, ϕ_0 belongs to the sector $[0, k]$). Now, absolute stability requires that the system (1)–(3) be stable with $\phi = \phi_0$. However, in the present case (2) becomes

$$\sigma = \mathbf{c}'\mathbf{x} + d\xi = \mathbf{c}'\mathbf{x} - k_0 d\sigma \tag{19}$$

$$(1 + k_0 d)\sigma = \mathbf{c}'\mathbf{x} \tag{20}$$

However, because $1 + k_0 d = 0$, (20) can hold only if $\mathbf{x} = \mathbf{0}$, which means that the feedback system is degenerate. The situation can also be explained using transfer functions. Define

$$\hat{g}(s) = \mathbf{c}(s\mathbf{I} - \mathbf{A})^{-1}\mathbf{b} + d \tag{21}$$

Then $\hat{g}$ is the open-loop transfer function. If we now apply a feedback gain of k_0, the closed-loop transfer function becomes

$$\hat{g}_c(s) = \frac{\hat{g}(s)}{1 + k_0\hat{g}(s)} \tag{22}$$

Now, because $\hat{g}$ is a rational function of s, so is $\hat{g}_c(s)$. If we let $|s| \to \infty$, we get

$$\hat{g}_c(\infty) = \frac{\hat{g}(\infty)}{1 + k_0\hat{g}(\infty)} = \frac{d}{1 + k_0 d} \tag{23}$$

However, note that $1 + k_0 d = 0$ (and that $d \neq 0$). Hence $\hat{g}(\infty) = \infty$, which means that $\hat{g}_c$ is not a *proper* rational function and so represents an unstable system (see Sec. 6.3).

The conclusion, therefore, is that if ϕ belongs to the sector $[0, k]$, then $1 + kd > 0$ is a *necessary* condition for absolute stability, so that we can safely assume that it holds.

Calculating the derivative of V along the solutions of (1)–(3), we get

$$\dot{V}(\mathbf{x}) = \dot{\mathbf{x}}'\mathbf{P}\mathbf{x} + \mathbf{x}'\mathbf{P}\dot{\mathbf{x}} = \mathbf{x}'(\mathbf{A}'\mathbf{P} + \mathbf{P}\mathbf{A})\mathbf{x} + 2\mathbf{x}'\mathbf{P}\mathbf{b}\xi \tag{24}$$

We would like to derive conditions on $\mathbf{A}$, $\mathbf{b}$, $\mathbf{c}$, d, and k that ensure that $-\dot{V}$ is an l.p.d.f. over all of R^n. To do this, we proceed as follows: First, it is easy to verify from (2) that

$$k\xi\mathbf{x}'\mathbf{c} + (1 + kd)\xi^2 - \xi(k\sigma + \xi) = 0 \tag{25}$$

Therefore, we may substract this quantity from $\dot{V}$ without affecting its value. This gives

$$\begin{aligned}\dot{V}(\mathbf{x}) = {} & \mathbf{x}'(\mathbf{A}'\mathbf{P} + \mathbf{P}\mathbf{A})\mathbf{x} + 2\xi\mathbf{x}'(\mathbf{P}\mathbf{b} - \tfrac{1}{2}k\mathbf{c}) \\ & - (1 + kd)\xi^2 + \xi(k\sigma + \xi)\end{aligned} \tag{26}$$

Now, because ϕ belongs to the sector $[0, k]$, we have

$$0 \leq \sigma\phi(t, \sigma) \leq k\sigma^2, \qquad \forall\, \sigma \in R \tag{27}$$

which can be rearranged as

$$0 \leq \frac{\phi(t, \sigma)}{\sigma} \leq k, \qquad \forall\, \sigma \neq 0 \tag{28}$$

This inequality (28) means (loosely speaking) that $\phi(t, \sigma)$ and $k\sigma - \phi(t, \sigma)$ always have the same sign; more precisely,

$$\phi(t, \sigma)[k\sigma - \phi(t, \sigma)] \geq 0, \qquad \forall\, \sigma \in R \tag{29}$$

In other words,

$$\xi(k\sigma + \xi) \leq 0 \tag{30}$$

Hence the last term in $\dot{V}$ is nonpositive. We now make use of Theorem (13) in order to *complete the square*, so as to make the rest of the terms in $\dot{V}$ negative. Specifically, apply Theorem (13), with $\mathbf{Q}$ any positive definite matrix, $\mathbf{A}$, $\mathbf{b}$ as at present, $\gamma = 1 + kd$, and $\mathbf{v} = k\mathbf{c}/2$. Then Theorem (13) assures us that if (17) holds (the implications of which are discussed later on), then there exist a positive definite matrix $\mathbf{P}$ and a vector $\mathbf{q}$ such that

$$\mathbf{A}'\mathbf{P} + \mathbf{P}\mathbf{A} = -\mathbf{q}\mathbf{q}' - \epsilon\mathbf{Q} \tag{31}$$

$$\mathbf{P}\mathbf{b} - \tfrac{1}{2}k\mathbf{c} = (1 + kd)^{1/2}\mathbf{q} \tag{32}$$

Thus $\dot{V}$ becomes

$$\begin{aligned}\dot{V}(\mathbf{x}) &= -\epsilon\mathbf{x}'\mathbf{Q}\mathbf{x} - \mathbf{x}'\mathbf{q}\mathbf{q}'\mathbf{x} + 2\xi\mathbf{x}'(1 + kd)^{1/2}\mathbf{q} - (1 + kd)\xi^2 \\ &\qquad + \xi(k\sigma + \xi) \\ &= -\epsilon\mathbf{x}'\mathbf{Q}\mathbf{x} - [\mathbf{x}'\mathbf{q} - (1 + kd)^{1/2}\xi]^2 + \xi(k\sigma + \xi) \\ &\leq -\epsilon\mathbf{x}'\mathbf{Q}\mathbf{x}\end{aligned} \tag{33}$$

where we make use of (30) and the obvious fact that

34 $$[\mathbf{x}'\mathbf{q} - (1 + kd)^{1/2}\xi]^2 \geq 0$$

Thus, clearly, V is a p.d.f., and $-\dot{V}$ is a p.d.f., whence by Theorem [5.2(58)] it follows that the system (1)–(3) is globally asymptotically stable.

Now, let us come back to condition (17), which, as has just been demonstrated, is a sufficient condition for global asymptotic stability. In the present case, we have

35 $$\hat{h}(s) = (1 + kd) + k\mathbf{c}'(s\mathbf{I} - \mathbf{A})^{-1}\mathbf{b} = 1 + k\hat{g}(s)$$

where

36 $$\hat{g}(s) \triangleq \mathbf{c}'(s\mathbf{I} - \mathbf{A})^{-1}\mathbf{b} + d$$

is the transfer function of the open-loop system (1)–(2). Therefore, a sufficient condition for the global asymptotic stability of the system (1)–(3) is

37 $$\operatorname{Re}[1 + k\hat{g}(j\omega)] > 0, \qquad \forall\, \omega \in R$$

This is stated formally as a fact.

38 **fact** If ϕ belongs to the sector $[0, k]$ and $1 + kd > 0$, then (37) is a sufficient condition for the absolute stability of the system (1)–(3).

If ϕ belongs to the sector $[0, \infty)$, then the corresponding absolute stability criterion can be obtained from (37) by dividing through by k and then letting $k \longrightarrow \infty$.

39 **fact** If ϕ belongs to the sector $[0, \infty)$ and $d > 0$, then a sufficient condition for absolute stability is

40 $$\operatorname{Re}\hat{g}(j\omega) > 0, \qquad \forall\, \omega \in R$$

Actually, it should be clear that we can drop the phrase "and $1 + kd > 0$" in the statement of fact (38), because if (37) holds, then automatically

41 $$1 + kd = \operatorname{Re}[1 + k\hat{g}(\infty)] > 0$$

Similar remarks apply to fact (39).

If ϕ belongs to some sector $[\alpha, \beta]$, then the system (1)–(3) can be made to satisfy the hypothesis of fact (38) by *loop transformations*, as shown in Fig. 5.9. In this case, the transformed forward path transfer function is

42 $$\hat{g}_t(s) = \frac{\hat{g}(s)}{1 + \alpha\hat{g}(s)}$$

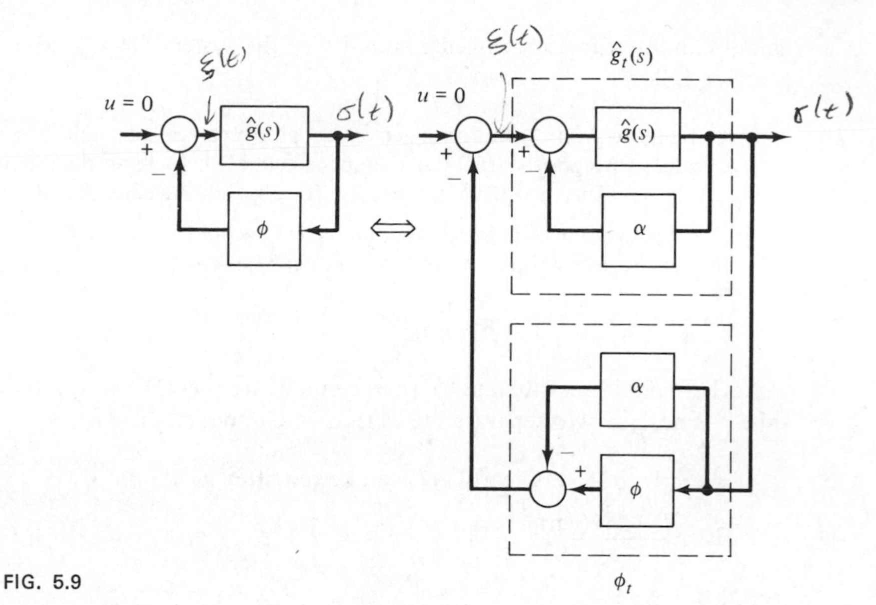

FIG. 5.9

while the transformed nonlinearity is

43 $$\phi_t(\sigma) = \phi(\sigma) - \alpha\sigma$$

Now, ϕ_t belongs to the sector $[0, \beta - \alpha]$. However, before we can apply fact (38), we must ensure that the transfer function $\hat{g}_t(s)$ can be represented in the form

44 $$\hat{g}_t(s) = \mathbf{c}_t(s\mathbf{I} - \mathbf{A}_t)^{-1}\mathbf{b}_t + d_t$$

where $\mathbf{A}_t$ *is a Hurwitz matrix*. Now, if the pair $(\mathbf{A}_t, \mathbf{b}_t)$ is controllable and the pair $(\mathbf{A}_t, \mathbf{c}_t)$ is observable, then the eigenvalues of $\mathbf{A}_t$ are the same as the poles of $\hat{g}_t(s)$. In turn, if $\alpha \neq 0$, then the poles of $\hat{g}_t(s)$ are the same as the zeroes of $1 + \alpha\hat{g}(s)$. Therefore, to apply fact (38) to the transformed system (42)–(43), we must first ensure that $1 + \alpha\hat{g}(s)$ has no zeroes in the closed right half-plane.

The last condition can be readily verified using the well-known Nyquist criterion. Suppose (1)–(2) is a realization of the untransformed transfer function $\hat{g}(s)$, and suppose (1) $\mathbf{A}$ has no eigenvalues on the imaginary axis, and (2) $\mathbf{A}$ has ν eigenvalues with positive real part. Then $\hat{g}(s)$ has no poles on the imaginary axis and has exactly ν poles in the open right half plane. In this context, the Nyquist criterion states that $1 + \alpha\hat{g}(s)$ has no zeroes in the closed right half-plane if and only if the Nyquist plot of $\hat{g}(j\omega)$ does not intersect the point $-1/\alpha + j0$ and encircles it in the counterclockwise sense exactly ν times.

We are now ready to extend fact (38) to the general case of a nonlinearity in the sector $[\alpha, \beta]$. Based on the foregoing discussion, a sufficient condition for the absolute stability of the system (1)–(3) can be given as follows.

45 **fact** Suppose **A** in (1) has no eigenvalues with zero real part and has ν eigenvalues with positive real part; suppose ϕ in (3) belongs to the sector $[\alpha, \beta]$, $\alpha \neq 0$. Then a sufficient condition for absolute stability is

46 The Nyquist plot of $\hat{g}(j\omega)$ does not intersect the point $-1/\alpha + j0$ and encircles it ν times in the counterclockwise sense.

$$\text{Re}\left[1 + (\beta - \alpha)\frac{\hat{g}(j\omega)}{1 + \alpha\hat{g}(j\omega)}\right] > 0, \qquad \forall\, \omega \in R \tag{47}$$

The conditions (46) and (47) can be made simpler. However, to do this, it is necessary to separate the cases $\alpha < 0$ and $\alpha > 0$.

Case 1 $\alpha > 0$: Clearly (47) can be rewritten as

$$\text{Re}\left[\frac{1 + \beta\hat{g}(j\omega)}{1 + \alpha\hat{g}(j\omega)}\right] > 0, \qquad \forall\, \omega \in R \tag{48}$$

Suppose

$$\hat{g}(j\omega) = u + jv \tag{49}$$

Then (48) becomes

$$\frac{(1 + \beta u)(1 + \alpha u) + \alpha\beta v^2}{(1 + \alpha u)^2 + v^2} > 0, \qquad \forall\, \omega \in R \tag{50}$$

If (46) holds, the denominator of (50) is never zero, so that (50) can be simplified to

$$(1 + \beta u)(1 + \alpha u) + \alpha\beta v^2 > 0, \qquad \forall\, \omega \in R \tag{51}$$

$$\left(u + \frac{1}{\alpha}\right)\left(u + \frac{1}{\beta}\right) + v^2 > 0, \qquad \forall\, \omega \in R \tag{52}$$

The condition (52) can be given a simple geometric interpretation. Let $D(\alpha, \beta)$ be the disk in the complex plane that is centered on the real axis and passes through $-1/\alpha + j0$ and $-1/\beta + j0$. This disk is shown in Fig. 5.10. The inequality (52) states that the Nyquist plot of $\hat{g}(j\omega)$ never enters the disk.

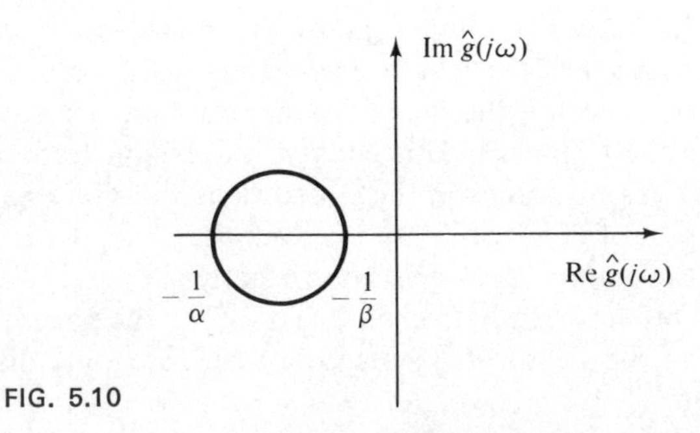

FIG. 5.10

The encirclement condition (46) can also be slightly simplified. If the Nyquist plot of $\hat{g}(j\omega)$ does not enter the disk $D(\alpha, \beta)$ and encircles the point $-1/\alpha + j0$ exactly ν times, then the plot must encircle *the disk* $D(\alpha, \beta)$ ν times and vice versa.

Combining these results, we get the following.

53 **fact** Suppose **A** in (1) has no eigenvalues with zero real part and has ν eigenvalues with positive real part; suppose ϕ in (3) belongs to the sector $[\alpha, \beta]$, with $\alpha > 0$. Then a sufficient condition for absolute stability is

54 The Nyquist plot of $\hat{g}(j\omega)$ does not enter the disk $D(\alpha, \beta)$ and encircles it in the counterclockwise sense exactly ν times.

Case 2a $\alpha < 0 < \beta$: In this case, it is clear, first, that a necessary condition for absolute stability is that **A** is Hurwitz. This is easily seen by setting $\phi(\sigma) \equiv 0$, which is an admissible nonlinearity. Thus (46) and (47) become, respectively,

55 The Nyquist plot of $\hat{g}(j\omega)$ neither encircles nor intersects the point $-1/\alpha + j0$.

56 $$\text{Re}\left[\frac{1 + \beta\hat{g}(j\omega)}{1 + \alpha\hat{g}(j\omega)}\right] > 0, \qquad \forall\, \omega \in R$$

Now (56) can be reduced to (51), as before. However, in going from (51) to (52), we divide by $\alpha\beta$, which is now negative. Hence the inequality is reversed, and (56) becomes

57 $$\left(u + \frac{1}{\alpha}\right)\left(u + \frac{1}{\beta}\right) + v^2 < 0, \qquad \forall\, \omega \in R$$

Thus the geometrical interpretation is that the Nyquist plot of $\hat{g}(j\omega)$ lies in the interior of the disk $D(\alpha, \beta)$. Moreover, if this is the case, (55) is automatically satisfied. Thus we have

58 **fact** Suppose **A** is Hurwitz, and suppose ϕ belongs to the sector $[\alpha, \beta]$, with $\alpha < 0 < \beta$. Then a sufficient condition for absolute stability is

59 The Nyquist plot of $\hat{g}(j\omega)$ lies in the interior of the disk $D(\alpha, \beta)$.

Case 2b $\alpha < \beta < 0$: In this case, replace $\hat{g}$ by $-\hat{g}$, α by $-\beta$, β by $-\alpha$, and apply fact (53).

The large number of criteria are now summarized in the following theorem, which is generally referred to as the *circle criterion.*

60 ***Theorem*** Suppose **A** has no eigenvalues on the imaginary axis and has ν eigenvalues with positive real parts; suppose ϕ belongs to the sector $[\alpha, \beta]$. Then a sufficient condition for absolute stability is one of the following, as appropriate:

(i) *If* $0 < \alpha < \beta$, the Nyquist plot of $\hat{g}(j\omega)$ does not enter the disk $D(\alpha, \beta)$ and encircles it ν times in the counterclockwise sense.

(ii) *If* $0 = \alpha < \beta$, the Nyquist plot of $\hat{g}(j\omega)$ lies in the half-plane

61 $$\left\{s\colon \operatorname{Re} s > -\frac{1}{\beta}\right\}$$

(iii) *If* $\alpha < 0 < \beta$, the Nyquist plot of $\hat{g}(j\omega)$ lies in the interior of the disk $D(\alpha, \beta)$.

(iv) *If* $\alpha < \beta < 0$, replace $\hat{g}$ by $-\hat{g}$, α by $-\beta$, β by $-\alpha$, and apply (i).

62 **Example.** We apply the circle criterion to study the stability of the damped Mathieu equation

63 $$\ddot{y} + 2\mu\dot{y} + (\mu^2 + a^2 - q\cos\omega_0 t)y = 0$$

It is easily verified that (63) can be put in the form (1)–(3) with

64 $$\mathbf{A} = \begin{bmatrix} 0 & 1 \\ -2\mu & \mu^2 + a^2 \end{bmatrix}; \quad \mathbf{b} = \begin{bmatrix} 0 \\ 1 \end{bmatrix}; \quad \mathbf{c} = \begin{bmatrix} 1 \\ 0 \end{bmatrix}; \quad d = 0$$

65 $$\phi(\sigma, t) = \sigma q \cos\omega_0 t$$

Our objective is to derive conditions on μ, a, q, and ω_0 that ensure the global asymptotic stability of the system (63). We assume that $\mu, a > 0$, so that $\mathbf{A}$ is Hurwitz and that $q > 0$. Next, note that

66 $$\hat{g}(s) = \frac{1}{s^2 + 2\mu s + \mu^2 + a^2}$$

67 $$\phi \text{ belongs to the sector } [-q, q]$$

Therefore, case (iii) of Theorem (60) applies here, with $\alpha = -q$ and $\beta = q$. Here the disk $D(\alpha, \beta)$ is the disk centered at the origin with radius $1/q$. Plotting $\hat{g}(j\omega)$, we note (see Fig. 5.11) that $\hat{g}(j\omega)$ attains its

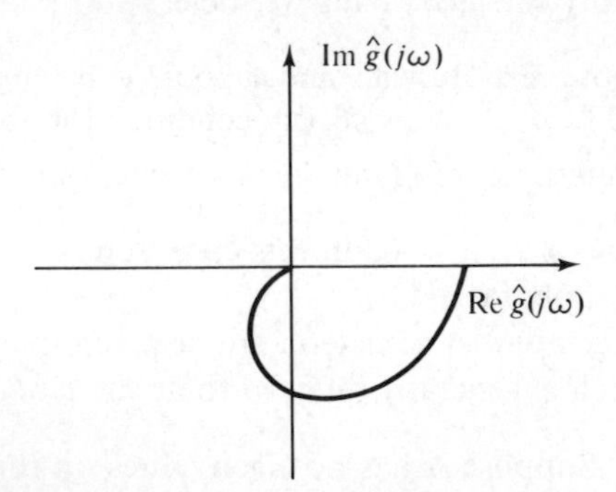

FIG. 5.11

maximum modulus at $\omega = \pm(a^2 - \mu^2)$ and that this maximum modulus is $1/(2\mu a)$. Thus the condition that the Nyquist plot of $\hat{g}(j\omega)$ lies in the interior of the disk $D(\alpha, \beta)$ reduces to

68 $$\frac{1}{2\mu a} < \frac{1}{q}$$

or, equivalently,

69 $$q < 2\mu a$$

Thus (69) is a sufficient condition for the global asymptotic stability of (63). Furthermore, because the circle criterion is a sufficient condition for *absolute* stability, (69) actually implies the global asymptotic stability of *all* systems of the form

70 $$\ddot{y} + 2\mu\dot{y} + (\mu^2 + a^2)y - qf(y) = 0$$

whenever f is any nonlinearity belonging to the sector $[-1, 1]$.

71 REMARKS: We shall conclude this subsection with a discussion of Theorem (60) (the circle criterion). One of the most appealing aspects of this result is its simple geometric interpretation, which is reminiscent of the Nyquist criterion.[17] Indeed, it is easy to see that if $\alpha = \beta$, then the *critical disk* $D(\alpha, \beta)$ shrinks to the *critical point* $-1/\alpha$, and the circle criterion reduces to the sufficiency portion of the Nyquist criterion.

Another appealing aspect of the circle criterion is that one needs only to plot the transfer function $\hat{g}(j\omega)$. This means that if we think of the forward path in Fig. 5.8 as a *black box*, then to apply Theorem (60) we need only to determine the transfer function of the black box and that there is no need to construct a state variable realization of the black box. Because the transfer function can usually be determined experimentally, based on input–output measurements, this feature greatly facilitates the use of the circle criterion.

In Chap. 6, we shall encounter another circle criterion which for all intents and purposes looks the same as Theorem (60). However, the one in Chap. 6 is applicable to a broader class of systems, e.g., distributed systems, systems with time delays, because the method of derivation in Chap. 6 is different. On the other hand, the present Theorem (60) is strictly limited to systems described by ordinary differential equations of finite order.

Finally, it is worth noting that the circle criterion is in some sense also a *necessary* condition for absolute stability. The appropriate setting for showing this is the functional analytic framework of Chap. 6, and a precise statement of this result can be found in Theorem [6.7(40)].

5.5.3 Lur'e–Postnikov Criterion

In this subsection, we shall once again consider the system (1)–(3). Throughout, we assume that $\mathbf{A}$ is Hurwitz, $d = 0$, and that ϕ is a *time-invariant* nonlinearity that belongs to the sector $[0, \infty)$.

---

[17] For a precise statement of this criterion, see Theorem [6.6(58)].

To prove the global asymptotic stability of the system (1)–(3), we employ a Liapunov function of the form

$$V(\mathbf{x}) = \mathbf{x}'\mathbf{P}\mathbf{x} + \beta\Phi(\sigma) \tag{72}$$

where

$$\Phi(\sigma) \triangleq \int_0^\sigma \phi(\zeta)\, d\zeta \tag{73}$$

P is a positive definite matrix, and $\beta \geq 0$. The function (72) is called a Liapunov function of the *Lur'e–Postnikov type*. Because ϕ belongs to the sector $[0, \infty)$, it is clear that $\Phi(\sigma)$ is always nonnegative, so that V is a p.d.f.

Calculating the derivative of V along the solutions of (1)–(3), we get

$$\begin{aligned} \dot{V}(\mathbf{x}) &= \mathbf{x}'(\mathbf{A}'\mathbf{P} + \mathbf{P}\mathbf{A})\mathbf{x} + 2\xi\mathbf{x}'\mathbf{P}\mathbf{b} + \beta\phi(\sigma)\dot{\sigma} \\ &= \mathbf{x}'(\mathbf{A}'\mathbf{P} + \mathbf{P}\mathbf{A})\mathbf{x} + 2\xi\mathbf{x}'\mathbf{P}\mathbf{b} - \beta\xi(\mathbf{c}'\mathbf{A}\mathbf{x} + \mathbf{c}'\mathbf{b}\xi) \end{aligned} \tag{74}$$

where we use the facts that $\xi = -\phi(\sigma)$ and

$$\dot{\sigma} = \mathbf{c}'\dot{\mathbf{x}} = \mathbf{c}'\mathbf{A}\mathbf{b} + \mathbf{c}'\mathbf{b}\xi \tag{75}$$

Defining

$$\mathbf{v} = \mathbf{P}\mathbf{b} - \frac{\beta}{2}\mathbf{A}'\mathbf{c} \tag{76}$$

(74) can be rewritten as

$$\dot{V}(\mathbf{x}) = \mathbf{x}'(\mathbf{A}'\mathbf{P} + \mathbf{P}\mathbf{A})\mathbf{x} + 2\xi\mathbf{x}'\mathbf{v} - \mathbf{c}'\mathbf{b}\xi^2 \tag{77}$$

Let us choose **P** to be the solution of the equation

$$\mathbf{A}'\mathbf{P} + \mathbf{P}\mathbf{A} = -\mathbf{Q} \tag{78}$$

where **Q** is a positive definite matrix of our choice. Then

$$\begin{aligned} \dot{V}(\mathbf{x}) &= -\mathbf{x}'\mathbf{Q}\mathbf{x} + 2\xi\mathbf{x}'\mathbf{v} - \mathbf{c}'\mathbf{b}\xi^2 \\ &= -(\mathbf{x} - \mathbf{Q}^{-1}\mathbf{v}\xi)'\mathbf{Q}(\mathbf{x} - \mathbf{Q}^{-1}\mathbf{v}\xi) - (\beta\mathbf{c}'\mathbf{b} - \mathbf{v}'\mathbf{Q}^{-1}\mathbf{v})\xi^2 \end{aligned} \tag{79}$$

We claim that if

$$\beta\mathbf{c}'\mathbf{b} - \mathbf{v}'\mathbf{Q}^{-1}\mathbf{v} > 0 \tag{80}$$

then $\dot{V}$ is a negative definite quadratic form in $\mathbf{x}$ and ξ. This is easily verified: Suppose (80) holds; then it is clear that $\dot{V}$ is at least negative semidefinite in $\mathbf{x}$ and ξ. Therefore it suffices to show that

$$\dot{V} = 0 \Longrightarrow \mathbf{x} = \mathbf{0}, \qquad \xi = 0 \tag{81}$$

Accordingly, suppose $\dot{V} = 0$; this implies, because of the positive definiteness of **Q** and (80), that

$$\mathbf{x} - \mathbf{Q}^{-1}\mathbf{v}\xi = \mathbf{0} \tag{82}$$

$$\xi = 0 \tag{83}$$

Clearly, (82) and (83) together establish (81). Next, because $\dot{V}$ is a negative definite quadratic form in $\mathbf{x}$ and ξ, there exist positive constants α_1 and α_2 such that

84 $$\dot{V}(\mathbf{x}) \leq -\alpha_1 \mathbf{x}'\mathbf{x} - \alpha_2 \xi^2$$

Hence

85 $$\dot{V}(\mathbf{x}) \leq -\alpha_1 \mathbf{x}'\mathbf{x}$$

which shows that $-\dot{V}$ is a p.d.f. and establishes the global asymptotic stability of (1)–(3).

The preceding discussion is summarized in the following.

86 ***Theorem*** Suppose that ϕ is a time-invariant nonlinearity belonging to the sector $[0, \infty)$ and that $d = 0$. Let $\mathbf{Q}$ be a positive definite matrix, and let $\mathbf{P}$ be the corresponding solution of (78). Define $\mathbf{v}$ by (76). Then (80) is a sufficient condition for the absolute stability of the system (1)–(3).

87 REMARKS: Theorem (86) is applicable to a more limited class of systems than Theorem (60), because Theorem (86) requires d to be zero and ϕ to be time-invariant; neither of these is needed in Theorem (60). Moreover, for a given $\mathbf{A}$, $\mathbf{b}$, $\mathbf{c}$, whether or not (80) holds depends strongly on the choice of $\mathbf{Q}$. This is a major drawback of this criterion, because there is no *a priori* method for choosing $\mathbf{Q}$ intelligently. At this stage, the Lur'e–Postnikov criterion is mainly of historical interest.

5.5.4 The Popov Criterion

In this subsection, we shall derive a well-known stability criterion for *autonomous* systems. This criterion was first derived by the Rumanian scientist V. M. Popov, who considered a class of systems slightly different from (1)–(3). Specifically, the class of systems studied by Popov are those of the form

88 $$\dot{\mathbf{x}}(t) = \mathbf{A}\mathbf{x}(t) - \mathbf{b}\phi[\sigma(t)]$$

89 $$\sigma(t) = \mathbf{c}'\mathbf{x}(t) + d\xi(t)$$

90 $$\dot{\xi}(t) = -\phi[\sigma(t)]$$

Clearly, the difference is between (90) and (3). By recasting (88)–(90) in the form

91 $$\begin{bmatrix} \dot{\mathbf{x}}(t) \\ \dot{\xi}(t) \end{bmatrix} = \begin{bmatrix} \mathbf{A} & \mathbf{0} \\ \mathbf{0}^T & 0 \end{bmatrix} \begin{bmatrix} \mathbf{x}(t) \\ \xi(t) \end{bmatrix} + \begin{bmatrix} \mathbf{b} \\ 1 \end{bmatrix} u(t)$$

92 $$\sigma(t) = [\mathbf{c}' \quad d]\begin{bmatrix}\mathbf{x}(t) \\ \xi(t)\end{bmatrix}$$

93 $$u(t) = -\phi[\sigma(t)]$$

We see that the present system is of the form (1)–(3), where the **A** matrix satisfies assumption (A2) instead of (A1).

Throughout this subsection, we shall make the following *Assumption*: ϕ belongs to the sector $(0, \infty)$, and $d \neq 0$.

The following result shows that $d > 0$ is actually a necessary condition for absolute stability.

94 **fact** Assuming that $d \neq 0$ and that ϕ belongs to the sector $(0, \infty)$, a necessary condition for the absolute stability of the system is $d > 0$.

***proof* (*outline of*)** If the system (88)–(90) is absolutely stable, then it is globally asymptotically stable for all ϕ of the form

95 $$\phi(\sigma) = \epsilon\sigma$$

for all $\epsilon > 0$. With this choice of ϕ, the system becomes

96 $$\begin{bmatrix}\dot{\mathbf{x}}(t) \\ \dot{\xi}(t)\end{bmatrix} = \begin{bmatrix}\mathbf{A} - \epsilon\mathbf{b}\mathbf{c}' & -\epsilon\mathbf{b}\mathbf{d} \\ -\epsilon\mathbf{c}' & -\epsilon d\end{bmatrix}\begin{bmatrix}\mathbf{x}(t) \\ \xi(t)\end{bmatrix}$$

which can be rewritten as

97 $$\begin{bmatrix}\dot{\mathbf{x}}(t) \\ \mu\dot{\xi}(t)\end{bmatrix} = \begin{bmatrix}\mathbf{A} - \epsilon\mathbf{b}\mathbf{c}' & -\epsilon\mathbf{b}d \\ -\mathbf{c}' & -d\end{bmatrix}\begin{bmatrix}\mathbf{x}(t) \\ \xi(t)\end{bmatrix}$$

where $\mu = 1/\epsilon$. The system (97) is similar to those studied in Sec. 4.3. If we neglect the terms involving ϵ on the right-hand side of (97), the system is qualitatively similar to

98 $$\begin{bmatrix}\dot{\mathbf{x}}(t) \\ \mu\dot{\xi}(t)\end{bmatrix} = \begin{bmatrix}\mathbf{A} & \mathbf{0} \\ -\mathbf{c}' & -d\end{bmatrix}\begin{bmatrix}\mathbf{x}(t) \\ \xi(t)\end{bmatrix}$$

Suppose now by way of contradiction that $d < 0$. Then, by letting $\epsilon \longrightarrow 0$ (i.e., $\mu \longrightarrow \infty$), we get the *simplified* system

99 $$\begin{bmatrix}0 \\ \mu\dot{\xi}(t)\end{bmatrix} = \begin{bmatrix}\mathbf{A} & \mathbf{0} \\ -\mathbf{c}' & -d\end{bmatrix}\begin{bmatrix}\mathbf{x}(t) \\ \xi(t)\end{bmatrix}$$

which has the solution

100 $$\mathbf{x}(t) \equiv 0$$

101 $$\xi(t) = \xi(0) \exp\frac{-dt}{\mu}$$

In deriving (100), we use the fact that **A** is Hurwitz and is therefore nonsingular. Clearly, by (101), the system (99) is unstable. Hence, by Theorem [4.3(44)], the system (98) is unstable for all sufficiently large values of μ.

Finally, by continuity, the system (97) is unstable for all sufficiently small values of ϵ. This shows that $d > 0$ is a necessary condition for absolute stability. ∎

To study the stability of the system (88)–(90), Popov uses a Liapunov function of the form

$$V = \mathbf{x}'\mathbf{P}\mathbf{x} + \alpha_1 \xi^2 + \beta \Phi(\sigma) \tag{102}$$

where Φ is defined in (73), $\mathbf{P}$ is a positive definite matrix, and α_1, β are nonnegative constants, not both of which are zero. Because $d > 0$, V can be rewritten as

$$V = \mathbf{x}'\mathbf{P}\mathbf{x} + \alpha(\sigma - \mathbf{c}'\mathbf{x})^2 + \beta \Phi(\sigma) \tag{103}$$

where $\alpha = \alpha_1/d^2$. Actually, Popov begins with a more general Liapunov function of the form

$$V = \mathbf{x}'\mathbf{P}\mathbf{x} + \alpha(\sigma - \mathbf{c}'\mathbf{x})^2 + \beta \Phi(\sigma) + \xi \mathbf{v}'\mathbf{x} \tag{104}$$

He then shows that if V is a p.d.f. and is a Liapunov function for *all* systems of the form (88)–(90), then $\mathbf{v}$ must be zero. Thus, in effect, there is no loss of generality in studying V-functions of the form (103) instead of (104). The details can be found in [16] Lefschetz [pp. 98–99].

Calculating the derivative of V along the trajectories of (88)–(90), we get

$$\begin{aligned} \dot{V} &= 2\mathbf{x}'\mathbf{P}\dot{\mathbf{x}} + 2\alpha(\sigma - \mathbf{c}'\mathbf{x})d\dot{\xi} + \beta\phi\dot{\sigma} \\ &= \mathbf{x}'(\mathbf{A}'\mathbf{P} + \mathbf{P}\mathbf{A})\mathbf{x} - 2\mathbf{x}'\mathbf{P}\mathbf{b}\phi - 2\alpha d(\sigma - \mathbf{c}'\mathbf{x})\phi \\ &\quad + \beta\phi(\mathbf{c}'\mathbf{A}\mathbf{x} - \mathbf{c}'\mathbf{b}\phi - d\phi) \end{aligned} \tag{105}$$

where ϕ is shorthand for $\phi(\sigma)$. Now, $\dot{V}$ can be rewritten as

$$\dot{V} = \mathbf{x}'(\mathbf{A}'\mathbf{P} + \mathbf{P}\mathbf{A})\mathbf{x} - 2\mathbf{x}'(\mathbf{P}\mathbf{b} - \mathbf{v})\phi - \gamma\phi^2 - 2\alpha d\sigma\phi \tag{106}$$

where

$$\mathbf{v} \triangleq \tfrac{1}{2}\beta\mathbf{A}'\mathbf{c} + \alpha d\mathbf{c} \tag{107}$$

$$\gamma \triangleq \beta(\mathbf{c}'\mathbf{b} + d) \tag{108}$$

We are now in a position to state Popov's criterion precisely.

109 ***Theorem*** The system (88)–(90) with $d \neq 0$ is absolutely stable for all nonlinearities ϕ belonging to the sector $(0, \infty)$ if there exists a nonnegative constant r such that

$$\operatorname{Re}\,[(1 + jr\omega)\hat{g}(j\omega)] > 0, \qquad \forall\, \omega \in R \tag{110}$$

where

$$\hat{g}(s) = \frac{d}{s} + \mathbf{c}'(s\mathbf{I} - \mathbf{A})^{-1}\mathbf{b} \tag{111}$$

proof We prove the theorem by once again using the Kalman–Yacubovitch lemma. Notice, first, that (110) is equivalent to

$$rd + \text{Re}\,[(1 + jr\omega)\mathbf{c}'(j\omega\mathbf{I} - \mathbf{A})^{-1}\mathbf{b}] > 0, \qquad \forall\ \omega \tag{112}$$

We now replace (112) by

$$\beta d + \text{Re}\,[(2\alpha d + j\omega\beta)\mathbf{c}'(j\omega\mathbf{I} - \mathbf{A})^{-1}\mathbf{b}] > 0, \qquad \forall\ \omega \tag{113}$$

where $\alpha > 0$ and $\beta \geq 0$ are chosen such that

$$r = \frac{\beta}{2\alpha d} \tag{114}$$

We next show that if (113) holds for a particular set of values α and β, then the corresponding V given by (103) is a Liapunov function for the system (88)–(90), provided of course that $\mathbf{P}$ is chosen properly. Specifically, Theorem (13) assures us that if $\mathbf{Q}$ is any positive definite matrix and

$$\gamma + 2\,\text{Re}\,\mathbf{v}'(j\omega\mathbf{I} - \mathbf{A})^{-1}\mathbf{b} > 0, \qquad \forall\ \omega \in R \tag{115}$$

then there exist a positive definite matrix $\mathbf{P}$ and a vector $\mathbf{q}$ such that

$$\mathbf{A}'\mathbf{P} + \mathbf{P}\mathbf{A} = -\epsilon\mathbf{Q} - \mathbf{q}\mathbf{q}' \tag{116}$$

$$\mathbf{P}\mathbf{b} - \mathbf{v} = \gamma^{1/2}\mathbf{q} \tag{117}$$

We shall return to condition (115) later on. Now, if (115) holds, then $\dot{V}$ becomes

$$\begin{aligned} \dot{V} &= -\epsilon\mathbf{x}'\mathbf{Q}\mathbf{x} - (\mathbf{x}'\mathbf{q})^2 - 2\mathbf{x}'\mathbf{q}\gamma^{1/2}\phi - \gamma\phi^2 - 2\alpha d\sigma\phi \\ &= -\epsilon\mathbf{x}'\mathbf{Q}\mathbf{x} - (\mathbf{x}'\mathbf{q} + \gamma^{1/2}\phi)^2 - 2\alpha d\sigma\phi \end{aligned} \tag{118}$$

Now, because $\mathbf{P}$ is positive definite and $\alpha > 0$, V is a p.d.f. in $\mathbf{x}$ and ξ. Next, it is clear from (118) that $\dot{V}$ is always nonpositive [recall that $\sigma\phi(\sigma) \geq 0$, $\forall\ \sigma$, because ϕ belongs to the sector $(0, \infty)$]. Thus, to show that $-\dot{V}$ is an l.p.d.f. over all of R^n, it is enough to show that $\dot{V} < 0$ whenever either $\mathbf{x}$ or ξ is nonzero. If $\mathbf{x} \neq \mathbf{0}$, $\dot{V} < 0$ because $\mathbf{Q}$ is positive definite. On the other hand, if $\mathbf{x} = \mathbf{0}$ and $\xi \neq 0$, then $\sigma \neq 0$; this shows that $\sigma\phi(\sigma) > 0$, because ϕ belongs to the sector $(0, \infty)$. Because $\alpha d > 0$, it follows that $\dot{V} < 0$ in this case also. Hence $-\dot{V}$ is an l.p.d.f. over all of R^n, and the system (88)–(90) is globally asymptotically stable.

We see, therefore, that (115) is a sufficient condition for absolute stability. Substituting for $\mathbf{v}$ and γ from (107) and (108), we get

$$\beta(\mathbf{c}'\mathbf{b} + d) + \text{Re}\,(\beta\mathbf{c}'\mathbf{A} + 2\alpha d\mathbf{c}')(j\omega\mathbf{I} - \mathbf{A})^{-1}\mathbf{b} > 0 \tag{119}$$

$$\beta d + \text{Re}\,\beta\mathbf{c}'[\mathbf{I} + \mathbf{A}(j\omega\mathbf{I} - \mathbf{A})^{-1}]\mathbf{b} + 2\alpha d\,\text{Re}\,\mathbf{c}'(j\omega\mathbf{I} - \mathbf{A})^{-1}\mathbf{b} > 0 \tag{120}$$

$$\beta d + \text{Re}\,[j\omega\beta\mathbf{c}'(j\omega\mathbf{I} - \mathbf{A})^{-1}\mathbf{b}] + 2\alpha d\,\text{Re}\,\mathbf{c}'(j\omega\mathbf{I} - \mathbf{A})^{-1}\mathbf{b} > 0 \tag{121}$$

$$\beta d + \text{Re}\,[(2\alpha d + j\omega\beta)\mathbf{c}'(j\omega\mathbf{I} - \mathbf{A})^{-1}\mathbf{b}] > 0, \qquad \forall\ \omega \in R \tag{122}$$

In going from (120) to (121), we use the obvious identity

$$\mathbf{I} + \mathbf{A}(j\omega\mathbf{I} - \mathbf{A})^{-1} = j\omega(j\omega\mathbf{I} - \mathbf{A})^{-1} \tag{123}$$

Finally, we see that (122) is the same as (113). This proves the theorem. ∎

As it stands, Theorem (109) shows that (110) is a *sufficient* condition for the existence of a Liapunov function of the form (103) for the system (88)–(90). However, Popov shows that (110) is also a *necessary* condition for the existence of a Liapunov function of the form (103). This should not be misinterpreted to mean that (110) is a necessary and sufficient condition for the absolute stability of the system (88)–(90). Rather, (110) is a necessary and sufficient condition for the *existence* of a Liapunov function of the type (103).

Geometric Interpretation

Like the circle criterion, the Popov criterion can also be given a geometric interpretation. Suppose we plot $\text{Re}\,\hat{g}(j\omega)$ vs. $\omega\,\text{Im}\,\hat{g}(j\omega)$, with ω as a parameter. Note that we only need to do the plotting for $\omega \geq 0$, because both $\text{Re}\,\hat{g}(j\omega)$ and $\omega\,\text{Im}\,\hat{g}(j\omega)$ are even functions of ω, and no new information is gained by plotting for $\omega < 0$. Such a plot is known as a *Popov plot*, in contrast with a Nyquist plot, which is a plot of $\text{Re}\,\hat{g}(j\omega)$ vs. $\text{Im}\,\hat{g}(j\omega)$. Now, the condition (110) simply states that there exists a nonnegative number r such that the Popov plot of $\hat{g}$ lies to the right of a straight line of slope $1/r$ passing through the origin. If $r = 0$, this straight line is taken as the vertical axis of the Popov plot. The situation is depicted in Fig. 5.12.

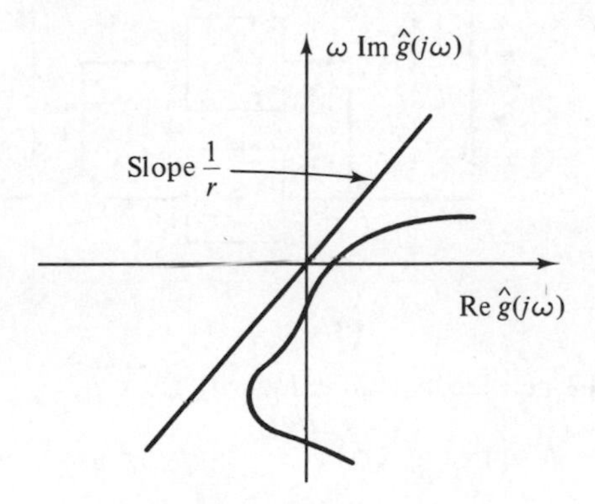

FIG. 5.12

Nonlinearity in a Sector (0, k)

Suppose now that we are interested in the absolute stability of the system (88)–(90) for all nonlinearities ϕ belonging to the sector $(0, k)$, where k is a given *finite* number. Because the class of allowed nonlinearities is smaller than before, one would expect to find a weaker sufficient condition for absolute stability than (110). And this is indeed what one finds.

124 ***Theorem*** A sufficient condition for the system (88)–(90) to be absolutely stable for all nonlinearities ϕ belonging to the sector $(0, k)$ is that there exists a nonnegative number r such that

$$\text{Re}\,[(1 + j\omega r)\hat{g}(j\omega)] + \frac{1}{k} > 0, \qquad \forall\, \omega \in R \tag{125}$$

proof The present problem is reduced to the problem covered by Theorem (109) by means of loop transformations. Suppose ϕ belongs to the sector $(0, k)$. Then the system (88)–(90) can be transformed to that in Fig. 5.13, where the new forward transfer function is

$$\hat{g}_1(s) = \hat{g}(s) + \frac{1}{k} \tag{126}$$

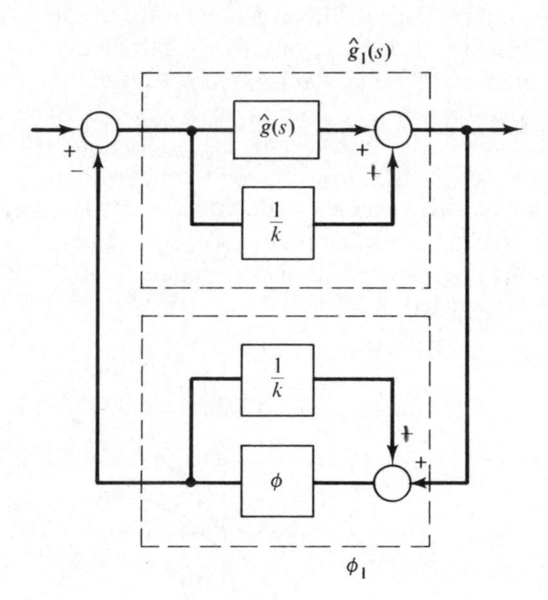

FIG. 5.13

while the new feedback nonlinearity is

$$\phi_1 = \phi \cdot \left(1 - \frac{1}{k} \cdot \phi\right)^{-1} \tag{127}$$

Note that, because ϕ belongs to the sector $(0, k)$, the inverse in (127) is well defined. Now, ϕ_1 belongs to the sector $(0, \infty)$. Therefore, by Theorem (109), a sufficient condition for the absolute stability of this system is that there exists a nonnegative constant r such that

$$\text{Re}\,[(1 + j\omega r)\hat{g}_1(j\omega)] > 0, \qquad \forall\, \omega \in R \tag{128}$$

It is easy to see that (128) and (125) are equivalent. ∎

REMARKS: In the case where ϕ belongs to the sector $(0, k)$, we use the sufficient condition (125), which is weaker than (110). Clearly, as $k \longrightarrow \infty$, (125) reduces to (110). The geometric interpretation of (125) is similar to that of (110): (125) states that the Popov plot of $\hat{g}$ lies to the right of a straight line of slope $1/r$ passing through the point $-1/k + j0$.

129 **Example.** Consider a system of the form (88)–(90), with

130 $$\hat{g}(s) = \frac{1}{s(s+1)^2}$$

In this case

131 $$\operatorname{Re} \hat{g}(j\omega) = \frac{-2}{(1+\omega^2)^2}$$

132 $$\omega \operatorname{Im} \hat{g}(j\omega) = \frac{\omega^2 - 1}{(1+\omega^2)^2}$$

The Popov plot of $\hat{g}$ is shown in Fig. 5.14. It is clear from this figure that (125) can never be satisfied unless $-1/k < -\frac{1}{2}$, i.e., $k < 2$. On the other hand, if $k < 2$, (125) can always be satisfied by a suitable choice of r. Hence this system is absolutely stable for all nonlinearities in the sector (0, 2).

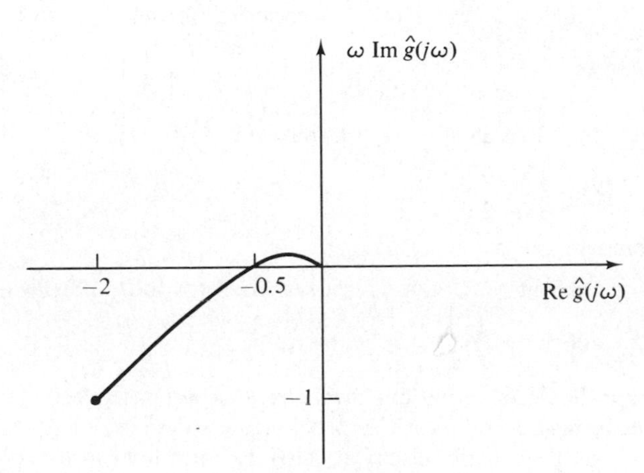

FIG. 5.14

5.5.5 Concluding Remarks

As is the case with the circle criterion, one of the main advantages of Popov's criterion is that it is applied directly to the transfer function $\hat{g}(s)$ and is thus independent of the state variable realization (88)–(90). In fact, there is no need to construct a state variable realization for a given $\hat{g}(s)$.

In Chap. 6, we shall encounter another Popov criterion, which is derived by using functional analytic methods. In contrast with the circle criterion (where the functional analytic methods yield a true generalization of the Liapunov methods), the Popov criterion of Chap. 6 differs in some details from the present one.

The noteworthy thing about Popov's criterion is the presence of the term $1 + j\omega r$, which is known as a *multiplier*. Other, more general multipliers are studied in [17] Narendra and Taylor.

Problem 5.19. Consider a feedback system with

$$\hat{g}(s) = \frac{1}{(s+1)(s+2)(s+3)}$$

Using the circle criterion, determine the largest range $[\alpha, \beta]$ such that the system is absolutely stable for all (possibly time-varying) nonlinearities ϕ in the sector $[\alpha, \beta]$.

Problem 5.20. Consider a feedback system with

$$\hat{g}(s) = \frac{1}{(s-1)(s+1)^2}$$

Using the circle criterion, determine the largest range $[\alpha, \beta]$ such that the system is absolutely stable for all nonlinearities in the sector $[\alpha, \beta]$.

Problem 5.21. Given a system of the form (1)–(3) with

$$\mathbf{A} = \begin{bmatrix} -3 & 4 \\ -1 & 1 \end{bmatrix}; \quad \mathbf{b} = \begin{bmatrix} 0 \\ 1 \end{bmatrix}; \quad \mathbf{c} = \begin{bmatrix} 1 \\ 0 \end{bmatrix}$$

apply the Lur'e–Postnikov criterion with (a) $\mathbf{Q} = \mathbf{I}$, (b)

$$\mathbf{Q} = \begin{bmatrix} 1 & 0 \\ 0 & 2 \end{bmatrix}$$

Compare.

Problem 5.22. Consider a system of the form (88)–(90), with

$$g(s) = \frac{1}{s(s+1)}$$

Using the Popov criterion, show that this system is absolutely stable for all time-invariant nonlinearities ϕ belonging to the sector $(0, k)$, where k is *any* finite number. Verify also that (110) does not hold in this particular case.

5.6 SLOWLY VARYING SYSTEMS

In this section, we shall prove a very general result on the stability of so-called *slowly varying* systems. The approach given here is due to [3] Barman.

Consider a general nonautonomous system described by

$$\dot{\mathbf{x}}(t) = \mathbf{f}[t, \mathbf{x}(t)], \quad t \geq 0 \tag{1}$$

where $\mathbf{f}: R \times R^n \longrightarrow R^n$. If we now consider the *autonomous* system

2 $$\dot{\mathbf{x}}(t) = \mathbf{f}[p, \mathbf{x}(t)], \qquad t \geq 0$$

where p is a particular nonnegative number, then we can think of (2) as a particular case of the system (1), with its time dependence "frozen" at time p. This concept can be clarified by considering linear systems. Suppose we are given the nonautonomous system

3 $$\dot{\mathbf{x}}(t) = \mathbf{A}(t)\mathbf{x}(t)$$

Then, for any $p \geq 0$, the autonomous system

4 $$\dot{\mathbf{x}}(t) = \mathbf{A}(p)\mathbf{x}(t)$$

can be thought of as describing the system (3), frozen at time p.

In this section, we shall address ourselves to the following question: Suppose that, for each $p \geq 0$, $\mathbf{0}$ is a globally asymptotically stable equilibrium point of the frozen system (1); under what additional conditions is it possible to conclude that $\mathbf{0}$ is a globally asymptotically stable equilibrium point of the original system (1)? The answer turns out to be that the additional requirements are (1) the global asymptotic stability of $\mathbf{0}$ for each of the frozen systems is in some sense uniform with respect to p, and (2) the rate of variation of $\mathbf{f}$ with respect to t is sufficiently small (hence the name slowly varying systems).

To streamline the presentation of the results, we shall employ the following notation: As in Sec. 5.1, $\mathbf{s}(t, \mathbf{x}, t_0)$ denotes the solution of (1) as evaluated at time t, starting from the initial state $\mathbf{x}$ at time t_0. Similarly, $\mathbf{s}_p(t, \mathbf{x}, t_0)$ denotes the solution of (2) as evaluated at time t, starting from the initial state $\mathbf{x}$ at time t_0. The symbol D is used to denote partial differentiation. For instance

5 $$D_t\mathbf{f}(t, \mathbf{x}) = \frac{\partial \mathbf{f}(t, \mathbf{x})}{\partial t}$$

and so on. Throughout, $\|\cdot\|$ denotes the l_2-norm.

The main result for nonlinear systems is given next.

6 ***Theorem*** Suppose the function $\mathbf{f}: R \times R^n \longrightarrow R^n$ satisfies the following conditions:

7 (i) $\mathbf{f}(t, 0) = \mathbf{0}$, $\forall\, t \geq 0$.

(ii) $\mathbf{f}$ is continuously differentiable.

(iii) $\mathbf{f}$ is globally Lipschitz-continuous; i.e., there exists a finite constant k such that

8 $$\|\mathbf{f}(t, \mathbf{x}) - \mathbf{f}(t, \mathbf{y})\| \leq k\|\mathbf{x} - \mathbf{y}\|, \qquad \forall\, t \geq 0, \forall\, \mathbf{x}, \mathbf{y} \in R^n$$

Suppose the family of frozen systems (2) satisfies

9 $$\|\mathbf{s}_p(t, \mathbf{x}, t_0)\| \leq m\|\mathbf{x}\| \exp[-\alpha(t - t_0)], \qquad \forall\, t \geq t_0, \forall\, t_0 \geq 0, \forall\, \mathbf{x} \in R^n, \forall\, p \geq 0$$

for some finite positive constants m and α. Under these conditions, there is a number δ, which depends only on k, m, and α, such that the equilibrium point $\mathbf{0}$ of (1) is globally asymptotically stable whenever

10 $$\|D_t\mathbf{f}(t, \mathbf{x})\| \leq \delta\|\mathbf{x}\|, \qquad \forall\, t \geq 0,\ \forall\, \mathbf{x} \in R^n$$

In fact, δ can be so chosen that the solution trajectories of (1) satisfy

11 $$\|\mathbf{s}(t, \mathbf{x}, t_0)\| \leq m_1\|\mathbf{x}\| \exp[-\beta(t - t_0)], \qquad \forall\, t \geq t_0,\ \forall\, t \geq 0,\ \forall\, \mathbf{x} \in R^n$$

proof The function $\mathbf{s}_p(t, \mathbf{x}, t_0)$ satisfies the integral equation

12 $$\mathbf{s}_p(t, \mathbf{x}, t_0) = \mathbf{x} + \int_{t_0}^{t} \mathbf{f}[p, \mathbf{s}_p(\tau, \mathbf{x}, t_0)]\, d\tau, \qquad \forall\, t \geq t_0$$

Because the frozen system (2) is autonomous, there is no loss of information in studying only $\mathbf{s}_p(t, \mathbf{x}, 0)$. From (12) we have

13 $$\mathbf{s}_p(t, \mathbf{x}, 0) = \mathbf{x} + \int_{0}^{t} \mathbf{f}[p, \mathbf{s}_p(\tau, \mathbf{x}, 0)]\, d\tau, \qquad \forall\, t \geq 0$$

Now, (8) and (9) together imply that

14 $$\|\mathbf{f}[p, \mathbf{s}_p(\tau, \mathbf{x}, 0)]\| \geq -k\|\mathbf{s}_p(\tau, \mathbf{x}, 0)\| \geq -km, \qquad \forall\, \tau \geq 0$$

Substituting from (14) in (13), we see that

15 $$\|\mathbf{s}_p(t, \mathbf{x}, 0)\| \geq \frac{\|\mathbf{x}\|}{2} \quad \text{whenever } t \in \left[0, \frac{1}{2\,km}\right]$$

Next, we differentiate both sides of (13) with respect to p to get

16 $$D_p\mathbf{s}_p(t, \mathbf{x}, 0) = \int_{0}^{t} \{D_{\mathbf{s}}\mathbf{f}[p, \mathbf{s}_p(\tau, \mathbf{x}, 0)] \cdot D_p\mathbf{s}_p(\tau, \mathbf{x}, 0) + D_p\mathbf{f}[p, \mathbf{s}_p(\tau, \mathbf{x}, 0)]\}\, d\tau$$

Now, because $\mathbf{f}$ satisfies the Lipschitz condition (8), it can be shown that the induced l_2-norm of the Jacobian matrix $D_{\mathbf{x}}\mathbf{f}(t, \mathbf{x})$ is always less than or equal to k. Moreover, $D_p\mathbf{f}(p, \mathbf{x})$ satisfies (10) (where δ is to be specified later). Utilizing these facts in (16), we get

17 $$\begin{aligned} \|D_p\mathbf{s}_p(t, \mathbf{x}, 0)\| &\leq \int_{0}^{t} [k\|D_p\mathbf{s}_p(\tau, \mathbf{x}, 0)\| + \delta\|\mathbf{s}_p(\tau, \mathbf{x}, 0)\|]\, d\tau \\ &\leq \delta \int_{0}^{t} m\|\mathbf{x}\| \exp(-\alpha\tau)\, d\tau + \int_{0}^{t} k\|D_p\mathbf{s}_p(\tau, \mathbf{x}, 0)\|\, d\tau \\ &\leq \frac{\delta m}{\alpha}\|\mathbf{x}\| + \int_{0}^{t} k\|D_p\mathbf{s}_p(\tau, \mathbf{x}, 0)\|\, d\tau \end{aligned}$$

Applying the Bellman–Gronwall inequality (Appendix I) to (17), we get

18 $$\|D_p\mathbf{s}_p(t, \mathbf{x}, 0)\| \leq \|\mathbf{x}\|\frac{\delta m}{\alpha} \exp kt$$

Next, we construct suitable Liapunov function candidates for the family of frozen systems (2) as well as for the original system (1). Choose $\gamma > 0$ such that

19 $$\gamma \geq \frac{k+\alpha}{2\alpha}$$

and choose

20 $$V_p(\mathbf{x}) = \int_0^\infty ||\mathbf{s}_p(\tau, \mathbf{x}, 0)]||^{2\gamma}\, d\tau = \int_0^\infty [\mathbf{s}_p'(\tau, \mathbf{x}, 0)\mathbf{s}_p(\tau, \mathbf{x}, 0)]^\gamma\, d\tau$$

as a Liapunov function candidate for the frozen system (2). Note that $V_p(\mathbf{x})$ is purely a "conceptual" Liapunov function candidate, because determining $V_p(\mathbf{x})$ requires explicit knowledge of the solution trajectories $\mathbf{s}_p(\tau, \mathbf{x}, 0)$. From (15) and (9), it is easy to show that V_p is both decrescent and a p.d.f. In fact, from (15) and (9), we have

21 $$\frac{||\mathbf{x}||^{2\gamma}}{2\,km\cdot 2^{2\gamma}} \leq V_p(\mathbf{x}) \leq ||\mathbf{x}||^{2\gamma}\frac{m^{2\gamma}}{2\gamma\alpha}, \quad \forall\, p \geq 0,\ \forall\, \mathbf{x} \in R^n$$

To calculate the derivative of $V_p(\cdot)$ along the solutions of (2), we observe that

22 $$\dot{V}_p(\mathbf{x}) = \lim_{h\to 0^+} \frac{V_p[\mathbf{s}_p(h, \mathbf{x}, 0)] - V_p(\mathbf{x})}{h}$$

Now

23 $$V_p[\mathbf{s}_p(h, \mathbf{x}, 0)] = \int_0^\infty ||\mathbf{s}_p[\tau, \mathbf{s}_p(h, \mathbf{x}, 0), 0]||^{2\gamma}\, d\tau$$

However, because the system (2) is autonomous, we have

24 $$\mathbf{s}_p[\tau, \mathbf{s}_p(h, \mathbf{x}, 0), 0] = \mathbf{s}_p(\tau + h, \mathbf{x}, 0)$$

Therefore

25 $$V_p[\mathbf{s}_p(h, \mathbf{x}, 0)] = \int_0^\infty ||\mathbf{s}_p(\tau + h, \mathbf{x}, 0)||^{2\gamma}\, d\tau = \int_h^\infty ||\mathbf{s}_p(\tau, \mathbf{x}, 0)||^{2\gamma}\, d\tau$$

26 $$V_p[\mathbf{s}(h, \mathbf{x}, 0)] - V_p(\mathbf{x}) = -\int_0^h ||\mathbf{s}_p(\tau, \mathbf{x}, 0)||^{2\gamma}\, d\tau$$

27 $$\dot{V}_p(\mathbf{x}) = -||\mathbf{x}||^{2\gamma}$$

because $\mathbf{s}_p(h, \mathbf{x}, 0) \longrightarrow \mathbf{x}$ as $h \longrightarrow 0$. Thus V_p satisfies all the hypotheses of Theorem [5.2(58)] and is a Liapunov function for the frozen system (2).

We shall now complete the proof of the theorem by demonstrating that the function $V(t, \mathbf{x})$ defined by

28 $$V(t, \mathbf{x}) = V_p(\mathbf{x})|_{p=t} = \int_0^\infty ||\mathbf{s}_t(\tau, \mathbf{x}, 0)||^{2\gamma}\, d\tau$$

is a Liapunov function for the *original* system (1), provided δ is sufficiently small. First, because (21) holds, we see that V is both decrescent and a p.d.f. Next, calculating the derivative of V along the trajectories of (1), we get

29 $$\dot{V}(t, \mathbf{x}) = D_{\mathbf{x}}V(t, \mathbf{x})\cdot\mathbf{f}(t, \mathbf{x}) + D_t V(t, \mathbf{x}) = -||\mathbf{x}||^{2\gamma} + D_t V(t, \mathbf{x})$$

where we use (27). Now,

30 $$D_t V(t, \mathbf{x}) = \int_0^\infty \gamma \| \mathbf{s}_t(\tau, \mathbf{x}, 0) \|^{2(\gamma-1)} \cdot [\mathbf{s}_t(\tau, \mathbf{x}, 0)]' \cdot D_t \mathbf{s}(\tau, \mathbf{x}, 0) \, d\tau$$

Hence

31 $$| D_t V(t, \mathbf{x}) | \leq \int_0^\infty \gamma \| \mathbf{s}_t(\tau, \mathbf{x}, 0) \|^{2(\gamma-1)} \| \mathbf{s}_t(\tau, \mathbf{x}, 0) \| \cdot \| D_t \mathbf{s}(\tau, \mathbf{x}, 0) \| \, d\tau$$

Substituting from (9) and (18) in (31), we get

32 $$\begin{aligned} | D_t V(t, \mathbf{x}) | &\leq \int_0^\infty \gamma \| \mathbf{x} \|^{2\gamma} m^{2\gamma-1} \exp [-(2\gamma - 1)\alpha\tau] \cdot \frac{\delta}{m\alpha} \exp k\tau \, d\tau \\ &= \| \mathbf{x} \|^{2\gamma} \delta \int_0^\infty \frac{m^{2\gamma-2}}{\alpha} \exp \{[k - (2\gamma - 1)\alpha]\tau\} \, d\tau \end{aligned}$$

From the manner in which γ was chosen [namely, (19)], we have

33 $$k - (2\gamma - 1)\alpha < 0$$

Hence the integral on the right-hand side of (32) is finite; let us denote it by M. Clearly M depends only on m, k, and α. Now if

34 $$\delta \cdot M < 1$$

it is clear from (29) that $-\dot{V}$ is a p.d.f. Hence V satisfies all the hypotheses of Theorem [5.2(58)]. Therefore we conclude that V is a Liapunov function for the system (1) and that $\mathbf{0}$ is a globally asymptotically stable equilibrium point of (1).

To prove that the solution trajectories of (1) actually satisfy (11), we proceed as follows: From (21), (29), and (34), we see that there exist finite positive constants c_1, c_2, and c_3 such that

35 $$c_1 \| \mathbf{x} \|^{2\gamma} \leq V(t, \mathbf{x}) \leq c_2 \| \mathbf{x} \|^{2\gamma}$$

36 $$\dot{V}(t, \mathbf{x}) \leq -c_3 \| \mathbf{x} \|^{2\gamma}$$

Therefore, along solution trajectories of (1), we have

37 $$\begin{aligned} \frac{d}{dt} V[t, \mathbf{s}(t, \mathbf{x}, t_0)] &\leq -c_3 \| \mathbf{s}(t, \mathbf{x}, t_0) \|^{2\gamma} \\ &\leq \frac{-c_3}{c_2} V[t, \mathbf{s}(t, \mathbf{x}, t_0)] \end{aligned}$$

From (37), we get, after using the integrating factor $\exp [-c_3(t - t_0)/c_2]$,

38 $$V[t, \mathbf{s}(t, \mathbf{x}, t_0)] \leq V(t, \mathbf{x}) \exp \left[\frac{-c_3(t - t_0)}{c_2} \right]$$

Using the bounds (35) in (38) gives

39 $$c_1 \| \mathbf{s}(t, \mathbf{x}, t_0) \|^{2\gamma} \leq c_2 \| \mathbf{x} \|^{2\gamma} \exp \left[\frac{-c_3(t - t_0)}{c_2} \right]$$

One can now obtain (11) from (39) by routine manipulations. ∎

40 REMARKS: What Theorem (6) shows is that *there is a* δ such that whenever (10) holds $\mathbf{0}$ is a globally asymptotically stable equilibrium

point of (1). In other words, if **0** is a globally asymptotically stable equilibrium point of each of the *frozen* systems (2) and if the rate of variation of f with respect to t is sufficiently slow, then **0** is also a globally asymptotically stable equilibrium point of the original system (1).

The specialization of Theorem (6) to linear systems is perhaps more easy to visualize.

41 **corollary** Consider the differential equation

42 $$\dot{\mathbf{x}}(t) = \mathbf{A}(t)\mathbf{x}(t)$$

Suppose

(i) $\mathbf{A}(\cdot)$ is continuously differentiable.
(ii) $\mathbf{A}(\cdot)$ is bounded.
(iii) There exists a positive number α such that, for each $t \geq 0$, the eigenvalues of $\mathbf{A}(t)$ all have real parts less than or equal to $-\alpha$.

Under these conditions, there exists a number $\delta > 0$ such that **0** is a uniformly asymptotically stable equilibrium point of (42) whenever

43 $$\|\dot{\mathbf{A}}(t)\|_i \leq \delta, \qquad \forall\, t \geq 0$$

where $\|\cdot\|_i$ denotes the induced l_2-norm.

6 *Input–output stability*

In this chapter, we shall present some basic results concerning input–output stability. Like Liapunov stability, input–output stability is a subject on which there exists a large amount of literature, and the contents of this chapter do not fully reflect the breadth of this subject. A thorough treatment of this subject can be found in [10] Desoer and Vidyasagar.

Before proceeding to the study of input–output *stability*, it is necessary to reconcile the input–output *approach* to system analysis with the state variable methods. The preceding four chapters are predicated on the system under study being described by a set of *differential equations* that describe the time evolution of the system *state variables*. In contrast, the systems encountered in this chapter are assumed to be described by an *input–output mapping* that assigns, to each *input*, a corresponding *output*. In view of this seeming disparity, it is important to realize that an input–output representation and a state representation are two different ways of looking at the *same system*—the two types of representation are used because they each give a different kind of insight into how the system works. It is now known that not only does there exist a close relationship between the input–output representation and the state representation of a given system but that there also exists a very close relationship between the kinds of *stability results* that one can get using the input–output approach and the state variable approach.

At this stage, one may well ask, why not simply use one of the approaches—why use both? The answer is that while the two approaches are *related*, they do not give identical results. Because, in analyzing a system, it is advantageous to have as many answers as we can, it is good to have, at our disposal, both the approaches, each yielding its own set of insights and information.

This chapter is organized as follows: In Sec. 6.1 we shall give a very brief introduction to L_p-spaces and their extensions and introduce the concepts of truncations and causality. In Sec. 6.2 we shall explore the relationship between the input–output representation and the state representation of a system. In Sec. 6.3 we shall present the basic definitions of input–output stability. In Sec. 6.4 we shall give some of the relationships between input–output stability and Liapunov stability. In Sec. 6.5 we shall discuss the open-loop stability of linear systems, while Sec. 6.6 contains some results on linear time-invariant feedback systems. The material in these two sections serves as a stepping stone for Sec. 6.7, where we shall state several criteria for the stability of nonlinear time-varying feedback systems.

6.1 INTRODUCTION TO L_p-SPACES AND THEIR EXTENSIONS

In this section, we shall give a brief introduction to L_p-spaces, extended L_p-spaces, truncations, and causality. Much of input–output stability, including the problem statement and the stability theorems, is couched in these terms, so that a certain degree of familiarity with these concepts is necessary to appreciate input–output theory. On the other hand, most of the input–output results given here do not require any deep results from Lebesgue theory, so that the pedestrian treatment given below is sufficient for the present purposes.

6.1.1 L_p-Spaces

A subset S of the real number system R is said to be of *measure zero* if S contains either a finite or a countably infinite number of elements. That is, S is of measure zero if the elements of S can be placed in one-to-one correspondence with some subset of the natural numbers $N = \{1, 2, 3, \ldots\}$. In this case, we say that S is *atmost countable*. A function $f(\cdot)$: $R \rightarrow R$ is said to be *measurable* if it is continuous everywhere except on a set of measure zero, i.e., except at an atmost countable number of points. Thus a reader who has no prior acquaintance with the Lebesgue theory of measure and integration can simply think of all functions encountered below as continuous functions and of all integrals as Riemann integrals. This would lead to no conceptual difficulties and no loss of insight, except that occasionally some results from Lebesgue theory would have to be accepted on faith.

1 *[definition]* For all $p \in [1, \infty)$, we label as $L_p[0, \infty)$ the set consisting of all measurable functions $f(\cdot): [0, \infty) \longrightarrow R$ such that

$$\int_0^\infty |f(t)|^p \, dt < \infty \tag{2}$$

The label $L_\infty[0, \infty)$ denotes the set of all measurable functions $f(\cdot): [0, \infty) \longrightarrow R$ that are essentially bounded[1] on $[0, \infty)$.

Thus, for $1 \leq p < \infty$, $L_p[0, \infty)$ denotes the set of measurable functions whose pth powers are absolutely integrable over $[0, \infty)$, whereas $L_\infty[0, \infty)$ denotes the set of essentially bounded measurable functions.

3 **Example.** The function $f(t) = e^{-\alpha t}$, $\alpha > 0$, belongs to $L_p[0, \infty)$ for all p. The function $g(t) = 1/(t + 1)$ belongs to $L_p[0, \infty)$ for all $p > 1$ but not to $L_1[0, \infty)$. The function

$$f_p(t) = \left[\frac{1}{t^{1/2}(1 + \log t)}\right]^{2/p} \tag{4}$$

where $p \in [1, \infty)$, belongs to the set $L_p[0, \infty)$ but does *not* belong to $L_q[0, \infty)$ for any $q \neq p$.

5 **fact** Whenever $p \in [1, \infty]$, the set $L_p[0, \infty)$ is a real linear vector space, in the sense of definition [3.1(1)].

Thus, whenever $f(\cdot), g(\cdot)$ belong to $L_p[0, \infty)$, we have that $(f + g)(\cdot)$ also belongs to $L_p[0, \infty)$, as does $(\alpha f)(\cdot)$, for all real numbers α. This fact follows from what is known as Minkowski's inequality (see [20] Royden).

6 *[definition]* For $p \in [1, \infty)$, the function $\|\cdot\|_p: L_p[0, \infty) \longrightarrow [0, \infty)$ is defined by

$$\|f(\cdot)\|_p = \left[\int_0^\infty |f(t)|^p \, dt\right]^{1/p} \tag{7}$$

The function $\|\cdot\|_\infty: L_\infty[0, \infty) \longrightarrow [0, \infty)$ is defined by[2]

$$\|f(\cdot)\| = \operatorname*{ess.\,sup.}_{t \in [0, \infty)} |f(t)| \tag{8}$$

---

[1] "Essentially bounded" means "bounded except on a set of measure zero." The reader need not worry about the distinction between "essentially bounded" and "bounded."

[2] For "ess. sup." (= essential supremum), one can safely read "supremum."

Definition (6) introduces functions $\|\cdot\|_p$, for $1 \leq p \leq \infty$, that map the linear space $L_p[0, \infty)$ into the interval $[0, \infty)$. Note that, by virtue of definition (1), the right-hand sides of (7) and (8) are well defined and finite.

9 **fact** (*Minkowski's inequality*) Let $p \in [1, \infty]$, and let $f(\cdot)$, $g(\cdot) \in L_p[0, \infty)$. Then

10 $$\|(f+g)(\cdot)\|_p \leq \|f(\cdot)\|_p + \|g(\cdot)\|_p$$

Thus fact (9) shows that $\|\cdot\|_p$ satisfies the triangle inequality. Because axioms (N1) and (N2) of definition [3.1(25)] are easily verified, we see that the pair $(L_p[0, \infty), \|\cdot\|_p)$ is a normed linear space, for each $p \in [1, \infty]$.

11 **fact** For each $p \in [1, \infty]$, the normed linear space $(L_p[0, \infty), \|\cdot\|_p)$ is complete and is hence a Banach space. For $p = 2$, the norm $\|\cdot\|_2$ corresponds to the inner product

12 $$\langle f(\cdot), g(\cdot)\rangle_2 = \int_0^\infty f(t)g(t)\, dt$$

Thus $(L_2[0, \infty), \langle\cdot, \cdot\rangle_2)$ is a complete inner product space, i.e., a Hilbert space.

Fact (11) brings out the main reason for dealing with L_p-spaces in studying input–output stability. We *could* instead study the set $C_p[0, \infty)$, which consists of all *continuous* functions $f(\cdot)$ satisfying (2). Then $C_p[0, \infty)$ is a subset of $L_p[0, \infty)$, and the pair $(C_p[0, \infty), \|\cdot\|_p)$ is also a normed linear space. However, it is *not* complete. Hence, if we require a Banach space as the setting for the stability theory, it is better to work with the spaces $L_p[0, \infty)$. Note that $(L_p[0, \infty), \|\cdot\|_p)$ is the completion of the normed linear space $(C_p[0, \infty), \|\cdot\|_p)$.

13 **fact** (*Holder's inequality*) Let $p, q \in [1, \infty]$, and suppose

14 $$\frac{1}{p} + \frac{1}{q} = 1$$

(Note that we take $p = 1$ if $q = \infty$ and vice versa.) Suppose $f(\cdot) \in L_p[0, \infty)$ and $g(\cdot) \in L_q[0, \infty)$. Then the function

15 $$h(t) \triangleq f(t)g(t)$$

belongs to $L_1[0, \infty)$. Moreover

16 $$\int_0^\infty |f(t)| \cdot |g(t)|\, dt \leq \left[\int_0^\infty |f(t)|^p\, dt\right]^{1/p} \left[\int_0^\infty |g(t)|^q\, dt\right]^{1/q}$$

The inequality (16) can be concisely expressed as

17 $$\|f(\cdot)g(\cdot)\|_1 \leq \|f(\cdot)\|_p \|g(\cdot)\|_q$$

6.1.2 Extended L_p-Spaces

Once the concept of L_p-spaces is understood, we can define extended L_p-spaces. First, we require the notion of truncated functions.

18 *[definition]* Let $x(\cdot)$: $[0, \infty) \rightarrow R$ be measurable. Then for all $T \in [0, \infty)$, the function $x_T(\cdot)$ defined by

19
$$x_T(t) = \begin{cases} x(t), & 0 \leq t \leq T \\ 0, & T < t \end{cases}$$

is called the *truncation* of $x(\cdot)$ to the interval $[0, T]$.

20 *[definition]* The set of all measurable functions $f(\cdot)$: $[0, \infty) \rightarrow R$, such that $f_T(\cdot) \in L_p[0, \infty)$ for all T, is denoted by $L_{pe}[0, \infty)$ and is called the *extension* of $L_p[0, \infty)$.

Thus the set $L_{pe}[0, \infty)$ (which is also sometimes called the *extended L_p-space*) consists of all measurable functions $f(\cdot)$ which have the property that all truncations $f_T(\cdot)$ of $f(\cdot)$ belong to $L_p[0, \infty)$, although $f(\cdot)$ itself may or may not belong to $L_p[0, \infty)$.

21 **Example.** Let $f(\cdot)$ be defined by

22
$$f(t) = t$$

Then for each $T < \infty$, the truncated function $f_T(\cdot)$ is the *triangular pulse* function defined by

23
$$f_T(t) = \begin{cases} t, & 0 \leq t \leq T \\ 0, & T < t \end{cases}$$

It is clear that, for each *finite* value of T, the function $f_T(\cdot)$ belongs to *all* the spaces $L_p[0, \infty)$. Hence the original function $f(\cdot)$ belongs to each of the *extended* spaces $L_{pe}[0, \infty)$, for each $p \in [1, \infty]$. However, $f(\cdot)$ does *not* belong to any of the *unextended* spaces $L_p[0, \infty)$.

The relationships between the unextended spaces $L_p[0, \infty)$ and the extended spaces $L_{pe}[0, \infty)$ are further brought out in the following.

24 ***lemma*** For each $p \in [1, \infty]$, the set $L_{pe}[0, \infty)$ is a real linear vector space. If $p \in [1, \infty]$ and $f(\cdot) \in L_{pe}[0, \infty)$, then

(i) $\|f_T(\cdot)\|_p$ is a nondecreasing function of T.

(ii) $f(\cdot) \in L_p[0, \infty)$ (the *unextended* space) if and only if there exists a finite constant m such that

25
$$\|f_T(\cdot)\|_p \leq m, \qquad \forall\, T < \infty$$

In this case

26
$$\|f(\cdot)\|_p = \lim_{T \to \infty} \|f_T(\cdot)\|_p$$

The proof is almost self-evident and is left as an exercise.

Thus, in summary, the extended space $L_{pe}[0, \infty)$ is a linear space that contains the unextended space $L_p[0, \infty)$ as a subset. Notice, however, that while $L_p[0, \infty)$ is a normed linear space, $L_{pe}[0, \infty)$ is only a linear vector space and cannot be made into a normed linear space.

27 *[definition]* The symbol $L_p^n[0, \infty)$ denotes the set of all n-tuples $\mathbf{f} = [f_1 \ldots f_n]'$, where $f_i \in L_p[0, \infty)$ for $i = 1, \ldots, n$. $L_{pe}^n[0, \infty)$ is defined similarly. The norm on $L_p^n[0, \infty)$ is defined by

28 $$||\mathbf{f}(\cdot)||_p \triangleq \left[\sum_{i=1}^{n} ||f_i(\cdot)||_p^2\right]^{1/2}$$

In other words, the norm of a vector-valued function $\mathbf{f}(\cdot)$ is the square root of the sum of the squares of the norms of the component functions $f_i(\cdot)$. This choice is to some extent arbitrary but has the advantage that, with this definition, $L_2^n[0, \infty)$ is a Hilbert space.

REMARKS ON NOTATION: Hereafter, we shall use L_p to denote $L_p[0, \infty)$; L_{pe}, L_p^n, and L_{pe}^n are defined similarly.

6.1.3 Causality

We shall close this section by introducing the notion of causality. If we think of a mapping A as representing a system and of $(Af)(\cdot)$ as the output of the system corresponding to the input $f(\cdot)$, then a causal system is one where the value of the output at any time t depends only on the values of the input up to time t. This is made precise next.

29 *[definition]* A mapping $A: L_{pe}^n \longrightarrow L_{pe}^m$ is said to be *causal* if

30 $$(Af)_T = (Af_T)_T, \qquad \forall\, T < \infty, \forall\, f \in L_{pe}^n$$

An alternative formulation of causality is provided by the following.

31 ***lemma*** Let $A: L_{pe}^n \longrightarrow L_{pe}^m$. Then A is causal in the sense of definition (29) if and only if, whenever $f, g \in L_{pe}^n$ and

32 $$f_T = g_T \qquad \text{for some } T < \infty$$

we have

33 $$(Af)_T = (Ag)_T$$

proof

(i) "*If*" Suppose A satisfies (32) and (33); we must show that A satisfies (30). Accordingly, let $f \in L_{pe}$ and $T < \infty$ be arbitrary. Then clearly $(f)_T = (f_T)_T$, so that by (32) and (33) we have

34 $$(Af)_T = (Af_T)_T$$

Because (34) holds for every T and for every $f \in L^n_{pe}$, (30) is established, and A is causal in the sense of definition (29).

(ii) "*Only If*" Suppose A satisfies (30). Let $f, q \in L^n_{pe}$, and suppose $f_T = g_T$ for some T; we must show that (33) holds. Now, by (30), we have

35 $$(Af)_T = (Af_T)_T$$

36 $$(Ag)_T = (Ag_T)_T$$

Moreover, because $f_T = g_T$, (35) and (36) together imply that

37 $$(Af)_T = (Ag)_T$$

which is the same as (33). ■

Definition (29) and lemma (31) provide two alternative (but entirely equivalent) interpretations of causality. Definition (29) states that a mapping A is causal if any truncation of Af to an interval $[0, T]$ depends only on the corresponding truncation f_T of f. To put it another way, the values of Af over the interval $[0, T]$ depend only on the values of f over $[0, T]$. Lemma (31) states that A is causal if and only if, whenever f and g are equal over an interval $[0, T]$, we have that Af and Ag are also equal over $[0, T]$.

Problem 6.1. Determine whether each of the functions below belongs to any of the unextended spaces L_p and to any of the extended spaces L_{pe}.

(a) $f(t) = te^{-t}$

(b) $f(t) = \tan t$

(c) $f(t) = 1/(1 + t)^{1/2}$

(d) $f(t) = e^{t^2}$

(e) $f(t) = \begin{cases} 0, & \text{if } t = 0 \\ 1/t^{1/2}, & \text{if } 0 < t < 1 \\ 1/t, & \text{if } 1 \leq t \end{cases}$

Problem 6.2. Prove lemma (24).

Problem 6.3. Suppose $H: L_{pe} \longrightarrow L_{pe}$ is one of the form

$$(Hf)(t) = \int_0^\infty h(t, \tau)f(\tau)\, d\tau$$

Show that H is causal if and only if

$$h(t, \tau) = 0 \quad \text{whenever } t < \tau$$

Problem 6.4. Suppose $A: L_{pe} \longrightarrow L_{pe}$ is of the form

$$(Af)(t) = \int_0^\infty a(t, \tau)n[f(\tau)]\, d\tau$$

where $n: h \longrightarrow h$ is not identically zero. Show that A is causal if and only if

$$a(t, \tau) = 0 \quad \text{whenever } t < \tau$$

6.2 INPUT–OUTPUT AND STATE REPRESENTATION OF SYSTEMS

In this section, we shall give a brief introduction to some results on constructing a state representation of a given input–output relationship. The purpose in presenting these results is to once again emphasize that the input–output representation and the state representation are two different ways of describing the *same* system.

Given is a system described in state variable form by the set of equations

$$\dot{\mathbf{x}}(t) = \mathbf{f}[t, \mathbf{x}(t), \mathbf{u}(t)]; \qquad \mathbf{x}(0) = \mathbf{x}_0 \quad \text{(fixed)} \tag{1}$$

$$\mathbf{y}(t) = \mathbf{g}[t, \mathbf{x}(t), \mathbf{u}(t)] \tag{2}$$

Suppose (1) has a unique solution for $\mathbf{x}(\cdot)$, for each $\mathbf{u}(\cdot)$ belonging to a prespecified class of inputs. [This is the case, for example, if $\mathbf{f}$ is continuous and satisfies a global Lipschitz condition in $\mathbf{x}$ for each fixed $\mathbf{u}(\cdot)$.] Then, once we know the $\mathbf{x}(\cdot)$ corresponding to a particular input $\mathbf{u}(\cdot)$, we can use (2) to calculate the output $\mathbf{y}(\cdot)$. Thus, conceptually at least, it is possible to construct the input–output relationship of a system given the state representation of the system.

For this reason, in this section we shall concentrate on the problem of constructing a state representation given the input–output relationship. There are many results available on this subject, and the brief discussion below presents only the simpler results.

6.2.1 Linear Time-Invariant Systems

Suppose that a system has the input–output relationship

$$\mathbf{y}(t) = \int_0^t \mathbf{H}(t - \tau)\mathbf{u}(\tau)\, d\tau \tag{3}$$

In words, the output $\mathbf{y}(\cdot)$ corresponding to the input $\mathbf{u}(\cdot)$ is given by the *convolution* of $\mathbf{u}(\cdot)$ and another function $\mathbf{H}(\cdot)$. The function $\mathbf{H}(\cdot)$ is characteristic of the system under study and is usually known as the *impulse response* of the system. Assuming that the functions $\mathbf{u}(\cdot)$, $\mathbf{y}(\cdot)$, and $\mathbf{H}(\cdot)$ are all Laplace-transformable, one can express (3) in the equivalent form

$$\hat{\mathbf{y}}(s) = \hat{\mathbf{H}}(s)\hat{\mathbf{u}}(s) \tag{4}$$

where a caret denotes the Laplace transform.

Our objective is to find matrices **A**, **B**, **C**, **D**, if possible, such that the system of equations

5 $$\dot{\mathbf{x}}(t) = \mathbf{A}\mathbf{x}(t) + \mathbf{B}\mathbf{u}(t); \qquad \mathbf{x}(0) = \mathbf{0}$$

6 $$\mathbf{y}(t) = \mathbf{C}\mathbf{x}(t) + \mathbf{D}\mathbf{u}(t)$$

has the input–output relationship (3) [or, equivalently, (4)]. In this case, (5)–(6) is a *state representation* of the *input–output relationship* (3). The conditions under which this can be done, as well as some properties of the resulting system of equations (5)–(6), are given next.

7 **fact** Given a Laplace-transformable distribution[3] $\mathbf{H}(\cdot)$, one can find **A**, **B**, **C**, **D** such that the system of equations (5)–(6) has the input–output relationship (3) if and only if $\hat{\mathbf{H}}(s)$ is a proper rational matrix in s; i.e., every element of the matrix $\hat{\mathbf{H}}(s)$ is a proper rational function of **s**.

8 **fact** Given a proper rational matrix $\hat{\mathbf{H}}(s)$, one can always find **A**, **B**, **C**, **D** such that

9 $$\hat{\mathbf{H}}(s) = \mathbf{C}(s\mathbf{I} - \mathbf{A})^{-1}\mathbf{B} + \mathbf{D}, \qquad \forall\, s$$

and such that

10 $$\text{rank}\,[\mathbf{B} \quad \mathbf{AB} \quad \ldots \quad \mathbf{A}^{n-1}\mathbf{B}] = n$$

11 $$\text{rank}\,[\mathbf{C}' \quad \mathbf{A}'\mathbf{C}' \quad \ldots \quad (\mathbf{A}')^{n-1}\mathbf{C}'] = n$$

where n is the order of the matrix **A**. In this case, the set of eigenvalues of **A** is the same as the set of poles of $\hat{\mathbf{H}}(s)$; i.e., a complex number λ is an eigenvalue of **A** if and only if $\hat{\mathbf{H}}(\cdot)$ has a pole at λ.

The above results can be found in any standard book on linear systems, e.g., [7] Chen.

In summary, facts (7) and (8) state that if $\hat{\mathbf{H}}(s)$ is a proper rational function of s, then one can always construct a system of equations of the form (5)–(6) which has the input–output relationship (3). Further, we can ensure that the system (5)–(6) is controllable and observable and that the set of eigenvalues of **A** coincides with the set of poles of $\hat{\mathbf{H}}(s)$.

6.2.2 Linear Time-Varying Systems

Consider now a linear system which has the input–output relationship

12 $$\mathbf{y}(t) = \int_0^t \mathbf{H}(t, \tau)\mathbf{u}(\tau)\, d\tau$$

---

[3] Note that if $\mathbf{D} \neq \mathbf{0}$, then the impulse response of the system (5)–(6) contains an impulse distribution at $t = 0$.

We would like, if possible, to construct a system of equations of the form[4]

13 $$\dot{\mathbf{x}}(t) = \mathbf{A}(t)\mathbf{x}(t) + \mathbf{B}(t)\mathbf{u}(t); \qquad \mathbf{x}(0) = \mathbf{0}$$

14 $$\mathbf{y}(t) = \mathbf{C}(t)\mathbf{x}(t)$$

such that the system of equations (13)–(14) has the input–output relationship (12). The conditions under which this can be done are given next.

15 **fact** Given $\mathbf{H}(\cdot, \cdot)$, one can find $\mathbf{A}(\cdot)$, $\mathbf{B}(\cdot)$, $\mathbf{C}(\cdot)$ such that the system of equations (13)–(14) has the input–output relationship (12) if and only if $\mathbf{H}(t, \tau)$ can be expressed in the form

16 $$\mathbf{H}(t, \tau) = \mathbf{\Phi}(t)\mathbf{\Psi}(\tau)$$

proof See [7] Chen.

6.3 DEFINITIONS OF INPUT–OUTPUT STABILITY

In this section, we shall introduce the basic definitions of input–output stability.

1 *[definition]* Let $\mathbf{A}: L_{pe}^n \longrightarrow L_{pe}^m$. We say that the mapping $\mathbf{A}$ (or the system represented by the mapping $\mathbf{A}$) is *L_p-stable* if (1) $\mathbf{Af} \in L_p^m$ whenever $\mathbf{f} \in L_p^n$, and (2) there exist finite constants k, b such that[5]

2 $$\|\mathbf{Af}\|_p \leq k\|\mathbf{f}\|_p + b, \qquad \forall\, \mathbf{f} \in L_p$$

Definition (1) can be interpreted as follows: Suppose that, from an input–output point of view, a system with n inputs and m outputs can be represented by the mapping $\mathbf{A}: L_{pe}^n \longrightarrow L_{pe}^m$. That is, given an input $\mathbf{f} \in L_{pe}^n$, the output of the system is $\mathbf{Af} \in L_{pe}^m$. Now, we say that the system is *L_p-stable* if, whenever the input $\mathbf{f}$ belongs to the *unextended* space L_p^n, the resulting output $\mathbf{Af}$ belongs to L_p^m and moreover the norm of the output $\mathbf{Af}$ is no larger than k times the norm of the input $\mathbf{f}$ plus the *offset* constant b.

For example, if we put $p = \infty$, then the concept of L_∞-stability becomes what is commonly referred to as BIBO (bounded-input/bounded-output) stability; namely, a system is L_∞-stable if, whenever

[4]Note that, in (14), there is no term of the form $\mathbf{D}(t)\mathbf{u}(t)$. This is in the interests of simplicity.

[5]Note that we use $\|\cdot\|_p$ to denote the norms on L_p^n as well as L_p^m.

the input $f \in L_\infty$ (is bounded), the output $Af \in L_\infty$ (is bounded) and moreover (2) holds.

Note that some authors use the adjective "L_p-stable" to describe a system **A** that satisfies only condition 1 of definition (1) (i.e., inputs in L_p produce outputs in L_p but with no norm restriction).

3 **Example.** Consider the mapping A defined by

4 $$(Af)(t) = \int_0^t e^{-\alpha(t-\tau)} f(\tau)\, d\tau$$

where $\alpha > 0$ is a given constant, and suppose we wish to study the L_∞-stability of this system. We first establish that $A: L_{\infty e} \to L_{\infty e}$, so that A is in the class of mappings covered by definition (1). Accordingly, suppose $f \in L_{\infty e}$. Then for each finite T, we have that $\|f_T\|_\infty$ is finite (see definition [6.1(20)]. Hence, for each finite T, there exists a finite constant m_T such that

5 $$|f(t)| \leq m_T \quad \text{a.e.,}^6 \quad \forall\, t \in [0, T]$$

To show that $Af \in L_{\infty e}$, let g denote Af. Then

6 $$g(t) = \int_0^t e^{-\alpha(t-\tau)} f(\tau)\, d\tau$$

Hence, for $t \leq T$, we have

7 $$\begin{aligned} |g(t)| &\leq \int_0^t |e^{-\alpha(t-\tau)}| \cdot |f(\tau)|\, d\tau \\ &\leq \int_0^t m_T |e^{-\alpha(t-\tau)}|\, d\tau = \frac{m_T(1 - e^{-\alpha t})}{\alpha} \\ &\leq \frac{m_T}{\alpha} \quad \text{a.e.} \end{aligned}$$

The inequality (7) shows that $g(\cdot)$ is bounded over $[0, T]$. Because this reasoning is valid for every finite T, it follows that $g(\cdot) \in L_{\infty e}$, i.e., that $A: L_{\infty e} \to L_{\infty e}$.

Next, we show that the mapping A is L_∞-stable in the sense of definition (1). Suppose $f \in L_\infty$; then there exists a finite constant m such that

8 $$|f(t)| \leq m \quad \text{a.e.,} \quad \forall\, t \geq 0$$

By using exactly the same reasoning as before, we can show that[7]

9 $$|g(t)| \leq \frac{m}{\alpha}, \quad \forall\, t \geq 0$$

---

[6]"a.e." stands for "almost everywhere," which means "except on a set of measure zero."

[7]Hereafter we shall drop the phrase "almost everywhere," it being implicitly understood.

Hence (2) is satisfied with $k = 1/\alpha$, $b = 0$, which shows that A is L_∞-stable.

10 **Example.** To illustrate the difference between conditions 1 and 2 of definition (1), let A be the mapping defined by

11 $$(Af)(t) = [f(t)]^2$$

First, it is easy to show that $A\colon L_{\infty e} \longrightarrow L_{\infty e}$, because if

12 $$|f(t)| \leq m, \qquad \forall\, t \in [0, T]$$

then

13 $$|(Af)(t)| \leq m^2, \qquad \forall\, t \in [0, T]$$

Also, by the same reasoning, it is clear that $Af \in L_\infty$ whenever $f \in L_\infty$. Hence A satisfies condition 1 of definition (1). However, it is clear that no k, b exist such that (2) is satisfied, because the function $x \mapsto x^2$ cannot be bounded by a straight line of the form $x \mapsto kx + b$. This is illustrated in Fig. 6.1.

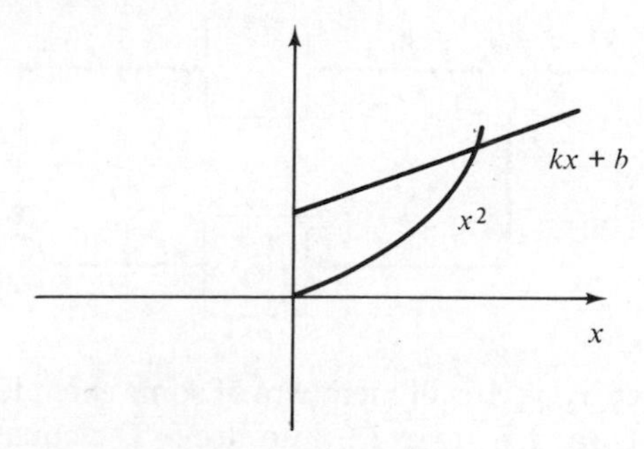

FIG. 6.1

14 **Example.** Consider the system whose input–output relationship is

15 $$(Af)(t) = \int e^{t-\tau} f(\tau)\, d\tau$$

This mapping A also maps $L_{\infty e}$ into itself. To see this, let $f \in L_{\infty e}$. Then, for each finite T, there exists a finite constant m_T such that

16 $$|f(t)| \leq m_T, \qquad \forall\, t \in [0, T]$$

Thus, whenever $t \in [0, T]$, we have

17 $$|(Af)(t)| \leq m_T \int_0^t |e^{(t-\tau)}|\, d\tau = m_T(e^T - 1)$$

Now, for each *finite* T, the right-hand side of (17) is a finite number. Hence $Af \in L_{\infty e}$.

On the other hand, A is *not* L_∞-stable, as can be seen by setting $f(t) = 1$, $\forall\, t$. Then $f(\cdot) \in L_\infty$, but

18 $$(Af)(t) = \int_0^t e^{(t-\tau)}\, d\tau = e^t - 1$$

which does *not* belong to L_∞ (although it does belong to $L_{\infty e}$). Hence there is at least one input in L_∞ whose corresponding output does not belong to L_∞, which shows that A is not L_∞-stable.

REMARKS: Example (14) illustrates the advantages of setting up the input–output stability problem in extended L_p-spaces. If we deal exclusively with L_p-stable systems, then we can represent such a system as a mapping from L_p (the unextended space) into itself, rather than from L_{pe} into itself. However, if we are interested in studying unstable systems (for example, the feedback stability of systems containing unstable subsystems), then we must have a way of mathematically describing such a system. This is accomplished by treating such a system as a mapping from L_{pe} into itself.

We shall now turn to the definitions of feedback stability. Consider the feedback system shown in Fig. 6.2, where the various quantities

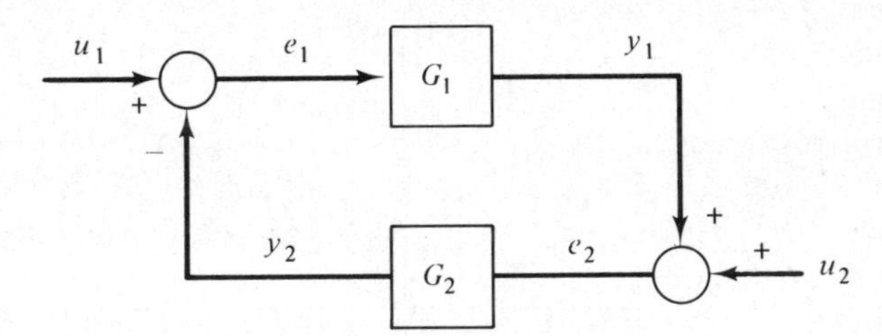

FIG. 6.2

$\mathbf{u}_1, \mathbf{u}_2, \mathbf{e}_1, \mathbf{e}_2, \mathbf{y}_1, \mathbf{y}_2$ are all members of some extended space L^n_{pe}, and the operators $\mathbf{G}_1$ and $\mathbf{G}_2$ map L^n_{pe} into itself. The equations describing this feedback system are

19 $$\mathbf{e}_1 = \mathbf{u}_1 - \mathbf{y}_2$$

20 $$\mathbf{e}_2 = \mathbf{u}_2 + \mathbf{y}_1$$

21 $$\mathbf{y}_1 = \mathbf{G}_1\mathbf{e}_1$$

22 $$\mathbf{y}_2 = \mathbf{G}_2\mathbf{e}_2$$

The idea is that $\mathbf{G}_1$ and $\mathbf{G}_2$ represent two given subsystems which are interconnected in the manner shown, $\mathbf{u}_1$ and $\mathbf{u}_2$ represent the (known) inputs, $\mathbf{e}_1$ and $\mathbf{e}_2$ are the (unknown) "errors," and $\mathbf{y}_1$ and $\mathbf{y}_2$ are the (unknown) outputs.

In analyzing the system described by (19)–(22), two distinct types of questions arise. It is clear that we can eliminate $\mathbf{y}_1, \mathbf{y}_2$ from (19)–(22) and rewrite the system equations as

23 $$\mathbf{e}_1 = \mathbf{u}_1 - \mathbf{G}_2\mathbf{e}_2$$

24 $$\mathbf{e}_2 = \mathbf{u}_2 + \mathbf{G}_1\mathbf{e}_1$$

Alternatively, we can eliminate $\mathbf{e}_1, \mathbf{e}_2$ and rewrite the system equations as

25 $$\mathbf{y}_1 = \mathbf{G}_1(\mathbf{u}_1 - \mathbf{y}_2)$$

26 $$\mathbf{y}_2 = \mathbf{G}_2(\mathbf{u}_2 + \mathbf{y}_1)$$

The system descriptions (19)–(22), (23)–(24), and (25)–(26) are all equivalent. With regard to this system, the first type of question that one can ask is the following: Given inputs $\mathbf{u}_1$ and $\mathbf{u}_2$ in L_{pe}^n, does there exist a set of solutions for $\mathbf{e}_1, \mathbf{e}_2, \mathbf{y}_1, \mathbf{y}_2$ in L_{pe}^n such that the system equations are satisfied, and if so, is this set of solutions unique? This is usually called the existence and uniqueness question. The second type of question takes the following form: Given inputs $\mathbf{u}_1, \mathbf{u}_2$ in L_p^n, assuming that solutions in L_{pe}^n exist for $\mathbf{e}_1, \mathbf{e}_2, \mathbf{y}_1, \mathbf{y}_2$, do these solutions actually belong to L_p^n? This is called the stability question. The reason for making the distinction between the two types of questions is that, in practice, different types of conditions arise in answering the two types of questions. For instance, existence and uniqueness of solutions can be assured under very mild conditions [usually some type of global Lipschitz conditions; see Theorem (27).] On the other hand, stability requires stronger conditions, which are the main subject of this chapter.

Accordingly, we shall state below a theorem regarding the existence and uniqueness of solutions to (19)–(22) and then proceed to a study of stability conditions. Theorem (27) is not the most general of its kind, but it is easy to prove and is adequate for most situations.

27 ***Theorem*** Consider the system described by (19)–(22), and suppose the operators $\mathbf{G}_1$ and $\mathbf{G}_2$ are of the form

28 $$(\mathbf{G}_1\mathbf{x})(t) = \int_0^t \mathbf{G}(t, \tau)\mathbf{n}_1[\tau, \mathbf{x}(\tau)]\, d\tau$$

29 $$(\mathbf{G}_2\mathbf{x})(t) = \mathbf{n}_2[\tau, \mathbf{x}(\tau)]$$

where $\mathbf{G}(\cdot, \cdot)$ is continuous, and the continuous functions $\mathbf{n}_1, \mathbf{n}_2$: $R_+ \times R^n \longrightarrow R^n$

(i) Are unbiased, i.e.,

30 $$\mathbf{n}_1(t, \mathbf{0}) = \mathbf{0} \qquad \forall\, t \geq 0$$

31 $$\mathbf{n}_2(t, \mathbf{0}) = \mathbf{0} \qquad \forall\, t \geq 0$$

(ii) Satisfy global Lipschitz conditions; i.e., there exist finite constants k_1 and k_2 such that

32 $$\|\mathbf{n}_1(t, \mathbf{x}) - \mathbf{n}_1(t, \mathbf{y})\| \leq k_1 \|\mathbf{x} - \mathbf{y}\|, \qquad \forall\, t \geq 0, \quad \forall\, \mathbf{x}, \mathbf{y} \in R^n$$

33 $$\|\mathbf{n}_2(t, \mathbf{x}) - \mathbf{n}_2(t, \mathbf{y})\| \leq k_2 \|\mathbf{x} - \mathbf{y}\|, \qquad \forall\, t \geq 0, \quad \forall\, \mathbf{x}, \mathbf{y} \in R^n$$

Under these conditions, $\mathbf{G}_1$ and $\mathbf{G}_2$ map L^n_{pe} into itself. Further, given any $\mathbf{u}_1, \mathbf{u}_2 \in L^n_{pe}$, there exists exactly one set of $\mathbf{e}_1, \mathbf{e}_2, \mathbf{y}_1, \mathbf{y}_2$, each belonging to L^n_{pe}, such that (19)–(22) are satisfied.

The proof is omitted here, because although it is quite straightforward, it is also rather messy. The interested reader is referred to [10] Desoer and Vidyasagar [Chapter III], where a more general version of this theorem is proved. The basis of the proof is the observation that, under the stated assumptions, the set of equations (23)–(24) [which is equivalent to (19)–(22)] reduces to a nonlinear integral equation of the Volterra type. Basically, Theorem (27) assures us that existence and uniqueness are guaranteed provided $\mathbf{G}_1$ is a causal integral operator (possibly nonlinear) and $\mathbf{G}_2$ is a (causal) memoryless nonlinear operator. Actually, the roles of $\mathbf{G}_1$ and $\mathbf{G}_2$ in Theorem (27) can be interchanged, and the conclusions still hold. In addition, even if both $\mathbf{G}_1$ and $\mathbf{G}_2$ are integral operators of the form (28), existence and uniqueness still follow. The key is that at least one of the operators $\mathbf{G}_1$ and $\mathbf{G}_2$ is an integral operator and as such introduces an element of *smoothing* in the closed loop. The situation that is *not* covered by Theorem (27) is the case where both $\mathbf{G}_1$ and $\mathbf{G}_2$ are memoryless. In this case, existence and uniqueness can fail in some seemingly innocent systems. See [10] Desoer and Vidyasagar [Chapter I].

With this cursory treatment of existence and uniqueness results, we shall turn now to the definitions of stability for the system described by (19)–(22).

34 *[definition]* The system represented by (25)–(26) is said to be *L_p-stable* if, whenever $\mathbf{u}_1, \mathbf{u}_2$ belong to L^n_p, we have that any corresponding $\mathbf{y}_1, \mathbf{y}_2$ in L^n_{pe} that satisfy (25)–(26) actually belong to L^n_p, and in addition, there exist finite constants k and b such that

$$\|\mathbf{y}_1\|_p \leq k(\|\mathbf{u}_1\|_p + \|\mathbf{u}_2\|_p) + b \tag{35}$$

$$\|\mathbf{y}_2\|_p \leq k(\|\mathbf{u}_1\|_p + \|\mathbf{u}_2\|_p) + b \tag{36}$$

whenever $\mathbf{u}_1, \mathbf{u}_2, \mathbf{y}_1, \mathbf{y}_2$ satisfy (25)–(26).

REMARKS:

1. Notice that, according to definition (34), the question of the *L_p-stability* of the system (25)–(26) is quite divorced from the question of existence and uniqueness of solutions to (25)–(26). If $\mathbf{u}_1, \mathbf{u}_2$ belong to L^n_p, then in order for the system (25)–(26) to be L_p-stable, one of two alternatives must occur: (i) either there do *not* exist any $\mathbf{y}_1, \mathbf{y}_2$, both belonging to

L_{pe}^n, such that (25)–(26) are satisfied, or (ii) there *do* exist $\mathbf{y}_1, \mathbf{y}_2$ in L_{pe}^n such that (25)–(26) are satisfied, and all such solutions in fact belong to the *unextended* space L_p^n and satisfy (35)–(36). This, and only this, is what is meant by L_p-stability. There is no a priori assumption that, corresponding to a given set of inputs $\mathbf{u}_1, \mathbf{u}_2$ in L_p^n, there actually exist $\mathbf{y}_1, \mathbf{y}_2$ in L_{pe}^n that satisfy (25)–(26). However, if such $\mathbf{y}_1, \mathbf{y}_2$ *do* exist, then they must actually belong to L_p^n and satisfy (35)–(36).

2. Definition (34) mentions only $\mathbf{y}_1, \mathbf{y}_2$ and does not say anything about $\mathbf{e}_1, \mathbf{e}_2$. However, the requirements on $\mathbf{e}_1, \mathbf{e}_2$ are implicit in definition (34). First, recall that the system descriptions (19)–(22) and (25)–(26) are equivalent and that L_p^n is a linear space. Hence, if $\mathbf{u}_1, \mathbf{u}_2, \mathbf{y}_1, \mathbf{y}_2$ satisfy (19) and (20), we see that $\mathbf{e}_1, \mathbf{e}_2$ also belong to L_p^n. Next, if $\mathbf{y}_1, \mathbf{y}_2$ satisfy (35)–(36), then clearly, from (19)–(20), we have

37 $$\|\mathbf{e}_1\|_p \leq (k+1)(\|\mathbf{u}_1\|_p + \|\mathbf{u}_2\|_p) + b$$

38 $$\|\mathbf{e}_2\|_p \leq (k+1)(\|\mathbf{u}_1\|_p + \|\mathbf{u}_2\|_p) + b$$

Hence $\mathbf{e}_1$, $\mathbf{e}_2$ satisfy conditions similar to (35)–(36). Thus, even though definition (34) does not explicitly mention $\mathbf{e}_1$ and $\mathbf{e}_2$, the conditions imposed on $\mathbf{y}_1$ and $\mathbf{y}_2$ imply that $\mathbf{e}_1$ and $\mathbf{e}_2$ satisfy similar conditions.

Actually, in definition (34), instead of imposing conditions on $\mathbf{y}_1, \mathbf{y}_2$, we could have imposed similar conditions on $\mathbf{e}_1$, $\mathbf{e}_2$, and the resulting definition would be entirely equivalent to definition (34), because (19) and (20) can be rewritten as

39 $$\mathbf{y}_1 = \mathbf{e}_2 - \mathbf{u}_2$$

40 $$\mathbf{y}_2 = \mathbf{u}_1 - \mathbf{e}_1$$

Hence, if $\mathbf{u}_1, \mathbf{u}_2, \mathbf{e}_1, \mathbf{e}_2$ belong to L_p^n and if $\mathbf{e}_1, \mathbf{e}_2$ satisfy conditions such as (37)–(38), then $\mathbf{y}_1, \mathbf{y}_2$ also belong to L_p^n and satisfy conditions similar to (35)–(36).

3. Comparing definitions (1) and (34), one may feel that there are two distinct concepts of L_p-stability, one for open-loop stability and one for feedback stability. However, this is not the case. It can be shown that if we take the system equations (25)–(26) as defining a *relation* between the input pair $(\mathbf{u}_1, \mathbf{u}_2)$ and the output pair $(\mathbf{y}_1, \mathbf{y}_2)$, then definition (34) requires the same properties of this relation as definition (1) does of the operator $\mathbf{A}$.
4. In definition (34), there is no requirement that the *subsystem* operators $\mathbf{G}_1$ and $\mathbf{G}_2$ be themselves L_p-stable.

Because we have not as yet presented any means of testing for stability, other than solving the system equations by brute force, the following example is of necessity rather simple.

41 **Example.** Consider a system described by (19)–(22), with

42 $$(G_1x)(t) \triangleq \int_0^t e^{-a(t-\tau)}x(\tau)\,d\tau$$

43 $$(G_2x)(t) \triangleq kx(t)$$

with k, a being given constants. Let $u_2(t) = 0$ for simplicity. The subsystem represented by G_1 has a *transfer function* of $1/(s + a)$, while G_2 represents a constant feedback gain of k. Thus, using elementary control theory, we see that the closed-loop transfer function is $1/(s + a + k)$, which corresponds to an *impulse response* of $e^{-(a+k)t}$. Thus, corresponding to an input $u_1(\cdot)$ in $L_{\infty e}$, there exists a unique set of solutions to (19)–(22),[8] given by

44 $$y_1(t) = \int_0^t e^{-(a+k)(t-\tau)}u_1(\tau)\,d\tau$$

45 $$y_2(t) = ky_1(t)$$

First, we see from Examples (3) and (14) that $y_1, y_2 \in L_{\infty e}$ whenever $u_1 \in L_{\infty e}$. Next, based on the same examples, it follows that $y_1, y_2 \in L_\infty$ whenever $u_1 \in L_\infty$, provided $a + k > 0$. Thus the system is L_∞-stable, in the sense of definition (34), if $a + k > 0$. On the other hand, if $a + k \leq 0$, then one can find inputs in L_∞ [e.g., $u_1(t) \equiv 1$] such that the corresponding y_1, y_2 do not belong to L_∞. Thus the system is L_∞-unstable if $a + k \leq 0$.

6.4 RELATIONSHIPS BETWEEN I/O STABILITY AND LIAPUNOV STABILITY

In some cases, input–output stability methods can be used to establish Liapunov stability. In this section, we shall present some results that state such results precisely.

1 ***Theorem*** Consider a system described by the vector differential equation

2 $$\dot{\mathbf{x}}(t) = \mathbf{A}\mathbf{x}(t) - \mathbf{f}[t, \mathbf{x}(t)]; \qquad \mathbf{x}(0) = \mathbf{x}_0$$

---

[8] The existence and uniqueness of solutions can also be established using Theorem (27).

where all eigenvalues of the $n \times n$ matrix $\mathbf{A}$ have negative real parts, and the mapping $f: R_+ \times R^n \longrightarrow R^n$ is continuous. Associated with system (2), define a corresponding nonlinear feedback system (shown in Fig. 6.3) described by

3 $$\mathbf{e}(t) = \mathbf{u}(t) - \int_0^t e^{\mathbf{A}(t-\tau)} y(\tau)\, d\tau$$

4 $$\mathbf{y}(t) = \mathbf{f}[t, \mathbf{e}(t)]$$

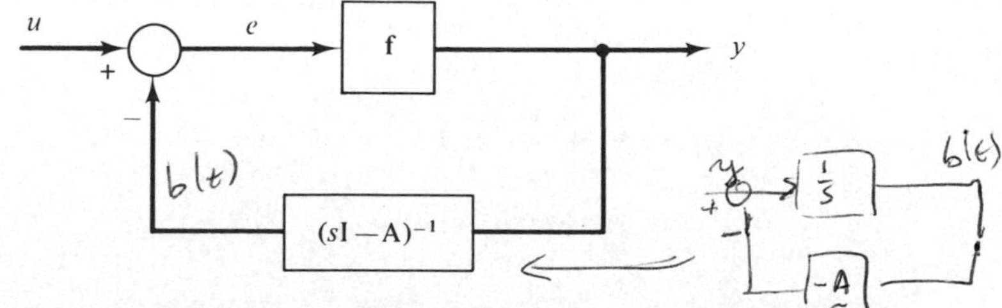

FIG. 6.3

With the stated hypotheses, if the feedback system (3)–(4) is L_2-stable in the sense of definition [6.3(34)], then the equilibrium point $\mathbf{0}$ of system (2) is globally asymptotically stable.

REMARKS: Theorem (1) states that, under certain conditions, the global asymptotic stability of system (2) can be established by proving the L_2-stability of the associated feedback system (3)–(4). As yet, we have not given any methods for proving L_2-stability—this will be done later.

proof We wish to show that, for all $\mathbf{x}_0$, the corresponding solution $\mathbf{x}(\cdot)$ of (2) has the property that $\mathbf{x}(t) \longrightarrow 0$ as $t \longrightarrow \infty$. We begin by expressing (2) as the equivalent nonlinear integral equation

5 $$\mathbf{x}(t) = e^{\mathbf{A}t}\mathbf{x}_0 - \int_0^t e^{\mathbf{A}(t-\tau)} \mathbf{f}[t, \mathbf{x}(\tau)]\, d\tau$$

Next, we observe that if we let

6 $$\mathbf{u}(t) = e^{\mathbf{A}t}\mathbf{x}_0$$

in the system (3)–(4) (i.e., the system of Fig. 6.3), then the "error" $\mathbf{e}(\cdot)$ is governed by (5). Now, because all eigenvalues of $\mathbf{A}$ have negative real parts, it is clear that the function $t \mapsto \|e^{\mathbf{A}t}\mathbf{x}_0\|$ is dominated by a decaying exponential; i.e., there exist constants $m > 0$, $a > 0$ such that

7 $$\|e^{\mathbf{A}t}\mathbf{x}_0\| \leq me^{-at}, \qquad \forall\, t \geq 0$$

where $\|\cdot\|$ denotes the Euclidean norm. Thus the particular input $\mathbf{u}(\cdot)$ defined by (6) belongs to L_2^n. Therefore, if the system (3)–(4) is L_2-stable (in the sense of definition [6.3(34)]), then both $\mathbf{e}(\cdot)$ and $\mathbf{y}(\cdot)$ belong to L_2^n. Let us define

8 $$\mathbf{z}(t) = \int_0^t e^{\mathbf{A}(t-\tau)} \mathbf{y}(\tau)\, d\tau$$

Then, by (3),

9 $$\mathbf{e}(t) = \mathbf{u}(t) - \mathbf{z}(t)$$

As mentioned previously, $\mathbf{e}(\cdot)$ in this case is the solution $\mathbf{x}(\cdot)$ of (2). Also, because of (7), we have that $\mathbf{u}(t) \longrightarrow \mathbf{0}$ as $t \longrightarrow \infty$. Thus if we can show that $\mathbf{z}(t) \longrightarrow \mathbf{0}$ as $t \longrightarrow \infty$, it will follow that $\mathbf{e}(t) \longrightarrow \mathbf{0}$ as $t \longrightarrow \infty$, and the global asymptotic stability of the equilibrium point $\mathbf{0}$ of (2) will be established. We shall now show that $\mathbf{z}(t) \longrightarrow \mathbf{0}$ as $t \longrightarrow \infty$, and this will conclude the proof of the theorem (readers disinterested in the details can skip to the next paragraph).

We have $\mathbf{y}(\cdot) \in L_2^n$ and

10 $$\begin{aligned} \mathbf{z}(t) &= \int_0^t e^{\mathbf{A}(t-\tau)}\mathbf{y}(\tau)\,d\tau \\ &= \int_0^{t/2} e^{\mathbf{A}(t-\tau)}\mathbf{y}(\tau)\,d\tau + \int_{t/2}^t e^{\mathbf{A}(t-\tau)}\mathbf{y}(\tau)\,d\tau \\ &= \int_{t/2}^t e^{\mathbf{A}\tau}\mathbf{y}(t-\tau)\,d\tau + \int_{t/2}^t e^{\mathbf{A}(t-\tau)}\mathbf{y}(\tau)\,d\tau \end{aligned}$$

after changing the dummy variable of integration in the first integral. Thus by the triangle inequality, we get

11 $$\|\mathbf{z}(t)\| \leq \int_{t/2}^t \|e^{\mathbf{A}\tau}\|_i \, \|\mathbf{y}(t-\tau)\|\,d\tau + \int_{t/2}^t \|e^{\mathbf{A}(t-\tau)}\|_i \, \|\mathbf{y}(\tau)\|\,d\tau$$

where $\|\cdot\|_i$ denotes the induced l_2-norm on $R^{n\times n}$ (see Sec. 3.2). Next, by Schwarz's inequality on L_2, we get

12 $$\begin{aligned} \|z(t)\| &\leq \left[\int_{t/2}^t (\|e^{\mathbf{A}\tau}\|_i)^2\,d\tau\right]^{1/2} \cdot \left[\int_{t/2}^t \|y(t-\tau)\|^2\,d\tau\right]^{1/2} \\ &\quad + \left[\int_{t/2}^t (\|e^{\mathbf{A}(t-\tau)}\|_i)^2\,d\tau\right]^{1/2} \cdot \left[\int_{t/2}^t \|\mathbf{y}(\tau)\|^2\,d\tau\right]^{1/2} \\ &\leq \left[\int_{t/2}^\infty (\|e^{\mathbf{A}\tau}\|_i)^2\,d\tau\right]^{1/2} \cdot \left[\int_0^\infty \|\mathbf{y}(\tau)\|\,d\tau\right]^{1/2} \\ &\quad + \left[\int_0^\infty (\|e^{\mathbf{A}\tau}\|_i^2\,d\tau\right]^{1/2} \cdot \left[\int_{t/2}^\infty \|\mathbf{y}(\tau)\|^2\,d\tau\right]^{1/2} \end{aligned}$$

Now we use the following well-known result from real analysis: If $f(\cdot) \in L_2$, then

13 $$\int_t^\infty f^2(\tau)\,d\tau \longrightarrow 0 \quad \text{as} \quad t \longrightarrow \infty$$

This result, coupled with the fact that both $\|e^{\mathbf{A}\tau}\|_i$ and $\|\mathbf{y}(\cdot)\|$ belong to L_2, implies that the right-hand side of (12) approaches zero as $t \longrightarrow \infty$. This shows that $\mathbf{z}(t) \longrightarrow \mathbf{0}$ as $t \longrightarrow \infty$. ∎

In some instances, the nonlinearity in the differential equation (2) enters only via a scalar nonlinear function. In this case, the global asymptotic stability of (2) can be established by studying the L_2-stability of a single-input/single-output system.

14 ***Theorem*** Consider a system described by the vector differential equation

15 $$\dot{\mathbf{x}}(t) = \mathbf{A}\mathbf{x}(t) - \mathbf{b}\phi[t, \mathbf{c}'\mathbf{x}(t)]; \qquad \mathbf{x}(0) = \mathbf{x}_0$$

where all eigenvalues of the $n \times n$ matrix $\mathbf{A}$ have negative real parts, $\phi: R_+ \times R \longrightarrow R$ is continuous, and the pair $(\mathbf{A}, \mathbf{c})$ is observable, i.e.,

16 $$\text{rank}\,[\mathbf{c}' \,|\, \mathbf{A}'\mathbf{c}' \,|\, \dots \,|\, (\mathbf{A}')^{n-1}\mathbf{c}'] = n$$

Associated with the system (15), define the single-input/single-output feedback system (shown in Fig. 6.4) described by

17 $$e(t) = u(t) - \int_0^t \mathbf{c}' e^{\mathbf{A}(t-\tau)}\mathbf{b}y(\tau)\, d\tau$$

18 $$y(t) = \phi[t, e(t)]$$

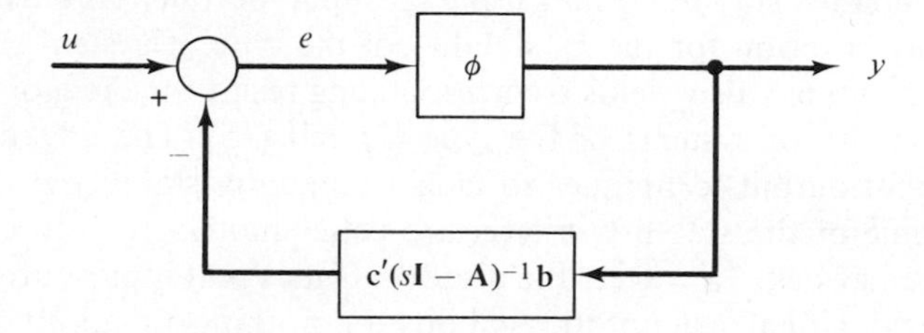

FIG. 6.4

With the stated hypotheses, if the system (17)–(18) is L_2-stable in the sense of definition [6.3(34)], then the equilibrium point $\mathbf{0}$ of (15) is globally asymptotically stable.

REMARKS: Note that (15) is the same as the system studied in Sec. 5.5 in connection with the Lur'e problem.

proof The proof closely parallels that of Theorem (1). First, (15) can be expressed as an equivalent nonlinear integral equation

19 $$\mathbf{x}(t) = e^{\mathbf{A}t}\mathbf{x}_0 - \int_0^t e^{\mathbf{A}(t-\tau)}\mathbf{b}\phi[\tau, \mathbf{c}'\mathbf{x}(\tau)]\, d\tau$$

Therefore

20 $$\mathbf{c}'\mathbf{x}(t) = \mathbf{c}'\mathbf{e}^{\mathbf{A}t}\mathbf{x}_0 - \int_0^t \mathbf{c}'\mathbf{e}^{\mathbf{A}(t-\tau)}\mathbf{b}\phi[\tau, \mathbf{c}'\mathbf{x}(\tau)]\, d\tau$$

If we define

21 $$e(t) = \mathbf{c}'\mathbf{x}(t)$$

22 $$u(t) = \mathbf{c}' e^{\mathbf{A}t}\mathbf{x}_0$$

then (20) becomes

23 $$e(t) = u(t) - \int_0^t \mathbf{c}' e^{\mathbf{A}(t-\tau)}\mathbf{b}\phi[\tau, e(\tau)]\, d\tau$$

which is the same as (17)–(18). Because $u(\cdot) \in L_2$, if the system (17)–(18) is L_2-stable, it follows that $e(\cdot), y(\cdot) \in L_2$. Next, as in the proof of Theorem (1), $y(\cdot) \in L_2$ implies that $z(t) \longrightarrow 0$ as $t \longrightarrow \infty$ and that $e(t) \longrightarrow 0$ as $t \longrightarrow \infty$. However, because $e(t)$ is a scalar and $\mathbf{x}(t)$ is a vector, we need to do some more work to show that $\mathbf{x}(t) \longrightarrow \mathbf{0}$ as $t \longrightarrow \infty$. If we define

$$\mathbf{v}(t) = \mathbf{b}y(t) = \mathbf{b}\phi[t, e(t)] \tag{24}$$

then the system description (15) becomes

$$\dot{\mathbf{x}}(t) = \mathbf{A}\mathbf{x}(t) + \mathbf{v}(t) \tag{25}$$

$$e(t) = \mathbf{c}'\mathbf{x}(t) \tag{26}$$

Now, using the facts that the pair $(\mathbf{A}, \mathbf{c})$ is observable and that $e(t) \longrightarrow 0$, $\mathbf{v}(t) \longrightarrow \mathbf{0}$ as $t \longrightarrow \infty$, we can show by standard arguments from linear system theory that $\mathbf{x}(t) \longrightarrow \mathbf{0}$ as $t \longrightarrow \infty$. ∎

Theorems (1) and (14) provide a glimpse of how one can prove Liapunov stability results using the input–output approach. Specifically, any criterion for the L_2-stability of the type of system shown in Figs. 6.3 and 6.4 also yields a corresponding result for the global asymptotic stability of systems of the type (2) and (15). The advantage of using input-output techniques to tackle Liapunov stability problems is that some of the stability criteria are better motivated in the input-output framework. However, the disadvantage is that input-output techniques yield global asymptotic stability or nothing at all. It is difficult to estimate a "region of attraction" using input-output techniques.

Problem 6.5. Prove the following generalization of Theorem (14): Consider a system described by the vector differential equation

$$\dot{\mathbf{x}}(t) = \mathbf{A}\mathbf{x}(t) + \mathbf{b}v(t); \;\; \mathbf{x}(0) = \mathbf{x}_0 \tag{27}$$

$$\sigma(t) = \mathbf{c}'\mathbf{x}(t) + dv(t) \tag{28}$$

$$v(t) = -\phi[t, \sigma(t)] \tag{29}$$

where all eigenvalues of $\mathbf{A}$ have negative real parts; $\phi: R_+ \times R \longrightarrow R$ is continuous; and the pair $(\mathbf{A}, \mathbf{c})$ is observable. Suppose $\phi(\cdot, \cdot)$ belongs to the sector $[\alpha, \beta]$, i.e.

$$\alpha\sigma^2 \leq \phi(t, \sigma) \leq \beta\sigma^2, \; \forall\, t \geq 0, \; \forall\, \sigma \in R \tag{30}$$

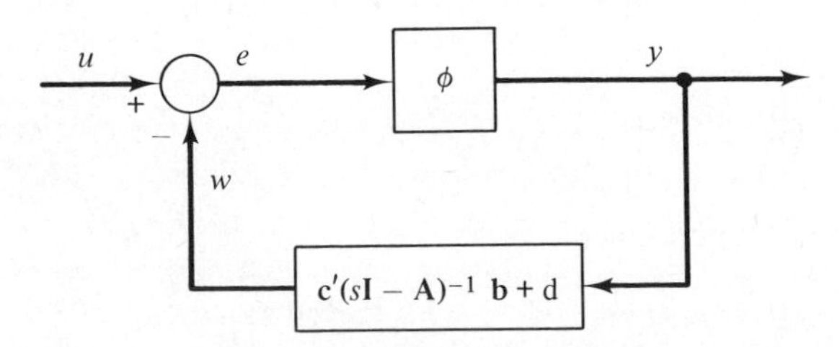

FIG. 6.5

and suppose $1 + kd \neq 0$, $\forall\, k \in [\alpha, \beta]$. Under these conditions, if the system shown in Fig. 6.5 is L_2-stable, then $\mathbf{0}$ is a globally asymptotically stable equilibrium point of the system (27)–(29) {Hint: Show that, for all $\mathbf{x}_0$, we have $y \in L_2$, $w(t) - dy(t) \longrightarrow 0$ as $t \longrightarrow \infty$ and hence $w(t) - d\phi[t, -w(t)] \longrightarrow 0$ as $t \longrightarrow \infty$. Thus conclude that $w(t) \longrightarrow 0$ and hence $\mathbf{x}(t) \longrightarrow \mathbf{0}$ as $t \longrightarrow \infty$.}

6.5 OPEN-LOOP STABILITY OF LINEAR SYSTEMS

Before attempting to study the stability of interconnected systems such as (19)–(22), it is helpful to first obtain conditions under which the operators $\mathbf{G}_1$ and $\mathbf{G}_2$ represent L_p-stable subsystems. In this section, we shall concentrate on linear systems for the most part and obtain necessary and sufficient conditions for a linear system to be L_p-stable. The term *open-loop stability* refers to the fact that we are studying the subsystems $\mathbf{G}_1$ and $\mathbf{G}_2$, rather than the overall *closed-loop* system, which is described by (19)–(22).

6.5.1 Time-Invariant Systems

Throughout most of this subsection, we shall study single-input/single-output systems. Consider a single-input/single-output time-invariant system, which is characterized by a scalar transfer function $\hat{h}(s)$. Suppose that, in addition, $\hat{h}(s)$ is a *rational* function of s. It is well known that such a system is L_∞-stable (BIBO stable) if and only if

1. $\hat{h}(s)$ is a *proper* rational function of s, and
2. All poles of $\hat{h}(\cdot)$ have negative real parts.

However, the situation is more complicated if $\hat{h}(s)$ is *not* rational. Such a situation arises whenever $\hat{h}(\cdot)$ is the transfer function of a distributed system, such as an RC transmission line (integrated circuit) or an LC transmission line (power line) or if $\hat{h}(\cdot)$ represents a simple delay, etc. In what follows, we shall give precise conditions under which a scalar $\hat{h}(s)$ (rational or irrational) represents an L_p-stable system. These conditions illustrate one of the chief advantages of the input–output approach to stability, namely, that it places lumped systems [rational $\hat{h}(s)$] and distributed systems [irrational $\hat{h}(s)$] in a *unified* framework. (This is much harder to achieve using Liapunov theory.) Then, once the single-input/single-output case is thoroughly analyzed, the results for the multi-input/multi-output case follow easily.

To do this, we introduce the sets $\mathcal{A}$ and $\hat{\mathcal{A}}$. Basically, (as shown later) $\mathcal{A}$ is the set of stable impulse responses, and $\hat{\mathcal{A}}$ is the set of stable transfer functions. The precise definitions are given next.

1 *[definition]* The symbol $\mathcal{A}$ denotes the set of generalized functions (distributions) of the form

2 $$f(t) = \begin{cases} 0, & t < 0 \\ \sum_{i=0}^{\infty} f_i \delta(t - t_i) + f_a(t), & t \geq 0 \end{cases}$$

where $\delta(\cdot)$ denotes the unit delta distribution, t_i are nonnegative constants, $f_a(\cdot)$ is a measurable function, and, further,

3 $$\sum_{i=0}^{\infty} |f_i| < \infty$$

4 $$\int_0^{\infty} |f_a(t)| \, dt < \infty$$

The *norm* $\|\cdot\|_{\mathcal{A}}$ of a distribution $f(\cdot)$ in $\mathcal{A}$ is defined by

5 $$\|f(\cdot)\|_{\mathcal{A}} = \sum_{i=0}^{\infty} |f_i| + \int_0^{\infty} |f_a(t)| \, dt$$

The *convolution* of two distributions $f(\cdot)$ and $g(\cdot)$ in $\mathcal{A}$, denoted by $f * g$, is defined by

6 $$(f * g)(t) = \int_0^t f(t - \tau) g(\tau) \, d\tau = \int_0^t f(\tau) g(t - \tau) \, d\tau$$

Thus, the set $\mathcal{A}$ consists of all distributions that vanish for $t < 0$, and for $t \geq 0$ consist of a sum of delayed impulses and a measurable function, with the additional property that the weights of the impulses form an absolutely summable sequence and the measurable function is absolutely integrable. One can think of $\mathcal{A}$ as the space L_1 augmented by delayed impulses.

Note that, in computing the convolution of two distributions, one should take

7 $$\delta(t - t_i) * \delta(t - t_j) = \delta(t - t_i - t_j)$$

8 $$\delta(t - t_i) * f_a(t) = f_a(t - t_i)$$

In other words, the convolution of two delayed unit impulses with delays of t_i and t_j, respectively, is another delayed unit impulse of delay $t_i + t_j$, and the convolution of a delayed unit impulse $\delta(t - t_i)$ and a measurable function $f_a(t)$ is the measurable function $f_a(t - t_i)$. Thus, given two elements $f(\cdot)$ and $g(\cdot)$ in $\mathcal{A}$, of the form

9 $$f(t) = \sum_{i=0}^{\infty} f_i \delta(t - t_i^{(f)}) + f_a(t)$$

10 $$g(t) = \sum_{i=0}^{\infty} g_i \delta(t - t_i^{(g)}) + g_a(t)$$

their convolution is given by

11
$$(f * g)(t) = \sum_{i=0}^{\infty} \sum_{j=0}^{\infty} f_i g_j \delta(t - t_i^{(f)} - t_j^{(g)}) + \sum_{i=0}^{\infty} f_i g_a(t - t_i^{(f)}) + \sum_{i=0}^{\infty} g_i f_a(t - t_i^{(g)}) + \int_0^t f_a(t - \tau) g_a(\tau)\, d\tau$$

12 **Example.** The function

13
$$f_1(t) = e^{-\alpha t}$$

belongs to $\mathcal{A}$, whenever $\alpha > 0$. The distribution

14
$$f_2(t) = \sum_{i=0}^{\infty} \frac{1}{(i+1)^2} \delta(t - iT), \qquad T > 0 \text{ a given number}$$

which is a sequence of evenly spaced delayed impulses, belongs to $\mathcal{A}$ because the sequence $[1/(i+1)^2]_{i=0}^{\infty}$ is absolutely summable. However, the distribution

15
$$f_3(t) = \sum_{i=0}^{\infty} \frac{1}{i+1} \delta(t - iT), \qquad T > 0 \text{ a given number}$$

does not belong to $\mathcal{A}$ because the sequence $[1/(i+1)]_0^{\infty}$ is not absolutely summable. The distribution

16
$$f_4(t) = \delta(t) + e^{-t}$$

belongs to $\mathcal{A}$ and has a norm of 2.

17 REMARKS:

1. Note that L_1 is a subset of $\mathcal{A}$; further, if $f(\cdot) \in L_1$, then

18
$$\|f(\cdot)\|_{\mathcal{A}} = \|f(\cdot)\|_1$$

2. If $f(\cdot)$ and $g(\cdot)$ are elements of $\mathcal{A}$, and at least one of them is in L_1 (i.e., does not contain any impulses), then $f * g$ does not contain any impulses. This is clear from (11).

As mentioned previously, the set $\mathcal{A}$ can be interpreted as the set of stable impulse responses; i.e., a system with impulse response $h(\cdot)$ is L_p-stable for all p if and only if $h(\cdot) \in \mathcal{A}$. To prove this important result, we shall first derive some useful properties of $\mathcal{A}$. The set $\mathcal{A}$ is an example of what is known as a *Banach algebra*—this is implied by the properties given below.

19 ***lemma*** The set $\mathcal{A}$, together with the function $\|\cdot\|_{\mathcal{A}}$ and the convolution $*$, has the following properties:

(i) $\|\cdot\|_{\mathcal{A}}$ is a norm on $\mathcal{A}$, and $\mathcal{A}$ is complete under this norm.

(ii) The convolution $*$ is distributive; i.e.,

20 $$f*(g+h) = f*g + f*h, \qquad \forall\, f, g, h \in \mathcal{A}$$

21 $$(f+g)*h = f*h + g*h, \qquad \forall\, f, g, h \in \mathcal{A}$$

(iii) The convolution $*$ is commutative; i.e.,

22 $$f*g = g*f, \qquad \forall\, f, g \in \mathcal{A}$$

(iv) Whenever $f, g \in \mathcal{A}$, we have that $f*g \in \mathcal{A}$, and in fact

23 $$\|f*g\|_{\mathcal{A}} \leq \|f\|_{\mathcal{A}} \cdot \|g\|_{\mathcal{A}}$$

(v) $\delta(\cdot)$ is the *unit* element of $\mathcal{A}$; i.e.,

24 $$\delta * f = f * \delta = f, \qquad \forall\, f \in \mathcal{A}$$

(vi) $\mathcal{A}$ has no divisors of the zero element; i.e.,

25 $$f*g = 0 \Longrightarrow f = 0 \quad \text{or} \quad g = 0$$

***proof* (*outline of*)**

(i) It is easy to verify that $\|\cdot\|_{\mathcal{A}}$ is a norm on $\mathcal{A}$. The completeness of $\mathcal{A}$ under this norm is more difficult and is stated here without proof.

(ii) and (iii) are obvious.

(iv) Suppose $f(\cdot)$ and $g(\cdot)$ are of the form (9) and (10), respectively. Then using the rules of convolution defined above,

26 $$\begin{aligned}(f*g)(t) &= \sum_{i=0}^{\infty}\sum_{j=0}^{\infty} f_i g_j \delta(t - t_i^{(f)} - t_j^{(g)}) \\ &\quad + \sum_{i=0}^{\infty} f_i g_a(t - t_i^{(f)}) + \sum_{i=0}^{\infty} g_i f_a(t - t_i^{(g)}) \\ &\quad + \int_0^t f_a(t-\tau) g_a(\tau)\, d\tau\end{aligned}$$

The first term on the right-hand side represents the distributional part of $f*g$, while the last three terms represent the measurable part of $f*g$. To calculate $\|f*g\|_{\mathcal{A}}$, we take each of the terms separately. First,

27 $$\sum_{i=0}^{\infty}\sum_{j=0}^{\infty} |f_i|\cdot|g_j| = \left(\sum_{i=0}^{\infty}|f_i|\right)\cdot\left(\sum_{j=0}^{\infty}|g_j|\right)$$

Next

28 $$\begin{aligned}\int_0^{\infty}\left|\sum_{i=0}^{\infty} f_i g_a(t - t_i^{(f)})\right| dt &\leq \sum_{i=0}^{\infty}|f_i|\int_0^{\infty}|g_a(t - t_i^{(f)})|\, dt \\ &= \left(\sum_{i=0}^{\infty}|f_i|\right)\cdot\left[\int_0^{\infty}|g_a(t)|\, dt\right]\end{aligned}$$

Similarly,

29 $$\int_0^{\infty}\left|\sum_{i=0}^{\infty} g_i f_a(t - t_i^{(g)})\right| dt \leq \left(\sum_{i=0}^{\infty}|g_i|\right)\cdot\left[\int_0^{\infty}|f_a(t)|\, dt\right]$$

Finally,

30
$$\int_0^\infty \left| \int_0^t f_a(t-\tau) g_a(\tau)\, d\tau \right| dt \leq \int_0^\infty \int_0^t |f_a(t-\tau)| \cdot |g_a(\tau)\, d\tau\, dt$$
$$= \int_0^\infty \int_\tau^\infty |f_a(t-\tau)| \cdot |g_a(\tau)|\, dt\, d\tau$$
(interchanging the order of integration)
$$= \int_0^\infty |g_a(\tau)| \cdot \left[\int_\tau^\infty |f_a(t-\tau)|\, dt \right] d\tau$$
$$= \left[\int_0^\infty |g_a(\tau)|\, d\tau \right] \cdot \left[\int_0^\infty |f_a(t)|\, dt \right]$$

Putting all these inequalities together proves (iv).

(v) is obvious.

(iv) The proof is beyond the scope of this book. ∎

Suppose $f(\cdot) \in \mathcal{A}$. Then, whenever $\text{Re}\, s \geq 0$, the integral

31
$$\hat{f}(s) = \int_0^\infty f(t) e^{-st}\, dt = \sum_{i=0}^\infty f_i e^{-st_i} + \hat{f}_a(s)$$

converges and is well defined. Therefore, all elements of $\mathcal{A}$ are *Laplace-transformable*, and the region of convergence of the Laplace transform includes the closed right half-plane

32
$$C_+ = \{s\colon \text{Re}\, s \geq 0\}$$

With this background, we define the set $\hat{\mathcal{A}}$.

33 *[definition]* The symbol $\hat{\mathcal{A}}$ denotes the set of all functions $\hat{f}\colon C_+ \to C$ that are Laplace transforms of elements of $\mathcal{A}$.

Thus, according to definition (33), $\hat{f}(\cdot) \in \hat{\mathcal{A}}$ is just another way of saying that the inverse Laplace transform of $\hat{f}(\cdot)$ belongs to $\mathcal{A}$. When we deal with feedback systems, the symbol $\hat{\mathcal{A}}$ comes in handy to keep the symbolism from proliferating.

34 **fact** Suppose $\hat{f}(s)$ is a rational function of s. Then $\hat{f}(\cdot) \in \hat{\mathcal{A}}$ if and only if

(i) $\hat{f}$ is proper, and

(ii) All poles of $\hat{f}$ have negative real parts.

proof If (i) and (ii) hold, it is clear that $f(\cdot)$ [the inverse Laplace transform of $\hat{f}(\cdot)$] consists of (i) a measurable function that is bounded by a decaying exponential, and (ii) possibly an impulse at $t = 0$. Hence $f(\cdot) \in \mathcal{A}$, i.e., $\hat{f}(\cdot) \in \hat{\mathcal{A}}$. On the other hand, if (i) does not hold, then $f(\cdot)$ contains higher-order impulses and hence does not belong to $\mathcal{A}$, while if (ii) does not hold, then the measurable part of $f(\cdot)$ is not absolutely integrable. ∎

Having defined $\mathcal{A}$, we can define its *extension* $\mathcal{A}_e$, in exactly the same way that we define L_{pe} from L_p.

35 *[definition]* The set consisting of all generalized functions $f(\cdot)$ which have the property that all truncations $f_T(\cdot)$ of $f(\cdot)$ belong to $\mathcal{A}$ for all finite T is denoted by $\mathcal{A}_e$ and is called the *extension* of $\mathcal{A}$.

The set $\mathcal{A}_e$ has some very useful properties. Most physical systems, even "unstable" systems, have impulse responses that belong to $\mathcal{A}_e$ [e.g., $h(t) = e^t$]. Moreover, it can be shown that if $h(\cdot)$ is any distribution that vanishes for $t < 0$, then the operator H is defined by

36 $$(Hf)(t) = \int_0^t h(t - \tau)f(\tau)\, d\tau$$

maps L_{pe} into L_{pe} for all $p \in [1, \infty]$ if and only if $h(\cdot) \in \mathcal{A}_e$.[9] In other words, the class of linear time-invariant systems that interest us (namely those that map L_{pe} into itself for all p) is exactly the same as the class of systems that have impulse responses in $\mathcal{A}_e$. Thus systems whose impulse responses lie in $\mathcal{A}_e$ are the most general (linear time-invariant) systems that we need to consider.

We shall now prove the main results of this subsection.

37 ***Theorem*** Consider the operator H defined by (36), where $h(\cdot) \in \mathcal{A}_e$. Then the following four statements are equivalent:

(i) H is L_1-stable.
(ii) H is L_∞-stable.
(iii) H is L_p-stable for all $p \in [1, \infty]$.
(iv) $h(\cdot) \in \mathcal{A}$.

REMARKS: Theorem (37) brings out fully the importance of the set $\mathcal{A}$. According to this theorem, a system with impulse response $h(\cdot)$ is (i) L_1-stable, or (ii) L_∞-stable, or (iii) L_p-stable for all $p \in [1, \infty]$ if and only if the impulse $h(\cdot)$ belongs to the set $\mathcal{A}$. This justifies the statement made earlier that $\mathcal{A}$ is the set of stable impulse responses and that $\hat{\mathcal{A}}$ is the set of stable transfer functions.

proof We shall first prove that (iv) implies each of (i), (ii), and (iii). Accordingly, suppose $h(\cdot) \in \mathcal{A}$.

(iv) → (i) If $f(\cdot) \in L_1$, then $f(\cdot) \in \mathcal{A}$, and in fact

38 $$\|f(\cdot)\|_{\mathcal{A}} = \|f(\cdot)\|_1$$

Hence, by (iv) of lemma (17), $h * f \in \mathcal{A}$, and

39 $$\|h * f\|_{\mathcal{A}} \leq \|h\|_{\mathcal{A}} \cdot \|f\|_{\mathcal{A}} = \|h\|_{\mathcal{A}} \|f\|_1$$

[9]The proof of this statement is very similar to that of Theorem (37) and is therefore left as an exercise.

Next, because $f(\cdot)$ contains no impulses, neither does $h*f$, which means that $h*f \in L_1$, and

40 $$\|h*f\|_{\alpha} = \|h*f\|_1$$

Now, (39) and (40) together imply that

41 $$\|h*f\|_1 \leq \|h\|_{\alpha}\cdot\|f\|_1$$

which shows that H is L_1-stable.

(iv) → (ii) Suppose that $f(\cdot) \in L_\infty$ and that $h(\cdot)$ is of the form

42 $$h(t) = \sum_{i=0}^{\infty} h_i\delta(t - t_i) + h_a(t)$$

Then

43 $$(h*f)(t) = \sum_{i=0}^{\infty} h_i f(t - t_i) + \int_0^t h_a(t-\tau)f(\tau)\,d\tau$$

44 $$\begin{aligned}|(h*f)(t)| &\leq \sum_{i=0}^{\infty} |h_i|\cdot|f(t-t_i)| + \int_0^t |h_a(t-\tau)|\cdot|f(\tau)|\,d\tau \\ &\leq \operatorname*{ess.\,sup.}_t |f(t)|\left[\sum_{i=0}^{\infty}|h_i| + \int_0^\infty |h_a(\tau)|\,d\tau\right] \\ &= \|f(\cdot)\|_\infty\cdot\|h(\cdot)\|_{\alpha}\end{aligned}$$

Because (44) holds for (almost) all t, we see that $h*f \in L_\infty$ and that

45 $$\|h*f\|_\infty \leq \|h\|_{\alpha}\cdot\|f\|_\infty$$

This shows that H is L_∞-stable.

(iv) → (iii) This part of the proof is rather involved and can be skipped without loss of continuity. Because we have already established (i) and (ii), we shall concentrate on L_p-stability for $1 < p < \infty$. Suppose $f(\cdot) \in L_p$. We first show that $h*f \in L_p$. Accordingly, suppose $h(\cdot)$ is of the form (42); then $h*f$ is of the form (43). Now, if $f(\cdot) \in L_p$, then clearly the function

46 $$g_1(t) \triangleq \sum_{i=0}^{\infty} h_i f(t-t_i)$$

belongs to L_p, and

47 $$\|g_1\|_p \leq \|f\|_p\cdot\left(\sum_{i=0}^{\infty}|h_i|\right)$$

Thus we concentrate on the integral

48 $$\int_0^\infty h_a(t-\tau)f(\tau)\,d\tau \triangleq g_2(t)$$

Let $q = p/(p-1)$ so that $(1/q) + (1/p) = 1$. Then

49 $$\begin{aligned}|g_2(t)| &\leq \int_0^t |h_a(t-\tau)|\cdot|f(\tau)|\,d\tau \\ &= \int_0^t |h_a(t-\tau)|^{1/q}\,|h_a(t-\tau)|^{1/p}|f(\tau)|\,d\tau \\ &\leq \left\{\int_0^t |h_a(t-\tau)|\,d\tau\right\}^{1/q}\cdot\left[\int_0^t |h_a(t-\tau)|\cdot|f(\tau)|^p\,d\tau\right]^{1/p}\end{aligned}$$

by Holder's inequality (fact [6.1(13)]). Hence

50
$$|g_2(t)|^p \leq \left[\int_0^t |h_a(t-\tau)|\,d\tau\right]^{p/q} \int_0^t |h_a(t-\tau)|\cdot|f(\tau)|^p\,d\tau$$
$$\leq \left[\int_0^\infty |h_a(\tau)|\,d\tau\right]^{p/q} \int_0^t |h_a(t-\tau)|\cdot|f(\tau)|^p\,d\tau$$

Now, if $f(\cdot) \in L_p$, then the function $\tau \mapsto |f(\tau)|^p \in L_1$. Thus, using the estimate (41), we get

51
$$\int_0^\infty |g_2(t)|^p\,dt \leq \left[\int_0^\infty |h_a(\tau)|\,d\tau\right]^{p/q} \cdot \int_0^\infty |h_a(\tau)|\,d\tau \cdot \int_0^\infty |f(\tau)|^p\,d\tau$$
$$= \left[\int_0^\infty |h_a(\tau)|\,d\tau\right]^p \int_0^\infty |f(\tau)|^p\,d\tau$$

because $(p/q) + 1 = p$. Raising both sides of (51) to the power $1/p$ gives

52
$$\|g_2\|_p \leq \int_0^\infty |h_a(\tau)|\,d\tau \cdot \|f\|_p$$

Because $h * f = g_1 + g_2$, (47) and (52) show that $h * f \in L_p$. Moreover,

53
$$\|h * f\|_p \leq \|g_1\|_p + \|g_2\|_p$$
$$\leq \|f\|_p \left[\sum_{i=0}^\infty |h_i| + \int_0^\infty |h_a(\tau)|\,d\tau\right]$$
$$= \|f\|_p \cdot \|h\|_{\mathcal{A}}$$

This shows that H is L_p-stable for all p.

Up to now, we have shown that (iv) implies (i), (ii), and (iii). We shall now prove the converse, namely, that (i), (ii), or (iii) each implies (iv).

(i) → (iv) Suppose (i) is true. Because H is linear, (i) implies that H is continuous. Now, because every element in $\mathcal{A}$ can be approximated arbitrarily closely (in the sense of distributions) by an element of L_1, this implies that H maps $\mathcal{A}$ into itself. Let $f(t) = \delta(t)$; then $h * f \in \mathcal{A}$, because H maps $\mathcal{A}$ into itself. However, $h * \delta = h$, which shows that $h \in \mathcal{A}$. Thus (i) implies (iv).

(ii) → (iv) This is a special case of a result contained in Theorem (70).

(iii) → (iv) Suppose (iii) is true, i.e., that H is L_p-stable for all $p \in [1, \infty]$. Then, in particular, H is L_1-stable. As shown above, this implies (iv). ∎

54 ***Theorem*** Consider the operator H defined by (36), where $h(\cdot) \in \mathcal{A}$. Then H is L_2-stable, and

55
$$\|h * f\|_2 \leq m \cdot \|f\|_2$$

where

56
$$m = \sup_\omega |\hat{h}(j\omega)|$$

REMARKS: The main purpose of Theorem (54) is to prove the bound (55). For an arbitrary $p \in [1, \infty]$, we have the inequality

(53). However, for $p = 2$, we can establish the tighter bound (55) (note that $m \leq \|h\|_{\mathcal{A}}$).

proof Let $g = h * f$. Then

57 $$\hat{g}(j\omega) = \hat{h}(j\omega)\hat{f}(j\omega)$$

Using Parseval's equality, we get

58 $$\|g\|_2^2 = \frac{1}{2\pi}\int_{-\infty}^{\infty} |\hat{g}(j\omega)|^2\, d\omega = \frac{1}{2\pi}\int_{-\infty}^{\infty} |\hat{h}(j\omega)|^2 |\hat{f}(j\omega)|^2\, d\omega$$

$$\leq \frac{m^2}{2\pi}\int_{-\infty}^{\infty} |\hat{f}(j\omega)|^2\, d\omega = m^2 \|f\|_2^2$$

Taking the square root of both sides of (58) establishes (55). ∎

We shall summarize the results of this subsection up to now:

1. We have introduced the sets $\mathcal{A}$ and $\hat{\mathcal{A}}$.
2. We have shown that a system with impulse response $h(\cdot)$ is (i) L_1-stable, (ii) L_∞-stable, (iii) L_p-stable for all $p \in [1, \infty]$ if and only if $h(\cdot) \in \mathcal{A}$.
3. We have established the useful bounds (53) and (55). These show that, given a system with impulse response $h(\cdot) \in \mathcal{A}$, the L_p-norm of the output $h * f$ is no larger than $\|h\|_{\mathcal{A}}$ times the L_p-norm of the input f; in the particular case where $f \in L_2$, a still tighter bound can be obtained.

Multi-Input/Multi-Output Systems

We shall turn now to systems with multiple inputs and outputs. The results in this case follow very easily from Theorem (37).

Consider a system with n inputs and m outputs, with the input/output relationship

59 $$(\mathbf{Hf})(t) \triangleq \int_0^t \mathbf{H}(t-\tau)\mathbf{f}(\tau)\, d\tau$$

where the impulse response matrix $\mathbf{H}(\cdot) \in \mathcal{A}_e^{m\times n}$;[10] i.e., the entries of the $m \times n$ matrix of distributions $\mathbf{H}$ all belong to $\mathcal{A}_e$. In analogy with the scalar (single-input/single-output) case, such types of operators $\mathbf{H}$ are the most general linear time-invariant operators that we need to consider.

The criteria for the L_p stability of systems of the form (59) are given next. Note that $\mathcal{A}^{m\times n}$ denotes the set of all $m \times n$ matrices of distributions in $\mathcal{A}$.

[10] To avoid confusion, we write $\mathbf{H}$ for the operator and $\mathbf{H}(\cdot)$ for the associated impulse response.

60 ***Theorem*** Consider an operator $\mathbf{H}$ of the form (59), where $\mathbf{H}(\cdot) \in \mathcal{A}_e^{m \times n}$. Under these conditions, the following statements are equivalent

(i) $\mathbf{H}$ is L_1-stable.

(ii) $\mathbf{H}$ is L_∞-stable.

(iii) $\mathbf{H}$ is L_p-stable for all $p \in [1, \infty]$.

(iv) $\mathbf{H}(\cdot) \in \mathcal{A}^{m \times n}$.

The proof is left as an exercise, because it is entirely parallel to that of Theorem (37).

Basically, Theorem (60) states that an n-input/m-output system with the impulse response matrix $\mathbf{H}(\cdot)$ is (i) L_1-stable, (ii) L_∞-stable, (iii) L_p-stable for all $p \in [1, \infty]$ if and only if each component of $\mathbf{H}(\cdot)$ belongs to $\mathcal{A}$.

In proving Theorem (39), we obtained the useful bound (53). A similar bound can be obtained for multivariable systems also and is given below without proof. In this context, it is worthwhile to recall the definition of the norm on L_p^n (definition [6.1(27)].

61 ***Theorem*** Consider an operator of the form (59), where $\mathbf{H}(\cdot) \in \mathcal{A}^{m \times n}$. Then, whenever $\mathbf{f}(\cdot) \in L_p^n$ for some $p \in [1, \infty]$ we have

62 $$||\mathbf{Hf}|| \leq \alpha_1 ||\mathbf{f}||_p$$

where

63 $$\alpha_1 = ||\mathbf{M}_1||_{i2}$$

$||\cdot||_{i2}$ indicates the l_2-induced matrix norm on $R^{m \times n}$, and $\mathbf{M}_1$ is the $m \times n$ matrix whose ijth entry is $||h_{ij}(\cdot)||_{\mathcal{A}}$. If $p = 2$, then

64 $$||\mathbf{Hf}||_2 \leq \alpha_2 ||\mathbf{f}||_2$$

where

65 $$\alpha_2 = ||\mathbf{M}_2||_{i2}$$

and $\mathbf{M}_2$ is the $m \times n$ matrix whose ijth entry is given by

66 $$(\mathbf{M}_2)_{ij} = \sup_{\omega} |\hat{h}_{ij}(j\omega)|, \qquad i = 1, \ldots, m, j = 1, \ldots, n$$

To explain in words, the bounding constants α_1 and α_2 are calculated as follows: Take the $m \times n$ matrix $\mathbf{H}(\cdot)$ and replace each element by its norm in $\mathcal{A}$. The resulting matrix is $\mathbf{M}_1$, and α_1 is the l_2-induced norm of $\mathbf{M}_2$. It is easy to verify that the bounds (62) and (64) reduce to (53) and (55), respectively, in the case of scalar systems.

6.5.2 Time-Varying Systems

In the previous subsection, we studied linear time-invariant operators of the form (36). The class of operators studied in this subsection represents a natural generalization of those of the form (36). Specifically, we shall consider operators of the form

67 $$(Gf)(t) = \sum_{i=0}^{\infty} g_i(t)f(t - t_i) + \int_0^t g_a(t, \tau)f(\tau)\, d\tau$$

Actually, because $f(t) = 0$ whenever $t < 0$, we can rewrite (67) as

68 $$(Gf)(t) = \sum_{i \in I(t)} g_i(t)f(t - t_i) + \int_0^t g_a(t, \tau)f(\tau)\, d\tau$$

where

69 $$I(t) = \{i\colon t_i \leq t\}$$

In other words, (68) is obtained from (67) by taking the summation only over those indices i such that $t_i \leq t$.

Theorem (70) gives necessary and sufficient conditions for an operator G of the form (68) to be L_∞-stable.

70 ***Theorem*** Consider an operator G of the form (67), where

71 $$t \mapsto \sum_{i \in I(t)} |g_i(t)| \in L_{\infty e}$$

72 $$\tau \mapsto g_a(t, \tau) \in L_1, \qquad \forall\, t \geq 0$$

73 $$t \mapsto \int_0^t |g_a(t, \tau)|\, d\tau \in L_{\infty e}$$

Then G maps $L_{\infty e}$ into itself. Further, G is L_∞-stable if and only if

74 $$\sup_t \left\{\sum_{i \in I(t)} |g_i(t)| + \int_0^t |g_a(t, \tau)|\, d\tau\right\} \triangleq c_\infty < \infty$$

proof We leave it as a problem to show that if (71)–(73) hold, then G maps $L_{\infty e}$ into itself.

To show that (74) is a necessary and sufficient condition for L_∞-stability, we shall first tackle the sufficiency part, which is simpler to prove.

(i) "*If*" Suppose that (74) holds and that $f(\cdot) \in L_\infty$. Then

75 $$\begin{aligned} |(Gf)(t)| &= \left|\sum_{i \in I(t)} g_i(t)f(t - t_i) + \int_0^t g_a(t, \tau)f(\tau)\, d\tau\right| \\ &\leq \left\{\sum_{i \in I(t)} |g_i(t)| + \int_0^t |g_a(t, \tau)|\, d\tau\right\} \cdot \|f(\cdot)\|_\infty \\ &\leq c_\infty \|f(\cdot)\|_\infty \end{aligned}$$

Because the right-hand side of (75) is independent of t, we see that $(Gf)(\cdot) \in L_\infty$ and that

76 $$\|(Gf)(\ )\|_\infty \leq c_\infty \|f(\cdot)\|_\infty$$

Hence G is L_∞-stable.

(ii) "*Only If*" We shall show the contrapositive, namely, that if (74) does not hold, then $\|(Gf)\|_\infty / \|f\|_\infty$ can be made arbitrarily large by suitably selecting $f(\cdot)$. Accordingly, suppose the function

77 $$r(t) \triangleq \sum_{i \in I(t)} |g_i(t)| + \int_0^t |g_a(t, \tau)|\, d\tau$$

is unbounded. Let k be an arbitrary constant; we show that a function $f_k(\cdot) \in L_\infty$ can be found such that $\|f_k(\cdot)\|_\infty = 1$ and $\|Gf_k\|_\infty \geq k$. The first step is to pick some $\epsilon > 0$ and to choose θ_k such that $r(\theta_k) \geq k + \epsilon$. This is always possible because $r(\cdot)$ is unbounded. Next, pick intervals δ_i of the form

78
$$\delta_i = [\theta_k - t_i - \sigma_i, \theta_k - t_i + \sigma_i], \qquad i \in I(\theta_k)$$

such that

79
$$\sum_{i \in I(\theta_k)} \int_{\tau \in \delta_i} |g_a(\theta_k, \tau)| \, d\tau \leq \frac{\epsilon}{2}$$

Clearly, this can always be accomplished by choosing the numbers σ_i sufficiently small. Finally, let $f_k(\cdot)$ be the function defined by

80
$$f_k(t) = \begin{cases} \text{sign } g_i(\theta_k), & \text{if } t \in \delta_i \\ \text{sign } g_a(\theta_k, t), & \text{if } t \notin \bigcup_{i \in I(\theta_k)} \delta_i \end{cases}$$

Then $f_k(\cdot) \in L_\infty$ and $\|f_k(\cdot)_\infty\| = 1$. Moreover, we have

81
$$(Gf_k)(\theta_k) = \sum_{i \in I(\theta_k)} g_i(\theta_k) f_k(\theta_k - t_i) + \int_0^{\theta_k} g_a(\theta_k, \tau) f_k(\tau) \, d\tau$$
$$= \sum_{i \in I(\theta_k)} g_i(\theta_k) f_k(\theta_k - t_i) + \sum_{i \in I(\theta_k)} \int_{\tau \in \delta_i} g_a(\theta_k, \tau) f_k(\tau) \, d\tau$$
$$+ \int_{\substack{\tau \in [0, \theta_k] \\ \tau \notin \bigcup_{i \in I(\theta_k)} \delta_i}} g_a(\theta_k, \tau) f_k(\tau) \, d\tau$$

Thus, rearranging, we get

82
$$(Gf_k)(\theta_k) \geq \sum_{i \in I(\theta_k)} |g_i(\theta_k)| + \int_0^{\theta_k} |g_a(\theta_k, \tau)| \, d\tau$$
$$- \sum_{i \in I(\theta_k)} \int_{\tau \in \delta_i} |g_a(\theta_k, \tau)| \, d\tau$$
$$- \sum_{i \in I(\theta_k)} \int_{\tau \in \delta_i} |g_a(\theta_k, \tau)| \cdot |f_k(\tau)| \, d\tau$$

However, because $|f_k(\tau)| \leq 1$ for all τ, the last inequality can be simplified to

83
$$(Gf_k)(\theta_k) \geq \left\{ \sum_{i \in I(\theta_k)} g_i(\theta_k) + \int_0^{\theta_k} |g_a(\theta_k, \tau)| \, d\tau \right\}$$
$$- \left\{ 2 \sum_{i \in I(\theta_k)} \int_{\tau \in \delta_i} |g_a(\theta_k, \tau)| \, d\tau \right\}$$
$$\geq k + \epsilon - \epsilon = k$$

because the first term in the braces is not less than $k + \epsilon$ by the way we chose θ_k, while the second term in the braces is not more than ϵ by (79). Finally,

84
$$\|Gf_k\| = \underset{t \geq 0}{\text{ess. sup.}} |(Gf_k)(t)| \geq |(Gf_k)(\theta_k)| \geq k$$

Because this process can be repeated for any arbitrary k, we see that G is not L_∞-stable. This shows that (74) is a necessary condition for G to be L_∞-stable. ∎

The next theorem gives necessary and sufficient conditions for G to be L_1-stable.

85 ***Theorem*** Consider an operator of the form (68), where

86 $$t \mapsto \sum_{i \in I(t)} |g_i(t + t_i)| \in L_{\infty e}$$

87 $$t \mapsto g_a(t, \tau) \in L_1, \qquad \forall\, \tau \geq 0$$

88 $$\tau \mapsto \int_\tau^\infty |g_a(t, \tau)|\, dt \in L_{\infty e}$$

Under these conditions, G maps L_{1e} into itself. Further, G is L_1-stable if and only if

89 $$\sup_\tau \left\{ \sum_{i=0}^\infty |g_i(\tau + t_i)| + \int_\tau^\infty |g_a(t, \tau)|\, dt \right\} \triangleq c_1 < \infty$$

proof We leave it as an exercise to show that if (86)–(88) hold, then G maps L_{1e} into itself.

We shall first prove the sufficiency of (89) for L_1-stability.

(i) "*If*" Suppose that (89) holds and that $f(\cdot) \in L_1$. Then

90 $$\begin{aligned}
\int_0^\infty |(Gf)(t)|\, dt &\leq \int_0^\infty \sum_{i \in I(t)} |g_i(t)| \cdot |f(t - t_i)|\, dt \\
&\quad + \int_0^\infty \int_0^t |g_a(t, \tau)| \cdot |f(\tau)|\, d\tau\, dt \\
&= \sum_{i=0}^\infty \int_0^\infty |g_i(t)| \cdot |f(t - t_i)|\, dt \\
&\quad + \int_0^\infty \int_\tau^\infty |g_a(t, \tau)| \cdot |f(\tau)|\, dt\, d\tau \\
&\leq \sum_{i=0}^\infty \int_0^\infty |g_i(\tau + t_i)| \cdot |f(\tau)|\, d\tau \\
&\quad + \int_0^\infty \int_\tau^\infty |g_a(t, \tau)| \cdot |f(\tau)|\, dt\, d\tau \\
&\leq \left[\sup_\tau \sum_{i=0}^\infty |g_i(\tau + t_i)| \right. \\
&\quad \left. + \int_\tau^\infty |g_a(t, \tau)|\, dt \right] \cdot \left\{ \int_0^\infty |f(\tau)|\, d\tau \right\} \\
&= c_1 \, \|f(\cdot)\|_1
\end{aligned}$$

This shows that G is L_1-stable.

(ii) "*Only If*" Suppose G is L_1-stable. Because G is linear, this implies that G is continuous on L_1 and hence that G maps $\mathcal{A}$ into $\mathcal{A}$. Let

$f(t) = \delta(t - \tau)$, where $\tau > 0$ is a given number. Then

91
$$(Gf)(t) = \sum_{i \in I(t)} g_i(t)\delta(t - \tau - t_i) + g_a(t, \tau)$$
$$= \sum_{i=0}^{\infty} g_i(\tau + t_i)\delta(t - \tau - t_i) + g_a(t, \tau)$$

By assumption, $(Gf)(\cdot) \in \mathfrak{A}$; i.e.,

92
$$\sum_{i=0}^{\infty} |g_i(\tau + t_i)| + \int_{\tau}^{\infty} |g_a(t, \tau)|\, dt < \infty$$

where we use the fact that $g_a(t, \tau) = 0$ for $t < \tau$. Now, the L_1-stability of G not only implies that (92) holds for each fixed τ but also that $\|Gf(\cdot)\|_{\mathfrak{A}}$ is bounded independently of τ; i.e.,

93
$$\sup_{\tau} \left\{ \sum_{i=0}^{\infty} |g_i(\tau + t_i)| + \int_{\tau}^{\infty} |g_a(t, \tau)|\, dt \right\} < \infty$$

which establishes (89). ∎

Finally, we shall show that if G of (68) is both L_1-stable and L_∞-stable, then it is L_p-stable for all $p \in [1, \infty]$.

94 ***Theorem*** Suppose an operator G of the form (68) satisfies both (74) and (89). Then G is L_p-stable for all $p \in [1, \infty]$. Further, whenever $f(\cdot) \in L_p$, we have

95
$$\|Gf\|_p \leq c_1^{1/p} c_\infty^{1/q} \|f\|_p$$

where $q = p/(p - 1)$ is the conjugate index of p.

proof We shall give the proof only in the case where $g_i \equiv 0$ for all i. The proof in the general case is similar.

If we assume that $g_i \equiv 0$ for all i, then (74) and (89) reduce, respectively, to

96
$$\sup_t \int_0^t |g_a(t, \tau)|\, d\tau \triangleq c_\infty < \infty$$

97
$$\sup_{\tau} \int_{\tau}^{\infty} |g_a(t, \tau)|\, dt \triangleq c_1 < \infty$$

Suppose $f(\cdot) \in L_p$. We can assume that $1 < p < \infty$, because (95) is clearly true if $p = 1$ or ∞. Then

98
$$|(Gf)(t)| \leq \int_0^t |g_a(t, \tau)| \cdot |f(\tau)|\, d\tau$$
$$= \int_0^t |g_a(t, \tau)|^{1/q} |g_a(t, \tau)|^{1/p} |f(\tau)|\, d\tau$$
$$\leq \left\{ \int_0^t |g_a(t, \tau)|\, d\tau \right\}^{1/q} \left\{ \int_0^t |g_a(t, \tau)| \cdot |f(\tau)|^p\, d\tau \right\}^{1/p}$$

by Holder's inequality. Next, we have

99
$$|(Gf)(t)|^p \le \left\{\int_0^t |g_a(t,\tau)|\, d\tau\right\}^{p/q} \left\{\int_0^t |g_a(t,\tau)|\cdot|f(\tau)|^p\, d\tau\right\}$$
$$\le c_\infty^{p/q} \cdot \int_0^t |g_a(t,\tau)|\cdot|f(\tau)|^p\, d\tau$$

100
$$\int_0^\infty |(Gf)(t)|^p\, dt \le c_\infty^{p/q} \int_0^\infty \int_0^t |g_a(t,\tau)|\cdot|f(\tau)|^p\, d\tau\, dt$$
$$= c_\infty^{p/q} \int_0^\infty \int_\tau^\infty |g_a(t,\tau)|\cdot|f(\tau)|^p\, dt\, d\tau$$
$$= c_\infty^{p/q} \int_0^\infty \left[\int_\tau^\infty |g_a(t,\tau)|\, dt\right] |f(\tau)|^p\, d\tau$$
$$\le c_\infty^{p/q} c_1 \|f\|_p^p$$

Raising both sides of (100) to the power $1/p$ gives (95). ∎

We shall summarize the results of this subsection up to now:

1. We have given a set of necessary and sufficient conditions for an operator G of the form (68) to be L_1-stable and to be L_∞-stable.
2. We have shown that if G is L_1-stable and L_∞-stable, then it is L_p-stable for all $p \in [1, \infty]$.
3. We have proved the useful inequality (95), which gives an upper bound for the L_p-norm of $(Gf)(\cdot)$.

We shall conclude this subsection by showing that if G is a *time-invariant* operator, then Theorems (70), (85), and (94) together reduce to Theorem (37). Accordingly, suppose

101
$$g_i(t) \equiv h_i \quad \text{(a constant)}, \qquad \forall\, t \ge 0$$

102
$$g_a(t,\tau) = h(t-\tau), \qquad \forall\, t, \tau \ge 0$$

so that G corresponds to a system with impulse response

103
$$h(t) = \sum_{i=0}^{\infty} h_i \delta(t - t_i) + h_a(t)$$

Then the condition (74) for L_∞-stability becomes

104
$$\sup_t \left\{\sum_{i \in I(t)} |h_i| + \int_0^t |h_a(t-\tau)|\, d\tau\right\} < \infty$$

However, as $t \to \infty$, $I(t)$ eventually includes all $i \ge 0$. Thus (104) is equivalent to requiring $h(\cdot)$ to belong to $\mathcal{A}$. Similarly, the condition (89) for L_1-stability becomes

105
$$\sup_\tau \left\{\sum_{i=0}^{\infty} |h_i| + \int_\tau^\infty |h_a(t-\tau)|\, d\tau\right\} < \infty$$

which also is equivalent to requiring $h(\cdot)$ to belong to $\mathcal{A}$. Finally, because we have

$$c_1 = c_\infty = \|h(\cdot)\|_{\mathcal{A}} \tag{106}$$

in this case, (95) reduces to (53).

Problem 6.6. Suppose $f(\cdot) \in \mathcal{A}$. Show that whenever $s \in C_+$ we have

$$|\hat{f}(s)| \leq \|f(\cdot)\|_{\mathcal{A}}$$

Problem 6.7. Show that if $\hat{f}(\cdot) \in \hat{\mathcal{A}}$, then the function $s \mapsto e^{-sT}\hat{f}(s)$ also belongs to $\hat{\mathcal{A}}$ for all positive T. [*Hint:* consider the inverse Laplace transform of $e^{-st}\hat{f}(s)$.]

Problem 6.8. Prove Theorem (60).

Problem 6.9. Determine whether each of the functions below belongs to $\mathcal{A}$ and/or to $\mathcal{A}_e$. If $f(\cdot) \in \mathcal{A}$, find $\|f(\cdot)\|_{\mathcal{A}}$.

(i) $f(t) = \sum_{i=0}^{\infty} \delta(t - iT),\ T > 0$

(ii) $f(t) = \delta(t - T) + e^{-at} \sin \omega t,\ T, a, \omega > 0.$

(iii) $f(t) = \sum_{i=0}^{\infty} e^{-it/T}$

Problem 6.10. Determine whether each of the functions below belongs to $\hat{\mathcal{A}}$.

(i) $\hat{f}(s) = e^{-sT}\dfrac{s^2 + 5s + 5}{s^2 + s + 10}$

(ii) $\hat{f}(s) = 1/\cosh\sqrt{s}$ (Hint: use partial fractions.)

Problem 6.11. Determine whether each of the operators below is L_1-stable and L_∞-stable, using Theorems (70) and (85).

(i) $(Hu)(t) = u(t - \delta) + \int_0^t \sin t\, e^{-2(t-\tau)} u(\tau)\, d\tau,\ \delta > 0$

(ii) $(Hu)(t) = \int_0^t e^{(-t+2\tau)} u(\tau)\, d\tau$

6.6 LINEAR TIME-INVARIANT FEEDBACK SYSTEMS

In this section, we study the conditions under which a feedback interconnection of linear time-invariant subsystems results in a stable system. Since the results for linear time-invariant systems are the ones most often used in practice, and since many of the stability results for nonlinear and/or time-varying systems use the results for linear time-invariant systems as a point of departure, it is important to have a good understanding of what makes a linear time-invariant feedback system stable or unstable.

This section is divided into two subsections. In Sec. 6.6.1, we deal exclusively with single-input single-output systems. The emphasis here is on the properties of the sets $\mathcal{A}$ and $\hat{\mathcal{A}}$, and how they can be exploited to obtain necessary and sufficient conditions for feedback stability. In Sec. 6.6.2, we tackle multi-input/multi-output systems, where some additional difficulties arise from dealing with matrix transfer functions.

6.6.1 Single-Input/Single-Output Systems

In this subsection, we study systems of the form shown in Fig. 6.6. Initially, we assume that $\hat{g}_1, \hat{g}_2 \in \hat{\mathcal{A}}$, i.e., that the system is "open-loop stable." We then consider the case where $\hat{g}_2$ is a constant, and $\hat{g}_1$ contains a finite number of right half-plane poles, but is otherwise stable. In both cases, we present necessary and sufficient conditions for stability that involve the behavior of $\hat{g}_1(s)$ and $\hat{g}_2(s)$ as s varies over the closed right half-plane. Next, we present a graphical procedure, whereby the stability conditions can be tested by examining only $\hat{g}_1(j\omega)$ and $\hat{g}_2(j\omega)$. This graphical procedure is a natural generalization of the standard Nyquist criterion. As an application, we show how the results presented here can be used to study the stability of networks containing uniform or nonuniform *RC* transmission lines and operational amplifiers.

The following result is the key tool used to derive all the stability criteria of this section.

1 ***lemma*** (*Paley–Wiener*) Suppose $\hat{p}(\cdot) \in \hat{\mathcal{A}}$. Then the function $\hat{q}(\cdot) = 1/\hat{p}(\cdot)$ belongs to $\hat{\mathcal{A}}$ if and only if

2 $$\inf_{s \in C_+} |\hat{p}(s)| > 0$$

where C_+ is the closed right half-plane, defined by

3 $$C_+ = \{s: \text{Re}\, s \geq 0\}$$

proof

(i) "*Only If*" Suppose $\hat{q} = 1/\hat{p} \in \hat{\mathcal{A}}$, and let $q(\cdot)$ denote the inverse Laplace transform of $\hat{q}(\cdot)$. Then by Problem 6.6, we have

4 $$|\hat{q}(s)| \leq \|q(\cdot)\|_{\mathcal{A}}, \qquad \forall\, s \in C_+$$

Therefore

5 $$\left|\frac{1}{\hat{p}(s)}\right| < \|q(\cdot)\|_{\mathcal{A}}, \qquad \forall\, s \in C_+$$

But this shows that

6 $$|\hat{p}(s)| \geq \frac{1}{\|q(\cdot)\|_{\mathcal{A}}} > 0, \qquad \forall\, s \in C_+$$

which establishes (2).

(ii) "*If*" The proof of the "if" part is far beyond the scope of this book and can be found in [14] Hille and Phillips [p. 150]. ▬

Lemma (1) states the following: Suppose $\hat{p}(\cdot)$ is the Laplace transform of an element of $\mathfrak{A}$. Then its reciprocal $1/\hat{p}(\cdot)$ is also the Laplace transform of an element of $\mathfrak{A}$ if and only if $\hat{p}(\cdot)$ is bounded away from zero over the right half-plane. In particular, this means that (1) $\hat{p}(\cdot)$ has no zeroes in C_+, and (2) no sequence $(s_i)_1^\infty$ in C_+ with $|s_i| \longrightarrow \infty$ can be found such that $\hat{p}(s_i) \longrightarrow 0$ as $i \longrightarrow \infty$.

Even though the "if" part of lemma (1) is very difficult to prove in the general case where $\hat{p}$ is an arbitrary element of $\hat{\mathfrak{A}}$, it is very easy to prove in the special case where $\hat{p}(s)$ is a rational function of s.

7 **fact** Suppose $\hat{p}(s)$ is a rational function of s and belongs to $\hat{\mathfrak{A}}$ [i.e., $\hat{p}(s)$ is proper and has no poles in C_+]. Then $1/\hat{p}(s)$ belongs to $\hat{\mathfrak{A}}$ if and only if (1) $\hat{p}$ has no zeroes in C_+, and (2) $\hat{p}(\infty) \neq 0$ [i.e., if (2) holds].

The proof is left as an exercise.

Notice that in proving the "only if" part of lemma (1) we never made use of the fact that $\hat{p} \in \hat{\mathfrak{A}}$. In view of this, we can state the following.

8 ***lemma*** Suppose $p(\cdot)$ is a distribution, and let $\hat{p}(\cdot)$ denote its Laplace transform. Under these conditions, if $1/\hat{p}(\cdot)$ belongs to $\hat{\mathfrak{A}}$, then (2) holds.

Lemma (8) states that if $\hat{p}$ is *any* Laplace transform, then (2) is a necessary condition for $1/\hat{p}$ to belong to $\hat{\mathfrak{A}}$ (although it may not be sufficient).

Consider now the feedback system shown in Fig. 6.6. By definition [6.3(34)], this system is L_p-stable if $y_1, y_2 \in L_p$ whenever $u_1, u_2 \in L_p$. Now, note that

9 $$\hat{y}_1(s) = \frac{\hat{g}_1(s)}{1 + \hat{g}_1(s)\hat{g}_2(s)}\hat{u}_1(s) - \frac{\hat{g}_1(s)\hat{g}_2(s)}{1 + \hat{g}_1(s)\hat{g}_2(s)}\hat{u}_2(s)$$

10 $$\hat{y}_2(s) = \frac{\hat{g}_1(s)\hat{g}_2(s)}{1 + \hat{g}_1(s)\hat{g}_2(s)}\hat{u}_1(s) - \frac{\hat{g}_2(s)}{1 + \hat{g}_1(s)\hat{g}_2(s)}\hat{u}_2(s)$$

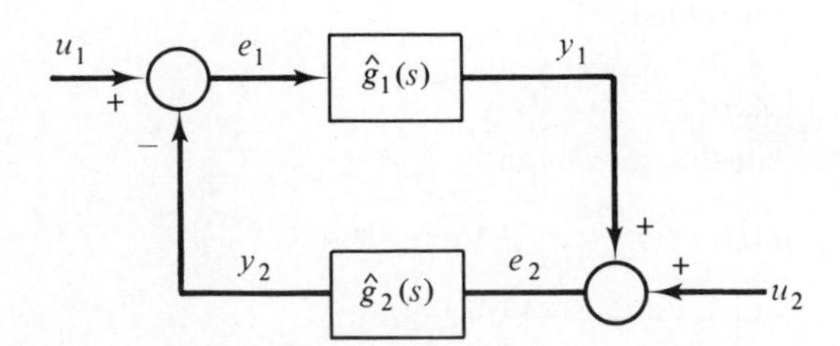

FIG. 6.6

We can express these equations in matrix form as

11 $$\begin{bmatrix} \hat{y}_1(s) \\ \hat{y}_2(s) \end{bmatrix} = \begin{bmatrix} \hat{h}_{11}(s) & \hat{h}_{21}(s) \\ \hat{h}_{12}(s) & \hat{h}_{22}(s) \end{bmatrix} \begin{bmatrix} \hat{u}_1(s) \\ \hat{u}_2(s) \end{bmatrix}$$

where the definition of $\hat{h}_{ij}(s)$ is obvious. Now, from Theorem [6.5(60)], we have the following result.

12 **fact** The system described by (9)–(10) is L_∞-stable[11] if and only if $\hat{h}_{ij}(\cdot) \in \hat{\mathfrak{A}}$, for $i, j = 1, 2$.

Thus the objective of this section is to derive necessary and sufficient conditions for $\hat{h}_{ij}(\cdot)$ to belong to $\hat{\mathfrak{A}}$, for $i, j = 1, 2$. We shall tackle first the so-called *open-loop stable* case, i.e., where $\hat{g}_1, \hat{g}_2$ belong to $\hat{\mathfrak{A}}$, because this is the easiest case.

13 ***lemma*** Suppose $\hat{g}_1, \hat{g}_2 \in \hat{\mathfrak{A}}$. Then $\hat{h}_{ij}(\cdot) \in \hat{\mathfrak{A}}$ for $i, j = 1, 2$ if and only if

14 $$\hat{r}(\cdot) \triangleq \frac{1}{1 + \hat{g}_1(\cdot)\hat{g}_2(\cdot)} \in \hat{\mathfrak{A}}$$

proof We shall do the proof in some detail, because the manipulations carried out below are repeated, in essentially the same form, in several later proofs.

(i) Suppose (14) is true. Clearly the function

15 $$\hat{f}(s) \equiv 1$$

belongs to $\hat{\mathfrak{A}}$, because it is the Laplace transform of the unit impulse. Therefore, if $\hat{r} \in \hat{\mathfrak{A}}$, then $1 - \hat{r} \in \hat{\mathfrak{A}}$, because $\hat{\mathfrak{A}}$ is a linear space. However,

16 $$1 - \hat{r}(s) = \frac{\hat{g}_1(s)\hat{g}_2(s)}{1 + \hat{g}_1(s)\hat{g}_2(s)}$$

This shows that $\hat{h}_{12}, \hat{h}_{21} \in \hat{\mathfrak{A}}$. Next, we have

17 $$\hat{h}_{11}(s) = \hat{g}_1(s)\hat{r}(s)$$

Because $\hat{g}_1$ and $\hat{r}$ both belong to $\hat{\mathfrak{A}}$, so does their product. (See lemma [6.5(19)].) Hence $\hat{h}_{11} \in \hat{\mathfrak{A}}$. Finally, $\hat{h}_{22} \in \mathfrak{A}$, because

18 $$\hat{h}_{22}(s) = \hat{g}_2(s)\hat{r}(s)$$

(ii) "*Only If*" Suppose $\hat{h}_{ij}(\cdot) \in \hat{\mathfrak{A}}$ for $i, j = 1, 2$. Then, in particular, $\hat{h}_{21}(\cdot) \in \hat{\mathfrak{A}}$. This means that $1 - \hat{h}_{21} \in \hat{\mathfrak{A}}$. However, $1 - \hat{h}_{21} = \hat{r}$. This shows that $\hat{r} \in \hat{\mathfrak{A}}$. ∎

The utility of fact (13) lies in that it leads to a simple necessary and sufficient condition for the stability of the system (9)–(10).

[11] Or, equivalently, (1) L_1-stable or (2) L_p-stable for all $p \in [1, \infty]$.

19 ***Theorem*** Suppose $\hat{g}_1, \hat{g}_2 \in \mathcal{A}$. Then the system (9)–(10) is L_∞-stable[12] if and only if

$$\inf_{s \in C_+} |1 + \hat{g}_1(s)\hat{g}_2(s)| > 0 \tag{20}$$

proof By fact (13), the system (9)–(10) is L_∞-stable if and only if $\hat{r} \in \hat{\mathcal{A}}$. Now, because $\hat{g}_1, \hat{g}_2 \in \hat{\mathcal{A}}$, we have that $1 + \hat{g}_1\hat{g}_2 \in \hat{\mathcal{A}}$. Hence, by lemma (1), $\hat{r} = 1/(1 + \hat{g}_1\hat{g}_2) \in \hat{\mathcal{A}}$ if and only if (20) holds. ∎

In engineering terms, the quantity $1 + \hat{g}_1(s)\hat{g}_2(s)$ is usually referred to as the *return difference*. Thus Theorem (19) states that the system described by (9)–(10) is L_∞-stable if and only if the magnitude of the return difference is bounded away from zero as s varies in the closed right half-plane.

In Theorem (19), we assumed that both $\hat{g}_1$ and $\hat{g}_2 \in \hat{\mathcal{A}}$. We next consider the situation where $\hat{g}_2$ is a nonzero constant, and $\hat{g}_1$ contains a finite number of right half-plane poles, but is otherwise stable.

21 ***lemma*** Suppose

$$\hat{g}_2(s) \equiv g_{20} \neq 0 \qquad \forall\, s \tag{22}$$

Then $\hat{h}_{ij} \in \mathcal{A}$ for $i, j = 1, 2$ if and only if $\hat{r} \in \hat{\mathcal{A}}$, where $\hat{r}$ is defined in (14).

REMARKS: Note that there are no assumptions about $\hat{g}_1$.

proof

(i) "*If*" Suppose $\hat{r} \in \hat{\mathcal{A}}$. Then,

$$1 - \hat{r}(s) = \frac{\hat{g}_1(s)g_{20}}{[1 + \hat{g}_1(s)g_{20}]} \tag{23}$$

also belongs to $\hat{\mathcal{A}}$. This shows that $\hat{h}_{12}, \hat{h}_{21} \in \hat{\mathcal{A}}$. Also, since g_{20} is a nonzero constant, we have that

$$\hat{h}_{11}(s) = [1 - \hat{r}(s)]/g_{20}$$

belongs to $\hat{\mathcal{A}}$. Finally,

$$\hat{h}_{22}(s) = g_{20}\hat{r}(s) \tag{24}$$

belongs to $\hat{\mathcal{A}}$.

(ii) "*Only If*" Exactly as in the proof of lemma (18). ∎

25 ***Theorem*** Suppose $\hat{g}_2$ is of the form (22), and suppose

$$\hat{g}_1(s) = \hat{g}_{1b}(s) + \hat{n}_1(s)/\hat{d}_1(s) \tag{26}$$

where $\hat{g}_{1b} \in \hat{\mathcal{A}}$, $\hat{n}_1$ and $\hat{d}_1$ are polynomials in s with no common zeroes, the degree of $\hat{n}_1$ is less than or equal to that of $\hat{d}_1$, and all

---

[12]Or equivalently, (i) L_1-stable, (ii) L_p-stable for all $p \in [1, \infty]$.

zeroes of $\hat{d}_1$ are in C_+. Under these conditions, the system (9)–(10) is L_∞-stable[12] if and only if

27 $$\inf_{s \in C_+} |1 + g_{20}\hat{g}_1(s)| > 0$$

REMARKS:

(i) The condition (27) has exactly the same interpretation as (20); namely, for the class of systems under study, a necessary and sufficient condition for stability is that the magnitude of the return difference is bounded away from zero as s varies over the closed right half-plane.

(ii) Note that there is no loss of generality in assuming that all zeroes of $\hat{d}_1$ are in C_+, since if $\hat{g}_1$ has some poles in the open left half-plane, the contribution of these poles can be removed by partial fraction expansion, and lumped in with $\hat{g}_{1b}$.

proof By lemma (21), we only need to show that (27) is a necessary and sufficient condition for $\hat{r}$ to belong to $\hat{\mathfrak{A}}$.

(i) "*Only If*" Since $\hat{r} = 1/(1 + g_{20}\hat{g}_1)$, the necessity of (27) follows from lemma (8).

(ii) "*If*" Let α denote the degree of the polynomial $\hat{d}_1$, and define

28 $$\hat{p}(s) = \frac{\hat{n}_1(s)}{(s+1)^\alpha}$$

29 $$\hat{q}(s) = \frac{\hat{d}_1(s)}{(s+1)^\alpha}$$

Since the degree of $\hat{n}_1$ is less than or equal to α, it is clear that $\hat{p}, \hat{q} \in \hat{\mathfrak{A}}$. Also, it is easy to verify that

30 $$\hat{g}_1(s) = \hat{g}_{1b}(s) + \frac{\hat{p}(s)}{\hat{q}(s)} = \frac{[\hat{p}(s) + \hat{q}(s)\hat{g}_{1b}(s)]}{\hat{q}(s)}$$

31 $$\hat{r}(s) = \frac{\hat{q}(s)}{\{\hat{q}(s) + g_{20}[\hat{p}(s) + \hat{q}(s)\hat{g}_{1b}(s)]\}}$$

We now claim that, if (27) is true, then

32 $$\inf_{s \in C_+} |\hat{q}(s) + g_{20}[\hat{p}(s) + \hat{q}(s)\hat{g}_{1b}(s)]| > 0$$

To show this, rewrite (32) equivalently as

33 $$\inf_{s \in C_+} |\hat{q}(s)[1 + g_{20}\hat{g}_1(s)]| > 0$$

Now, suppose (33) is false. Then there is a sequence $(s_i)_1^\infty$ in C_+ such that $\hat{q}(s_i)[1 + g_{20}\hat{g}_1(s_i)] \longrightarrow 0$ as $i \longrightarrow \infty$. There are two possibilities to consider: (i) $(s_i)_1^\infty$ is bounded, and (ii) $(s_i)_1^\infty$ is unbounded. Each of these possibilities is studied separately.

If $(s_i)_1^\infty$ is bounded, it has a convergent subsequence. Renumber it as $(s_i)_1^\infty$, and suppose that it converges to $s_0 \in C_+$. Since $\hat{q} + g_{20}(\hat{p} + \hat{q}\hat{g}_{1b}) \in \hat{\mathfrak{A}}$, it is continuous over C_+, which means that

34 $$\hat{q}(s_0) + g_{20}[\hat{p}(s_0) + \hat{q}(s_0)\hat{g}_{1b}(s_0)] = 0$$

However, (34) leads to a contradiction, as shown next. Either s_0 is a pole of $\hat{g}_1$, or it is not. If s_0 is a pole of $\hat{g}_1$, then it is a zero of $\hat{d}_1$, i.e., of $\hat{q}$. Hence, at $s = s_0$, we have

35 $$\hat{q}(s_0) + g_{20}[\hat{p}(s_0) + \hat{q}(s_0)\hat{g}_{1b}(s_0)] = g_{20}\hat{p}(s_0) \neq 0$$

because $\hat{n}_1$ and $\hat{d}_1$ have no common zeros; thus (34) is contradicted. If s_0 is not a pole of $\hat{g}_1$, then $\hat{q}_1(s_0)$ is a nonzero finite number, and so is $1 + g_{20}\hat{g}_1(s_0)$ [because of (27)]. Hence

36 $$\hat{q}(s_0) + g_{20}[\hat{p}(s_0) + \hat{q}(s_0)\hat{g}_{1b}(s_0)] = \hat{q}(s_0)[1 + g_{20}\hat{g}_1(s_0)] \neq 0$$

Thus (32) is once again contradicted.

Now consider the situation where $(s_i)_1^\infty$ is an unbounded sequence. Then we can pick a subsequence, which is again renumbered as $(s_i)_1^\infty$, such that $|s_i| \longrightarrow \infty$ as $i \longrightarrow \infty$. Now, since $\hat{q}$ is a rational function whose numerator and denominator are of equal degree, it is easy to see that $\hat{q}(s_i)$ approaches a nonzero constant as $|s_i| \longrightarrow \infty$. Also, by (27), $[1 + g_{20}\hat{g}_1(s_i)]$ remains bounded away from zero, which contradicts the assumption that $\hat{q}(s_i)[1 + g_{20}\hat{g}_1(s_i)] \longrightarrow 0$ as $i \longrightarrow \infty$.

Hence we conclude that no such sequence $(s_i)_1^\infty$ exists, i.e., that (32) is true. Next, by lemma (1) and the fact that $\hat{q} + g_{20}(\hat{p} + \hat{q}\hat{g}_{1b}) \in \hat{\mathfrak{a}}$, we have that $1/[\hat{q} + g_{20}(\hat{p} + \hat{q}\hat{g}_{1b})] \in \hat{\mathfrak{a}}$, whence $\hat{r} \in \hat{\mathfrak{a}}$, by (29). By lemma (21), this shows that the system (9)–(10) is L_∞-stable. ∎

Consider next the case where $\hat{g}_1$ and $\hat{g}_2$ are both of the form (26). In this case, the necessary and sufficient conditions for stability are slightly more complicated.

37 ***Theorem*** Suppose $\hat{g}_i$, $i = 1, 2$ is of the form

38 $$\hat{g}_i(s) = \hat{g}_{ib}(s) + \hat{n}_i(s)/\hat{d}_i(s)$$

where $\hat{g}_{ib} \in \hat{\mathfrak{a}}$, $\hat{n}_i$ and $\hat{d}_i$ are polynomials in s with no common zeros, the degree of $\hat{n}_i$ is less than or equal to that of $\hat{d}_i$, and all zeros of $\hat{d}_i$ are in C_+. Under these conditions, the system (9)–(10) is L_∞-stable[13] if and only if

39 (i) $\inf_{s \in C_+} |1 + \hat{g}_1(s)\,\hat{g}_2(s)| > 0$,

40 (ii) $\hat{d}_1(s)\,\hat{g}_1(s) \neq 0$ whenever $\hat{d}_2(s) = 0$,

41 (iii) $\hat{d}_2(s)\,\hat{g}_2(s) \neq 0$ whenever $\hat{d}_1(s) = 0$.

Remarks: Theorem (37) states that three conditions together are necessary and sufficient for stability:

(i) the return difference condition (39), which is the same as before;

(ii) the quantity $\hat{d}_1\hat{g}_1$ does not vanish at the poles of $\hat{g}_2$ in C_+; and

(iii) the symmetrical condition about $\hat{d}_2\hat{g}_2$ and $\hat{g}_1$.

---

[13] Or equivalently, (i) L_1-stable, (ii) L_p-stable for all $p \in [1, \infty]$.

Conditions (ii) and (iii) assure the complete *shifting* of the C_+ poles of $\hat{g}_1$ and $\hat{g}_2$, and can be rationalized as follows: suppose $s_0 \in C_+$ is a pole of $\hat{g}_1$, and that $\hat{d}_2(s_0)\,\hat{g}_2(s_0) = 0$; then in effect the feedback loop is open at the frequency $s = s_0$, so that there is nothing to stabilize the unstable subsystem $\hat{g}_1$. As a result, the closed-loop transfer function contains a pole at s_0. Similar reasoning applies to the C_+ poles of $\hat{g}_2$.

The proof of Theorem (37) is omitted in the interests of brevity, and can be found in [10] Desoer and Vidyasagar, together with some generalizations of Theorem (37).

6.6.2 Graphical Stability Criterion

In applying Theorems (19) or (25), we have to verify a condition of the form (20), which involves the behavior of the return difference $\hat{r}(s)$ as s varies over C_+. Similarly, in applying Theorem (37), conditions (40) and (41) can be checked easily, and only (39) is involved. The rest of this subsection is devoted to the presentation of a graphical stability criterion for verifying a condition of the form (20) [or (27) or (39)], by examining only $1 + \hat{g}_1(j\omega)\,g_2(j\omega)$ as ω varies over R. The criterion here is a natural generalization of the well known Nyquist criterion, and has the same advantages associated with the Nyquist criterion, namely: (i) the test can be applied directly to the experimentally measured data, i.e., $1 + \hat{g}_1(j\omega)\,\hat{g}_2(j\omega)$, and (ii) even if the stability criterion is not satisfied, it is easy to visualize the form of "compensation" needed to satisfy the criterion, i.e., to stabilize the system.

To present the graphical test for stability, we assume that the return difference $\hat{r}(\cdot)$ is of the form

42 $$\hat{r}(s) = 1 + \hat{g}_1(s)\hat{g}_2(s) = 1 + \sum_{i=0}^{\infty} r_i e^{-st_i} + \hat{r}_a(s) + \sum_{i=1}^{n_p} \sum_{j=1}^{m_i} \frac{r_{ij}}{(s - p_i)^j}$$

where t_i are nonnegative constants, $p_i \in C_+$; m_i is the multiplicity of the pole p_i; and n_p is the total number of poles of $\hat{r}$ in C_+; in addition,

43 $$\sum_{i=0}^{\infty} |r_i| < \infty$$

44 $$r_a(\cdot) \in L_1$$

Basically, in (42), we have split up $\hat{r}$ into a part that belongs to $\hat{\mathfrak{a}}$ (namely the first three terms on the right side), plus the contribution of the right half-plane poles of $\hat{r}$, expressed in partial fraction form. Now, if Theorem (19) or Theorem (25) is being applied, then the return difference $\hat{r}$ *is* of the form (42). However, if Theorem (37) is being applied, then $\hat{r}$ may not be of the form (42), since it may contain terms of the type

45 $$\frac{\hat{f}(s)}{(s-p)^k}$$

where $\hat{f} \in \hat{\mathfrak{a}}$ and $p \in C_+$. Thus the graphical stability criterion that we are about to present does not apply to *all* the systems covered by Theorem (37), but it does apply to all the systems covered by Theorems (19) and (25).

Thus the problem at hand is as follows: Given $\hat{r}$ of the form (42), determine conditions on $\hat{r}(j\omega)$, $\omega \in R$, which insure that

46 $$\inf_{s \in C_+} |\hat{r}(s)| > 0$$

In general, this problem is difficult to solve. However, if the "delays" t_i in (42) are rationally related, i.e., if there exist a number T and integers n_i such that

47 $$t_i = n_i T, \qquad \forall\ i$$

then a relatively simple criterion can be presented.

Before presenting the testing procedure formally, we introduce two preliminary notions: (i) the *indented* $j\omega$-axis, and (ii) the phase of $\hat{r}(j\omega)$. Suppose we study $\hat{r}(j\omega)$, where $\hat{r}$ is of the form (42). The first problem is that if $\text{Re}\, p_i = 0$ for some i, then $\hat{r}(j\omega)$ becomes unbounded as $j\omega \longrightarrow p_i$. To circumvent this problem, we *indent* the $j\omega$-axis as shown in Fig. 6.7, by *going around* the pole, p_i, via a semicircle of radius ϵ_i centered at p_i. Let δ_i denote the half-disk shaded in Fig. 6.7; i.e., let

48 $$\delta_i = \{s : |s - p_i| \leq \epsilon_i\}$$

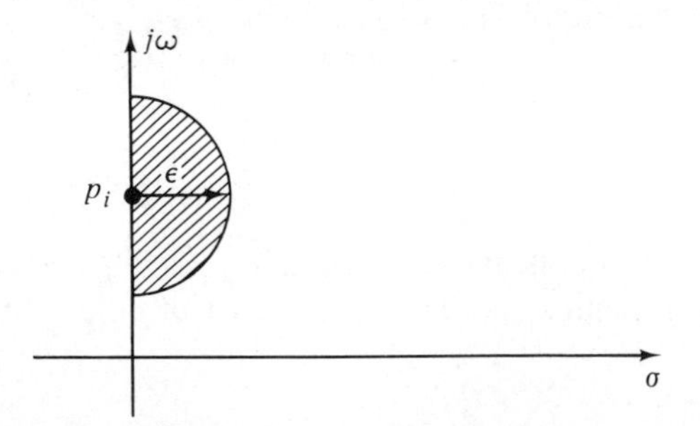

FIG. 6.7

Then, since $\hat{r}(s)$ is unbounded at p_i, clearly we can choose ϵ_i sufficiently small that δ_i contains no poles of $\hat{r}$ other than p_i; and such that

49 $$\inf_{s \in \delta_i} |\hat{r}(s)| > 0$$

If we repeat this procedure at each $j\omega$-axis pole of $\hat{r}$, we get a finite number of half-disk regions where (49) holds. Let us now delete these half-disks from C_+, and denote the resulting *indented* right half-plane as C_{i+} (see Fig. 6.8). Now, because of (49), we have

50 $$\inf_{s \in C_{i+}} |\hat{r}(s)| > 0$$

if and only if (46) holds.

After we indent the $j\omega$-axis, it is clear that, corresponding to each $\omega \in R$, there exists exactly one point on the indented $j\omega$-axis whose imaginary part is ω, but whose real part may or may not be zero. By a slight abuse of notation, let $\hat{r}(j\omega)$ denote the value of $\hat{r}$ at the point on the indented $j\omega$-axis whose imaginary part is ω (this too is illustrated in Fig. 6.8).

Once the $j\omega$-axis is indented, $\hat{r}(j\omega)$ is continuous in ω, and we can define the argument or phase of $\hat{r}(j\omega)$.

51 *[definition]* The *argument* of $\hat{r}(j\omega)$, denoted by $\phi(j\omega)$, is the complex number such that

52 $$\hat{r}(j\omega) = |\hat{r}(j\omega)| \exp [j\phi(j\omega)]$$

53 $$\phi(0) = 0 \quad \text{if} \quad \hat{r}(0) > 0,\ \pi \text{ if } \hat{r}(0) < 0$$

In applying (53), note that if $\hat{r}(0) = 0$, then (50) is immediately violated, and there is no need to test any further.

We now quote, without proof, the desired result.

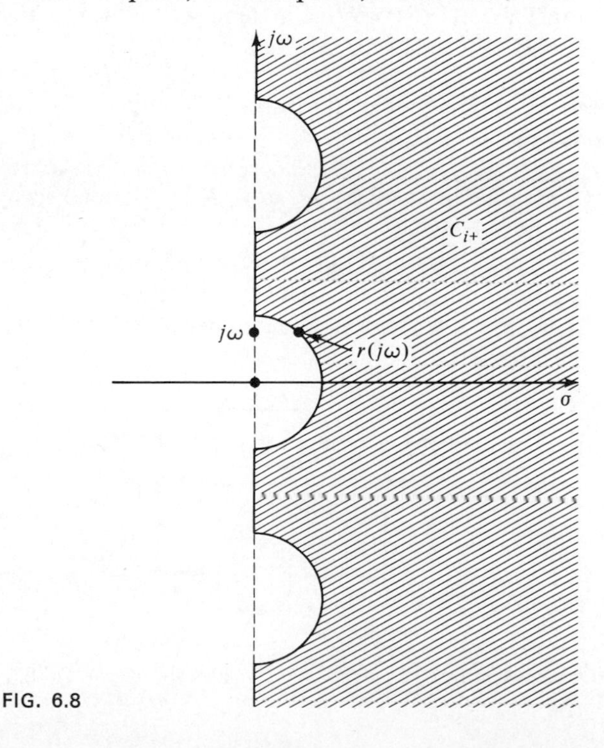

FIG. 6.8

54 ***Theorem*** (*Graphical Stability Test*) Given $r(\cdot)$ of the form (42), we have that (46) is true if and only if

55 (i) $\inf_{\omega \in R} |r(j\omega)| > 0$

56 (ii) $\lim_{n \to \infty} \phi(j2\pi n/T) - \phi(-j2\pi n/T) = 2\pi \nu_p$, where ν_p is the number of poles of $\hat{r}$ with positive real part.

REMARKS: Theorem (54) is a natural generalization of the well-known Nyquist criterion: (55) states that the polar (or Nyquist) plot of $\hat{r}(j\omega)$ is bounded away from the origin, while (56) states that the polar plot of $\hat{r}(j\omega)$ encircles the origin, in the counterclockwise sense, exactly as many times as there are poles of $\hat{r}$ with positive real part.

The proof of Theorem (54), as well as of its generalization to the case where the assumption (47) is dropped, can be found in [10] Desoer and Vidyasagar.

One of the common applications of Theorem (54) is in finding the range of values of the constant k for which the feedback system shown in Fig. 6.9 is stable. Suppose $\hat{g}(\cdot)$ is of the form (26); then by Theorem (25), we know that the closed-loop system is L_∞-stable if and only if

57 $$\inf_{s \in C_+} |1 + k\, g(s)| > 0$$

By applying Theorem (54) to this system, we get a graphical stability criterion that is well suited for applications.

58 ***Theorem*** Suppose $\hat{g}$ is of the form (26), and let $k \neq 0$. Then the system of Fig. 6.9 is L_∞-stable[14] if and only if

(i) The polar plot of $\hat{g}(j\omega)$, $\omega \in R$ is bounded away from the point $-1/k + j0$

(ii) The polar plot of $\hat{g}(j\omega)$ encircles the point $-1/k + j0$ exactly ν_p times in the counterclockwise sense as ω increases from $-\infty$ to ∞, where ν_p is the number of poles of $\hat{g}$ with positive real part.

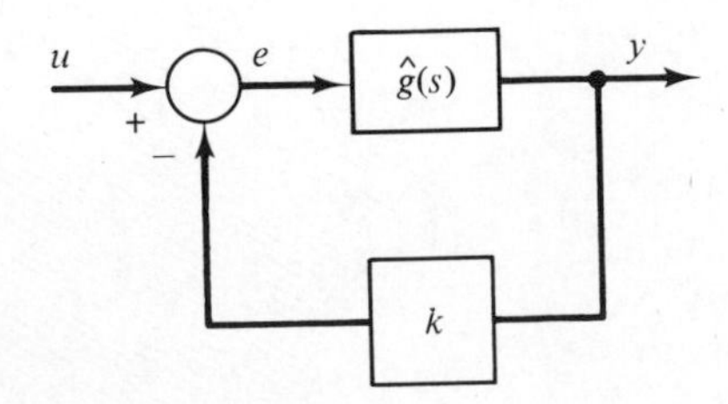

FIG. 6.9

[14] Or equivalently, (i) L_1-stable, (ii) L_p-stable for all $p \in [1, 0]$.

59 **Example.** Consider a system of the form shown in Fig. 6.9, with

60 $$\hat{g}_1(s) = e^{-s}\frac{s^2 + 4s + 2}{s^2 - 1}$$

In this case, $\hat{g}_1$ represents a system with a rational transfer function, followed by a transport delay. It is easy to verify that $\hat{g}_1 \notin \mathfrak{A}$ and hence that $\nu_p = 1$. The Nyquist plot of $\hat{g}_1(j\omega)$ is shown in Fig. 6.10. Note that, as ω increases, the Nyquist plot approaches the unit circle, because $\hat{g}_1(j\omega)$ approaches a periodic function. Applying Theorem (58), we see that this system is L_∞-stable whenever $0.5 < k < 0.568$.

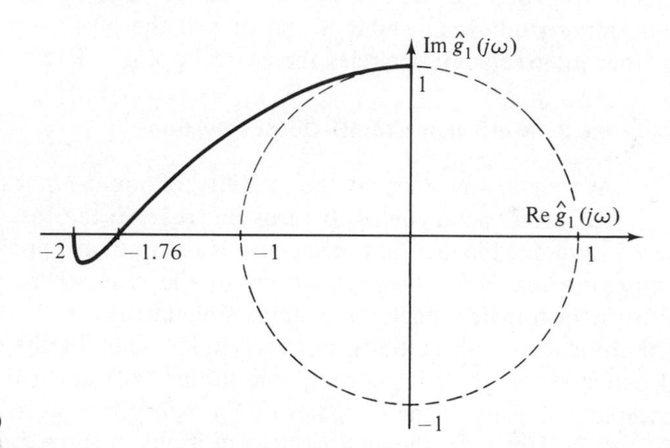

FIG. 6.10

61 **Example.** To illustrate the power of Theorem (58), we shall consider the stability of a nonuniform *RC* transmission line with an operational amplifier in the feedback, as shown in Fig. 6.11. This system is of the type shown in Fig. 6.9, where $\hat{g}_1$ is given by

62 $$\hat{g}_1(s) = \frac{1}{\hat{A}(s)}$$

where $\hat{A}(\cdot)$ is one of the so-called *chain parameters* of the transmission

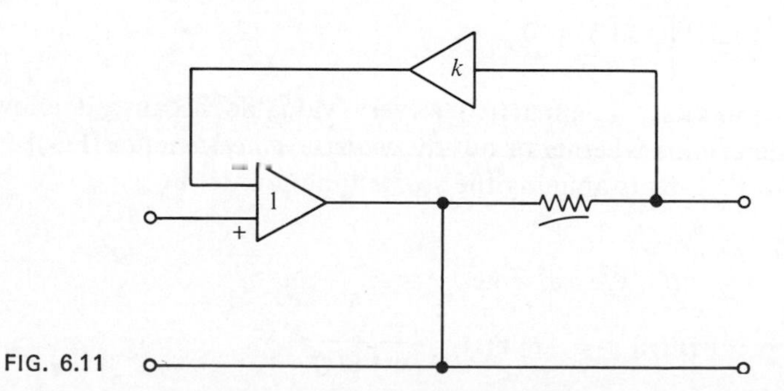

FIG. 6.11

line. It is known that, for a general nonuniform line, $\hat{A}(s)$ is of the form

63 $$\hat{A}(s) = \prod_{i=1}^{\infty} (s + p_i)$$

where p_i is asymptotically of the form

64 $$p_i \sim \alpha i^2$$

with α being a physical constant depending on the parameters of the line. [For an infinitely long uniform line, $\hat{A}(s) = \cosh \sqrt{\lambda s}$, where λ is a physical constant.] It is also known that $\hat{g}_1 \in \hat{\mathfrak{A}}$, and, in fact, $g_1 \in L_1$. Hence applying Theorem (97) to the present case, we see that the system under study is L_∞-stable if and only if the Nyquist plot of $\hat{g}_1(j\omega)$ neither intersects nor encircles the critical point $-1/k + j0$.

6.6.3 Multi-Input/Multi-Output Systems

We shall now turn to the stability of multi-input/multi-output (m.i.m.o.) feedback systems. It turns out that, in the case of *open-loop stable* systems, the fact that the system is m.i.m.o. as opposed to s.i.s.o. causes no real difficulties. However, in the case where some of the subsystem transfer functions contain singularities in C_+, the stability conditions are substantially more complex than in the s.i.s.o. case. Specifically, in the s.i.s.o. case, conditions (40) and (41) assure the complete shifting of the C_+ poles of the subsystem transfer functions $\hat{g}_1$ and $\hat{g}_2$. However, the analogous conditions in the m.i.m.o. case are more complex and involve what are known as coprime factorizations. We shall not develop the theory of coprime factorizations, because this would take us too far afield. Rather, we shall state the basic results for the open-loop stable case, and state refer the reader to [10] Desoer and Vidyasagar for further details.

We shall begin with a result that extends lemma (1) to the matrix case.

65 ***lemma*** Let $\hat{\mathbf{F}}(\cdot) \in \hat{\mathfrak{A}}^{n \times n}$. Then $[\hat{\mathbf{F}}(\cdot)]^{-1} \in \hat{\mathfrak{A}}^{n \times n}$ if and only if

66 $$\inf_{s \in C_+} |\det \hat{\mathbf{F}}(s)| > 0$$

REMARKS: Lemma (65) is very valuable because it allows us to determine whether or not the *matrix-valued function* $[\hat{\mathbf{F}}(\cdot)]^{-1}$ belongs to $\hat{\mathfrak{A}}^{n \times n}$ by examining the *scalar* function $\det \hat{\mathbf{F}}(\cdot)$.

proof

(i) "*If*" We can write

67 $$[\hat{\mathbf{F}}(s)]^{-1} = \text{Adj } \hat{\mathbf{F}}(s) \quad \frac{1}{\det \hat{\mathbf{F}}(s)}$$

where Adj $\hat{\mathbf{F}}(s)$ denotes the adjoint matrix of $\hat{\mathbf{F}}(s)$, i.e., the matrix consisting of the cofactors of $\hat{\mathbf{F}}(s)$. Because every cofactor of $\hat{\mathbf{F}}(s)$ consists of sums and products of the elements of $\hat{\mathbf{F}}(s)$ and because every element of $\hat{\mathbf{F}}(\cdot)$ belongs to $\hat{\mathfrak{A}}$, it follows that Adj $\hat{\mathbf{F}}(\cdot) \in \hat{\mathfrak{A}}^{n \times n}$. Similarly, det $\hat{\mathbf{F}}(\cdot) \in \hat{\mathfrak{A}}$. Now, if (66) holds, then by lemma (1), $1/\det \hat{\mathbf{F}}(\cdot) \in \hat{\mathfrak{A}}$. Hence $[\hat{\mathbf{F}}(\cdot)]^{-1}$ is the product of matrix in $\hat{\mathfrak{A}}^{n \times n}$ and a scalar in $\hat{\mathfrak{A}}$ and therefore belongs to $\hat{\mathfrak{A}}^{n \times n}$.

(ii) "*Only If*" Suppose $[\hat{\mathbf{F}}(\cdot)]^{-1} \in \hat{\mathfrak{A}}^{n \times n}$. Then, by the same reasoning as above, $\det [\hat{\mathbf{F}}(\cdot)]^{-1} \in \hat{\mathfrak{A}}$. However, $\det [\hat{\mathbf{F}}(\cdot)]^{-1} = 1/\det [\hat{\mathbf{F}}(\cdot)]$, and this belongs to $\hat{\mathfrak{A}}$. Hence, by lemma (8), we have that (66) holds. ∎

Lemma (8) can be similarly extended to the multivariable case, as follows.

68 ***lemma*** Suppose that $F(\cdot)$ is a matrix of distributions and that $[\hat{\mathbf{F}}(\cdot)]^{-1} \in \hat{\mathfrak{A}}^{n \times n}$. Then (66) holds.

The proof is left as an exercise.

Throughout the rest of this subsection, we shall study feedback systems described by

$$\hat{\mathbf{e}}_1 = \hat{\mathbf{u}}_1 - \hat{\mathbf{y}}_2 \tag{69}$$

$$\hat{\mathbf{e}}_2 = \hat{\mathbf{u}}_2 + \hat{\mathbf{y}}_1 \tag{70}$$

$$\hat{\mathbf{y}}_1 = \hat{\mathbf{G}}_1 \hat{\mathbf{e}}_1 \tag{71}$$

$$\hat{\mathbf{y}}_2 = \hat{\mathbf{G}}_2 \hat{\mathbf{e}}_2 \tag{72}$$

where $\hat{\mathbf{e}}_1(s), \hat{\mathbf{e}}_2(s), \hat{\mathbf{u}}_1(s), \hat{\mathbf{u}}_2(s), \hat{\mathbf{y}}_1(s), \hat{\mathbf{y}}_2(s) \in R^n$ and $\hat{\mathbf{G}}_1(s), \hat{\mathbf{G}}_2(s) \in R^{n \times n}$. The system (69)–(72) is the same as that in Fig. 6.5, except that $\hat{\mathbf{g}}_1$ and $\mathbf{g}_2$ are replaced by the matrix transfer functions $\hat{\mathbf{G}}_1$ and $\hat{\mathbf{G}}_2$. As in the s.i.s.o. case, (69)–(72) can be rewritten as

$$\hat{\mathbf{e}} = \hat{\mathbf{u}} - \mathbf{H}\hat{\mathbf{y}} \tag{73}$$

$$\hat{\mathbf{y}} = \hat{\mathbf{G}}\hat{\mathbf{e}} \tag{74}$$

where

$$\hat{\mathbf{e}} = [\hat{\mathbf{e}}_1' \quad \hat{\mathbf{e}}_2']' \tag{75}$$

$$\hat{\mathbf{u}} = [\hat{\mathbf{u}}_1' \quad \hat{\mathbf{u}}_2']' \tag{76}$$

$$\hat{\mathbf{y}} = [\hat{\mathbf{y}}_1' \quad \hat{\mathbf{y}}_2']' \tag{77}$$

$$\mathbf{G}(s) = \begin{bmatrix} \hat{\mathbf{G}}_1(s) & 0 \\ 0 & \mathbf{G}_2(s) \end{bmatrix} \tag{78}$$

$$\mathbf{H} = \begin{bmatrix} \mathbf{0} & \mathbf{I} \\ -\mathbf{I} & \mathbf{0} \end{bmatrix} \tag{79}$$

80 ***Theorem*** The system (73)–(74) is L_∞-stable[15] if and only if one of the following equivalent conditions holds: (1)

81 $$[\mathbf{I} + \mathbf{H}\hat{\mathbf{G}}(\cdot)]^{-1} \in \hat{\mathcal{A}}^{n \times n}$$

or (2)

82 $$\hat{\mathbf{G}}(\cdot)[\mathbf{I} + \mathbf{H}\hat{\mathbf{G}}(\cdot)]^{-1} \in \hat{\mathcal{A}}^{n \times n}$$

proof From (73)–(74), we can solve for $\hat{\mathbf{e}}$ and $\hat{\mathbf{y}}$ to get

83 $$\hat{\mathbf{e}} = (\mathbf{I} + \mathbf{H}\hat{\mathbf{G}})^{-1}\hat{\mathbf{u}}$$

84 $$\hat{\mathbf{y}} = \hat{\mathbf{G}}(\mathbf{I} + \mathbf{H}\hat{\mathbf{G}})^{-1}\hat{\mathbf{u}}$$

The conditions (81) and (82) now follow readily. ∎

Thus our objective is to find conditions on the subsystem transfer functions $\hat{\mathbf{G}}_1$ and $\hat{\mathbf{G}}_2$ which ensure that (83) or (84) holds. In the case where both $\hat{\mathbf{G}}_1$ and $\hat{\mathbf{G}}_2$ belong to $\hat{\mathcal{A}}^{n \times n}$ (i.e., the *open-loop stable* case), this is very easily done.

85 ***Theorem*** Suppose $\hat{\mathbf{G}}_1, \hat{\mathbf{G}}_2 \in \hat{\mathcal{A}}^{n \times n}$. Then the system (69)–(72) is L_∞-stable[16] if and only if

86 $$\inf_{s \in C_+} |\det[\mathbf{I} + \mathbf{H}\hat{\mathbf{G}}(s)]| > 0$$

proof Since $\hat{\mathbf{G}}_1, \hat{\mathbf{G}}_2 \in \hat{\mathcal{A}}^{n \times n}$, $\hat{\mathbf{G}} \in \hat{\mathcal{A}}^{2n \times 2n}$, whence $\mathbf{I} + \mathbf{H}\hat{\mathbf{G}} \in \hat{\mathcal{A}}^{2n \times 2n}$. Hence (81) holds if and only if (86) holds, by lemma (65). ∎

REMARKS: Theorem (85) states that, in the open-loop stable case, the feedback system (69)–(72) is L_∞-stable if and only if the determinant of the *return difference matrix* $\mathbf{I} + \mathbf{H}\hat{\mathbf{G}}$ is bounded away from zero over C_+. Note that

87 $$\det(\mathbf{I} + \mathbf{H}\hat{\mathbf{G}}) = \det\begin{bmatrix} \mathbf{I} & \hat{\mathbf{G}}_2 \\ -\hat{\mathbf{G}}_1 & \mathbf{I} \end{bmatrix} = \det(\mathbf{I} + \hat{\mathbf{G}}_2\hat{\mathbf{G}}_1)$$

so that the stability criterion (86) can be expressed in terms of the subsystem transfer matrices $\hat{\mathbf{G}}_1$ and $\hat{\mathbf{G}}_2$. Also, once the quantity $\det(\mathbf{I} + \mathbf{H}\hat{\mathbf{G}})$ is computed, the condition (86) can be verified graphically by using Theorem (54).

In the case where $\hat{\mathbf{G}}_1$ and $\hat{\mathbf{G}}_2$ have singularities in C_+, the situation is substantially more complicated in the m.i.m.o. case than it is in the s.i.s.o. case, because of the complexity of *poleshifting* in the m.i.m.o. case. To illustrate this point, consider the system of Fig. 6.6, and let

---

[15] Or, equivalently, (1) L_1-stable, (2) L_p-stable for $p \in [1, \infty]$.

[16] Or, equivalently, (1) L_1-stable, (2) L_p-stable for all $p \in [1, \infty]$.

$\hat{u}_2 = 0$, $\hat{g}_2 = k_1$, a nonzero constant. Then the closed-loop transfer function from $\hat{u}_1$ to $\hat{y}_1$ is

88 $$\hat{h}_{11}(s) = \frac{\hat{g}_1(s)}{1 + k\hat{g}_1(s)}$$

Now, suppose $s_0 \in C_+$ is a pole of $\hat{g}$. Because we can write

89 $$\hat{h}_{11}(s) = \frac{1}{k + 1/\hat{g}_1(s)}$$

it is clear that $\hat{h}_{11}$ is analytic at s_0 and that $\hat{h}_{11}(s_0) = 1/k$. Thus, if we place a nonzero feedback around $\hat{g}_1$, all poles of $\hat{g}_1$ in C_+ are shifted from their original locations. Now, if we can assure that the new pole locations are *not* in C_+, the system is stable (loosely speaking). This is the purpose of condition (39). Conditions (40) and (41) ensure that the effective feedback at the pole locations of $\hat{g}_1$ and $\hat{g}_2$ is nonzero, so that pole shifting does take place.

In contrast, in the multivariable case, it is quite possible for both the open-loop and closed-loop transfer functions to have poles at the same point in C. For example, let

$$\hat{\mathbf{G}}_1(s) = \begin{bmatrix} 1/(s-1) & 0 \\ 0 & 1/(s+1) \end{bmatrix}; \qquad \hat{\mathbf{G}}_2(s) \equiv \begin{bmatrix} 2 & 2 \\ 2 & 0 \end{bmatrix}$$

and suppose $\hat{\mathbf{u}}_2 \equiv 0$. Then $\hat{\mathbf{G}}_2$ represents a constant nonsingular feedback matrix, and the transfer function matrix from $\hat{\mathbf{u}}_1$ to $\hat{\mathbf{y}}_1$ is

90 $$\hat{\mathbf{H}}_{11}(s) = \hat{\mathbf{G}}_1(\mathbf{I} + \hat{\mathbf{G}}_2\hat{\mathbf{G}}_1)^{-1} = \begin{bmatrix} \dfrac{s+1}{(s-1)(s+3)} & \dfrac{-2}{(s-1)(s+3)} \\ \dfrac{-2}{(s-1)(s+3)} & \dfrac{s+1}{(s-1)(s+3)} \end{bmatrix}$$

Thus both $\hat{\mathbf{G}}_1$ and $\hat{\mathbf{H}}_{11}$ have poles at $s = 1$, even though the feedback matrix is nonsingular. This phenomenon cannot occur in s.i.s.o. systems and is the source of the complications in the m.i.m.o. case. This difficulty can be overcome by using the concept of coprime factorizations, and the details can be found in [10] Desoer and Vidyasagar.

Problem 6.12. Consider the feedback system shown in Fig. 6.8. Determine the range of values of the constant k for which the feedback system is L_∞-stable, with

(i) $\hat{g}(s) = \dfrac{e^{-s}(s^2 + 1)}{s^2 + 2s}$

(ii) $\hat{g}(s) = \dfrac{1 - e^{-2s}s}{s + 1}$

(iii) $\hat{g}(s) = \dfrac{e^{-.1s}(s^2 + 4s + 4)}{s^2 - 2s - 3}$

(iv) $\hat{g}(s) = 1 - e^{-s} + e^{-3s} + \dfrac{e^{-4s}}{s + 5}$

Problem 6.13. Consider the feedback system shown in Figure 6.5. Suppose

$$\hat{\mathbf{G}}_1(s) = \begin{bmatrix} \dfrac{e^{-s}s}{s+2} & 1 \\ \dfrac{-2}{s+1} & 1 - e^{-3s} \end{bmatrix}$$

Is the feedback system L_∞-stable with

(i) $\hat{\mathbf{G}}_2(s) = \begin{bmatrix} 1 & 1 \\ 0 & 2 \end{bmatrix}$ (ii) $\hat{\mathbf{G}}_2(s) = \begin{bmatrix} 1 & \dfrac{-1}{s+3} \\ 2 & 0 \end{bmatrix}$

(iii) $\hat{\mathbf{G}}_2(s) = \begin{bmatrix} 1 & 0 \\ 0 & e^{-s} \end{bmatrix}$

6.7 TIME-VARYING AND/OR NONLINEAR SYSTEMS

In the previous section, we studied the feedback stability of linear time-invariant systems. In this section, we study nonlinear and/or time-varying systems. Our attention is centered exclusively on single-input/single-output systems. The stability of multi-input/multi-output systems, which requires more elaborate techniques, is studied in detail in [10] Desoer and Vidyasagar.

In this section, we prove two well known results, known as the circle criterion and the (generalized) Popov criterion, respectively. The circle criterion provides a sufficient condition for the L_2-stability of a feedback system containing a linear time-invariant element in the forward path and a memoryless time-varying and/or nonlinear element in the feedback path. We show that the circle criterion is also a *necessary* condition for stability, in a sense to be made precise later on. The generalization of the Popov criterion that is presented here provides a sufficient condition for the L_2-stability of a feedback system containing a linear *time-invariant* system in the forward path and a *time-invariant* memoryless nonlinearity in the feedback path. Both criteria are graphical in nature, i.e., they are frequency domain criteria, and this feature makes them easy to apply.

6.7.1 The Circle Criterion

To develop the circle criterion, we shall first introduce a definition.

1 *[definition]* A continuous function $\phi: R_+ \times R \longrightarrow R$ is said to belong to the sector $[\alpha, \beta]$, $\alpha \leq \beta$, if

2 $$\alpha x^2 \leq x\phi(t, x) \leq \beta x^2, \qquad \forall\, t \geq 0, \forall\, x \in R$$

or, equivalently, if (1) $\phi(t, 0) = 0$, $\forall\, t \geq 0$, and (2)

3 $$\alpha \leq \frac{\phi(t, x)}{x} \leq \beta, \qquad \forall\, t \geq 0, \forall\, x \neq 0$$

REMARKS: A continuous function $\phi: R_+ \times R \longrightarrow R$ belongs to the sector $[\alpha, \beta]$ if, for each fixed t, the graph of $\phi(t, x)$ vs. x lies between the straight lines passing through the origin with slopes α and β, respectively. This is illustrated in Fig. 6.12.

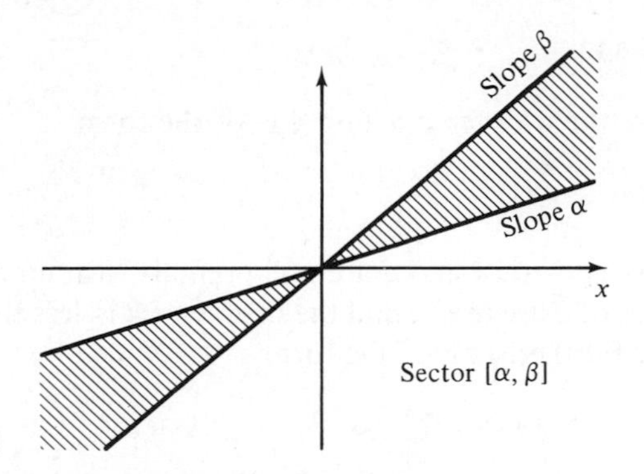

FIG. 6.12

We shall next prove a result very similar to Theorem [6.5(54)].

4 ***lemma*** Suppose that $h(\cdot) \in \mathcal{A}$ and that $u \in L_{2e}$. Then $h * u \in L_{2e}$, and, moreover,

5 $$\|(h * u)_T\|_2 \leq \sup_\omega |\hat{h}(j\omega)| \cdot \|u_T\|_2$$

proof Clearly, the map $u \mapsto h * u$ is causal. Hence by the definition of causality,

6 $$(h * u)_T = (h * u_T)_T$$

Thus

7 $$\|(h * u)_T\|_2 = \|(h * u_T)_T\|_2 \leq \|h * u_T\|_2 \leq \sup_\omega |\hat{h}(j\omega)| \cdot \|u_T\|_2$$

where the last step follows from Theorem [6.5(54)]. ▬

We shall now prove a result that is commonly known as the circle criterion.

8 ***Theorem*** (*Circle Criterion*) Consider a feedback system of the type shown in Fig. 6.13. The element G is linear, time-invariant, and described by

9 $$y_1(t) = (Ge_1)(t) = \int_0^t g(t-\tau)e_1(\tau)d\tau$$

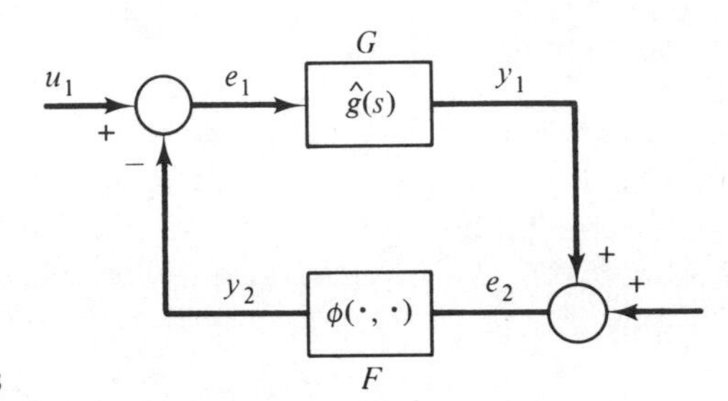

FIG. 6.13

where the transfer function $\hat{g}$ is of the form

10 $$\hat{g}(s) = \hat{g}_b(s) + \frac{\hat{n}(s)}{\hat{d}(s)}$$

where $\hat{g}_b \in \hat{\mathfrak{a}}$, $\hat{n}$ and $\hat{d}$ are polynomials with no common zeroes, all zeroes of $\hat{d}$ are in C_+, and the degree of $\hat{n}$ is less than or equal to that of $\hat{d}$. Further, $\hat{g}_b$ is of the form

11 $$\hat{g}_b(s) = \hat{g}_a(s) + \sum_{i=0}^{\infty} g_i e^{-isT}, \qquad g_a(\cdot) \epsilon L_1$$

(i.e., the "delays" are all equally spaced). The element F is memoryless and is described by

12 $$y_2(t) = (Fe_2)(t) = \phi[t, e_2(t)]$$

where $\phi(\cdot, \cdot)$ is continuous and belongs to the sector $[\alpha, \beta]$. Under these conditions, the system under study is L_2-stable if one of the following sets of conditions, as appropriate, holds: If $\alpha, \beta \neq 0$, let D be the disk in the complex plane centered on the real axis and passing through the points $-1/\alpha + j0$ and $-1/\beta + j0$.

(i) *If* $0 < \alpha < \beta$: (a) When the $j\omega$-axis is indented as in Sec. 6.6.1, the graph of $\hat{g}(j\omega)$ is bounded away from the disk D; i.e.,

13 $$\inf_{\substack{\omega \in R \\ z \in D}} |\hat{g}(j\omega) - z| > 0$$

(b) The graph of $g(j\omega)$ encircles the disk D exactly ν_p times, where ν_p is the number of poles of $\hat{g}$ with positive real part; i.e.,

14 $$\lim_{n\to\infty} \operatorname{Arg}\left(\hat{g}\frac{j2\pi n}{T} - z\right) - \operatorname{Arg}\left(\hat{g}\frac{-j2\pi n}{T} - z\right) = 2\pi\nu_p, \quad \forall z \in D$$

(ii) *If* $0 = \alpha < \beta$: (a) $\nu_p = 0$, i.e., $\hat{g}$ has no poles with positive real parts, and (b)

15 $$\inf_{\omega\in R} \operatorname{Re} \hat{g}(j\omega) + \frac{1}{\beta} > 0$$

(iii) *If* $\alpha < 0 < \beta$: (a) $\hat{g}$ has no poles in C_+, and (b) the graph of $\hat{g}(j\omega)$ is entirely contained in the disk D and is bounded away from the boundary of D.

(iv) *If* $\alpha < \beta < 0$: Replace $\hat{g}$ by $-\hat{g}$, α by $-\beta$, β by $-\alpha$, and apply (i), (ii), or (iii) above, as appropriate.

REMARKS:

1. Suppose

16 $$\hat{g}(s) = \mathbf{c}'(s\mathbf{I} - \mathbf{A})^{-1}\mathbf{b} + d$$

where the matrix $\mathbf{A}$ is Hurwitz (i.e., all eigenvalues of $\mathbf{A}$ have negative real parts), the pair $(\mathbf{A}, \mathbf{b})$ is controllable, and the pair $(\mathbf{A}, \mathbf{c})$ is observable. Then Theorem (8) gives a sufficient condition for the L_2-stability of the system of Fig. 6.5. By Problem 6.5, the same conditions also guarantee the global asymptotic stability of the equilibrium point $\mathbf{0}$ of the system [6.4 (27)]–[6.4(29)]; see also Problem 6.15. Thus Theorem (8) contains, as a special case, the circle criterion for Liapunov stability (Theorem [5.5(60)]). The details are left as an exercise. On the other hand, Theorem (8) is more general than Theorem [5.5(60)], because the former is applicable even to *distributed* systems, i.e., to systems where the transfer function $\hat{g}$ is irrational.

2. If $\beta - \alpha \to 0$ and β, α approach $k \neq 0$, then Theorem (8) reduces to the sufficiency part of Theorem [6.6(58)].

proof Define

17 $$c = \frac{\beta + \alpha}{2}$$

18 $$r = \frac{\beta - \alpha}{2}$$

so that

19 $$\alpha = c - r$$

20 $$\beta = c + r$$

Rearrange the system of Fig. 6.13 as in Fig. 6.14. In this way, the system under study is transformed into another equivalent system whose *forward* element G_t has the transfer function

21 $$\hat{g}_t(s) = \frac{\hat{g}(s)}{1 + c\hat{g}(s)}$$

and whose *feedback* element F_t is described by

22 $$(F_t y)(t) = \phi[t, y(t)] - cy(t) \triangleq \phi_t[t, y(t)]$$

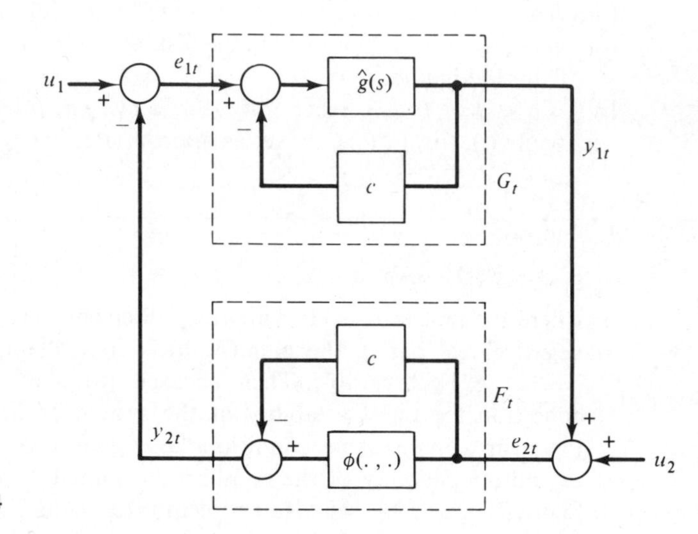

FIG. 6.14

Clearly, the new function $\phi_t(\cdot, \cdot)$ belongs to the sector $[-r, r]$. Now, regardless of which set of conditions (i)–(iv) is applicable, one can verify, using Theorem [6.6(58)], that $\hat{g}_t \in \hat{\mathcal{A}}$ and that

23 $$\sup_{\omega} |\hat{g}_t(j\omega)| \triangleq \gamma_0 < \frac{1}{r}$$

The details of the verification are left as a problem (see Problem 6.15).

Now, with regard to the rearranged system of Fig. 6.14, we have that $\hat{g}_t \in \mathcal{A}$ and that $\phi_t(\cdot, \cdot)$ belongs to the sector $[-r, r]$. The latter condition means that $y_{2t} \in L_{2e}$ whenever $e_{2t} \in L_{2e}$ and that

24 $$||(y_{2t})_T||_2 \leq r||(e_{2t})_T||_2, \qquad \forall\, T \geq 0$$

Also, by lemma (4), we have $G_t e_{1t} \in L_{2e}$ whenever $e_{1t} \in L_{2e}$ and

25 $$||(G_t e_{1t})_T||_2 \leq \gamma_0 ||(e_{1t})_T||_2, \qquad \forall\, T \geq 0$$

We can now complete the proof. Suppose that $u_1, u_2 \in L_2$ and that $e_{1t}, e_{2t} \in L_{2e}$ satisfy the system equations

26 $$e_{1t} = u_1 - F_t e_{2t}$$

27 $$e_{2t} = u_2 + G_t e_{1t}$$

Taking the truncations to $[0, T]$ of (26)–(27) and noting that F_t, G_t are causal, we get

28 $$(e_{1t})_T = u_{1T} - [F_t(e_{2t})_T]_T$$

29 $$(e_{2t})_T = u_{2T} - [G_t(e_{1t})_T]_T$$

Taking norms, we get

30 $$\begin{aligned}\|(e_{1t})_T\|_2 &\leq \|u_{1T}\|_2 + \|[F_t(e_{2t})_T]_T\|_2 \\ &\leq \|u_{1T}\|_2 + \|F_t(e_{2t})_T\|_2 \\ &\leq \|u_{1T}\|_2 + r\|(e_{2t})_T\|_2 \qquad \text{[by (24)]}\end{aligned}$$

31 $$\begin{aligned}\|(e_{2t})_T\|_2 &\leq \|u_{2T}\|_2 + \|[G_t(e_{1t})_T]_T\|_2 \\ &\leq \|u_{1T}\|_2 + \|G_t(e_{1t})_T\|_2 \\ &\leq \|u_{1T}\|_2 + \gamma_0\|(e_{1t})_T\|_2 \qquad \text{[by (25)]}\end{aligned}$$

Eliminating $\|(e_{2t})_T\|_2$ from (30) and (31) gives

32 $$\|(e_{1t})_T\|_2 \leq \|u_{1T}\|_2 + r[\|u_{2T}\|_2 + \gamma_0\|(e_{1t})_T\|_2]$$

33 $$(1 - r\gamma_0)\|(e_{1t})_T\|_2 \leq \|u_{1T}\|_2 + r\|u_{2T}\|_2 \leq \|u_1\|_2 + r\|u_2\|_2$$

Because $r\gamma_0 < 1$ [by (23)], we can divide both sides of (33) by $1 - r\gamma_0$, to obtain

34 $$\|(e_{1t})_T\|_2 \leq \frac{\|u_1\|_2 + r\|u_2\|_2}{1 - r\gamma_0}$$

Because the right-hand side of (34) is finite and independent T, we see that $e_{1t} \in L_2$. Similarly, if we eliminate $\|(e_{1t})_T\|_2$ from (30) and (31), we get

35 $$\|(e_{2t})_T\|_2 \leq \frac{\|u_2\|_2 + \gamma_0\|u_1\|_2}{1 - r\gamma_0}$$

which shows that $e_{2t} \in L_2$. Thus the system is L_2-stable, in the sense of definition [6.3(34)]. ∎

36 **Example.** Consider a system of the form shown in Fig. 6.13, with $\hat{g}(\cdot)$ as in Example [6.6(59)], i.e.,

$$\hat{g}(s) = e^{-s}\frac{s^2 + 4s + 2}{s^2 - 1}$$

Suppose it is desired to find the largest sector $[\alpha, \beta]$ such that the circle criterion is satisfied. Note that $\hat{g}$ has one pole with a positive real part, $\nu_p = 1$; therefore if $\alpha \leq 0$, the conditions for cases (ii)–(iv) of Theorem (8) can never be satisfied.[17] Thus, for the circle criterion to hold, we must have $\alpha > 0$.

The plot of $\hat{g}(j\omega)$ is shown in Fig. 6.10. For the system to be L_2-stable, the plot of $\hat{g}(j\omega)$ must encircle the *critical disk D* exactly once, in

[17]The circle criterion is a *sufficient* condition for L_2-stability. Hence if it is not satisfied, we *cannot* conclude that the system is unstable.

a counterclockwise sense. From Fig. 6.10, we see that this is the case provided $0.5 < \alpha \leq \beta < 0.55$. Hence the system is L_2-stable for *all* nonlinearities in the sector $[\alpha, \beta]$ with $\alpha > 0.5$ and $\beta < 0.55$.

We now show that the circle criterion of Theorem (8) is also a *necessary* condition for the stability of a certain class of systems. First of all, by applying Theorem (8), case (iii), we get the following result:

37 ***corollary*** Consider a feedback system of the form shown in Fig. 6.13. Let G, F be as in Theorem (8), and suppose $\phi(\cdot, \cdot)$ belongs to the sector $[-r, r]$ for some $r > 0$. Then the system under study is L_2-stable if $\hat{g}$ has no poles in C_+, and

38
$$\sup_{\omega} |\hat{g}(j\omega)| < r^{-1}$$

Now, consider the *linear time-invariant* feedback system shown in Fig. 6.15, where $\hat{g}$ is of the form

39
$$\hat{g}(s) = g_0 + \hat{g}_a(s) + \frac{\hat{n}(s)}{\hat{d}(s)}, \qquad g_a(\cdot) \in L_1$$

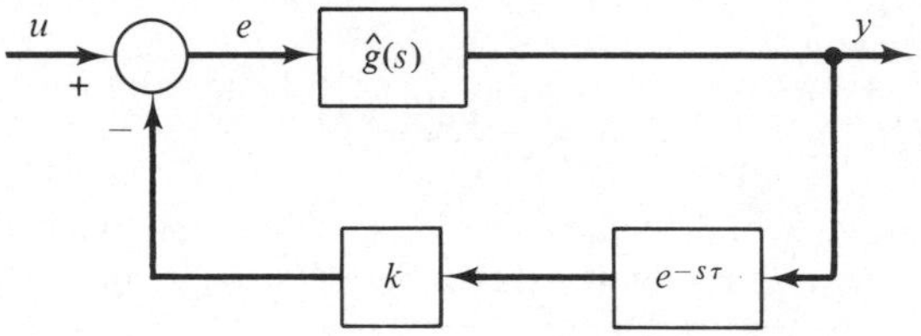

FIG. 6.15

Thus $\hat{g}$ is of the form (10)–(11), except that there are no delay terms. Suppose we ask: What conditions must $\hat{g}$ satisfy in order for the system of Fig. 6.15 to be L_∞-stable for *all* gains $k \in [-r, r]$ and for *all* delays $\tau \geq 0$? To put it another way, under what conditions does $\hat{g}(s)/[1 + k\, e^{-s\tau}\hat{g}(s)] \in \hat{\mathcal{A}}$ for *all* $k \in [-r, r]$ and for *all* $\tau \geq 0$? The answer is readily obtained from an application of the graphical stability criterion Theorem [6.6(58)], and is given next.

40 ***Theorem*** Suppose $\hat{g}$ is of form (39). Then $\hat{g}(s)/[1 + k\, e^{-s\tau}\hat{g}(s)] \in \hat{\mathcal{A}}$ $\forall\, k \in [-r, r]$ and $\forall\, \tau \geq 0$, if and only if $\hat{g} \in \hat{\mathcal{A}}$ and (38) is satisfied.

proof

(i) "*If*" If $\hat{g} \in \hat{\mathcal{A}}$ and (38) is satisfied, then the polar plot of $\hat{g}(j\omega)$ is contained within a circle of radius less than r^{-1}. Since $|e^{-j\omega\tau}| = 1$ $\forall\, \omega$ and $\forall\, \tau \geq 0$, we see that the polar plot of $e^{-j\omega\tau}\, \hat{g}(j\omega)$ is also contained within a circle of radius less than r^{-1}. Thus the polar plot of $e^{-j\omega\tau}\hat{g}(j\omega)$ is bounded away from, and does not encircle, the point $-1/(k + j0)$ whenever $k \in [-r, r]$. Also $e^{-s\tau}\hat{g}(s) \in \hat{\mathcal{A}}$,

$\forall\, \tau \geq 0$. Hence by Theorem [6.6(58)], we see that $\hat{g}(s)/[1 + ke^{-s\tau}\hat{g}(s)] \in \hat{\alpha}$, $\forall\, k \in [-r, r]$, $\forall\, \tau \geq 0$.

(ii) "*Only If*" Suppose $\hat{g}(s)/[1 + ke^{-s\tau}\hat{g}(s)] \in \hat{\alpha}$ $\forall\, k \in [-r, r]$, $\forall\, \tau \geq 0$. Then, by setting $k = 0$, $\tau = 0$, we see that $\hat{g} \in \hat{\alpha}$. Next, by the Riemann–Lebesgue lemma, $\hat{g}_a(j\omega) \longrightarrow 0$ as $|\omega| \longrightarrow \infty$, which means that $\hat{g}(j\omega) \longrightarrow g_0$ as $|\omega| \longrightarrow \infty$. The hypothesis implies that $|g_0| < r^{-1}$, because if $|g_0| \geq r^{-1}$, then the function $\hat{g}(s)/[1 + k\hat{g}(s)] \notin \hat{\alpha}$ with $k = 1/g_0$. Finally, suppose (38) is false; then there exist a complex number z with $|z| \geq r$ and a sequence $(\omega_i)_1^\infty$ such that $\hat{g}(j\omega_i) \longrightarrow z$ as $i \longrightarrow \infty$. Clearly the sequence $(\omega_i)_1^\infty$ is bounded, because $\hat{g}(j\omega) \longrightarrow g_0$ as $|\omega| \longrightarrow \infty$ and we have just shown that $|g_0| < r^{-1}$ (whereas $|z| \geq r^{-1}$). So there exists a *finite* frequency ω_0 such that $\hat{g}(j\omega_0) = z$. Now let $z = |z| \exp(j\phi)$, and let $\tau = (\phi + 2\pi)/\omega_0$. Then we have

41 $$[e^{-j\omega\tau}\hat{g}(j\omega)]_{\omega=\omega_0} = e^{-j\omega_0\tau}\hat{g}(j\omega_0) = |z| \geq r^{-1}$$

Hence the polar plot of $e^{-j\omega\tau}\hat{g}(j\omega)$ intersects the point $|z| + j0$, which shows that $\hat{g}(s)/[1 + ke^{-s\tau}\hat{g}(s)] \notin \hat{\alpha}$ with $k = |z|^{-1}$, because the stability criterion [6.6(57)] is violated. This contradicts the hypothesis that $\hat{g}(s)/[1 + ke^{-s\tau}\hat{g}(s)] \in \hat{\alpha}$ $\forall\, k \in [-r, r]$ and $\forall\, \tau \geq 0$. Hence we conclude that no such number z exists, i.e., that (38) holds. ∎

Thus Theorem (40) shows that the circle criterion is not at all conservative when applied to systems of the form shown in Fig. 6.15. Actually, we have restricted k to belong to the symmetrical interval $[-r, r]$ only to keep things simple, and we can prove similar results for the case where k in Fig. 6.15 varies over an arbitrary interval $[\alpha, \beta]$.

Combining Theorems (40) and (8), we get the following important result.

42 ***Theorem*** Suppose $\hat{g}$ is of the form (39) with $g_0 = 0$ and that the system of Fig. 6.15 is L_∞-stable for all $k \in [-r, r]$ and for all $\tau \geq 0$. Then the system of Fig. 6.13 is L_2-stable for all nonlinearities $\phi(\cdot, \cdot)$ in the sector $[-r, r]$.

Finally, combining Theorem (42) and Problem 6.5, we can prove the following result, which is reminiscent of Aizerman's conjecture—except that the result below is not a conjecture but a fact.

43 ***Theorem*** Suppose

44 $$\hat{g}(s) = \mathbf{c}'(s\mathbf{I} - \mathbf{A})^{-1}\mathbf{b} + d$$

where the matrix $\mathbf{A}$ is Hurwitz, the pair $(\mathbf{A}, \mathbf{b})$ is controllable, and the pair $(\mathbf{A}, \mathbf{c})$ is observable. Suppose the system of Fig. 6.15 is L_∞-

stable $\forall\, k \in [-r, r]$ and $\forall\, \tau \geq 0$. Then the equilibrium point $\mathbf{0}$ of the system

45 $$\dot{\mathbf{x}}(t) = \mathbf{A}\mathbf{x}(t) - \mathbf{b}u(t)$$

46 $$y(t) = \mathbf{c}'\mathbf{x}(t) + du(t)$$

47 $$u(t) = -\phi[t, y(t)]$$

is globally asymptotically stable whenever $\phi(\cdot, \cdot)$ belongs to the sector $[-r, r]$.

6.7.2 The Popov Criterion

In this subsection we shall state and prove a generalization of what is known as the Popov criterion. This criterion pertains to the system of Fig. 6.16. Note that this system is the same as that in Fig. 6.13, except that the nonlinear element $\phi(\cdot, \cdot)$ is replaced by the *time-invariant* nonlinearity $\phi(\cdot)$. For the purpose of proving the Popov criterion, we shall rearrange the system as shown in Fig. 6.17.

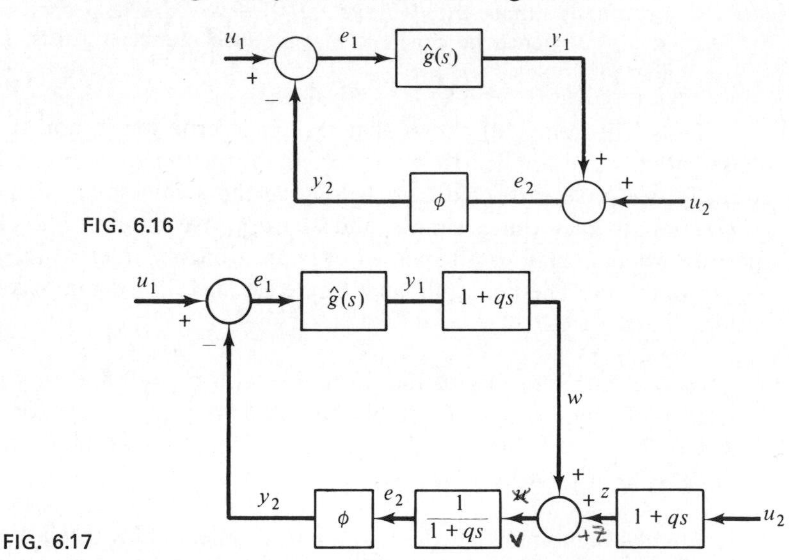

FIG. 6.16

FIG. 6.17

The following two lemmas pave the way to Popov's criterion.

48 ***lemma*** Suppose $\phi\colon R \longrightarrow R$ belongs to the sector $[0, k]$; i.e., suppose

49 $$0 \leq \sigma\phi(\sigma) \leq k\sigma^2, \qquad \forall\, \sigma \in R$$

With the symbols as defined in Fig. 6.17, whenever $q \geq 0$, we have

50 $$\int_0^T y_2(t)v(t)\, dt \geq \frac{1}{k}\int_0^T y_2^2(t)\, dt, \qquad \forall\, T \geq 0,\ \forall\, v \in L_{2e}$$

proof Define $\Phi: R \longrightarrow R$ by

51 $$\Phi(\sigma) = \int_0^\sigma \phi(\sigma)\, d\sigma$$

Then $\Phi(\sigma) \geq 0,\ \forall\ \sigma \in R$, because of the sector condition (49). From Fig. 6.17, we have

52 $$y_2(t) = \phi[e_2(t)]$$

53 $$v(t) = e_2(t) + q\dot{e}_2(t)$$

so that

54 $$\begin{aligned}\int_0^T y_2(t)v(t)\, dt &= \int_0^T e_2(t)\phi[e_2(t)]\, dt + q\int_0^T \dot{e}_2(t)\phi[e_2(t)]\, dt \\ &\geq \frac{1}{k}\int_0^T \{\phi[e_2(t)]\}^2\, dt + q\int_0^T \dot{e}_2(t)\phi[e_2(t)]\, dt \\ &= \frac{1}{k}\int_0^T y_2^2(t)\, dt + q\int_0^T \dot{e}_2(t)\phi[e_2(t)]\, dt\end{aligned}$$

where we use (49). Hence (50) is established if we show that

55 $$q\int_0^T \dot{e}_2(t)\phi[e_2(t)]\, dt \geq 0$$

If $q = 0$, (55) follows immediately, so suppose $q > 0$. Then

56 $$\begin{aligned}q\int_0^T \dot{e}_2(t)\phi[e_2(t)]\, dt &= q\int_{e_2(0)}^{e_2(T)} \phi(\sigma)\, d\sigma \\ &= q\{\Phi[e_2(T)] - \Phi[e_2(0)]\}\end{aligned}$$

However, because $e_2(\cdot)$ is the convolution of $v(\cdot)$ and the function $(1/q)$ $\exp(-t/q)$, we see that $e_2(0) = 0$, whence $\Phi[e_2(0)] = 0$. Hence

57 $$q\int_0^T \dot{e}_2(t)\phi\,[e_2(t)]\, dt = q\Phi\,[e_2(T)] \geq 0$$

This establishes (50). ∎

58 ***lemma*** Let $\hat{h} \in \hat{\mathfrak{A}}$, and define

59 $$\delta = \inf_{\omega \in R} \operatorname{Re} \hat{h}(j\omega)$$

Then, for all $f \in L_{2e}$, we have

60 $$\int_0^T r(t)f(t)\, dt \geq \delta \int_0^T f^2(t)\, dt, \qquad \forall\ T \geq 0$$

where $r = h * f$.

proof By causality, we have $r_T = (h * f_T)_T$. Now,

61 $$\begin{aligned}\int_0^T r(t)f(t)\, dt &= \int_0^\infty r_T(t)f_T(t)\, dt \\ &= \int_0^\infty (h * f_T)(t)f_T(t)\, dt\end{aligned}$$

By Parseval's equality, we now have

62
$$\int_0^T r(t)f(t)\,dt = \frac{1}{2\pi}\,\text{Re}\int_{-\infty}^{\infty} (h * f_T)(j\omega)f_T^*(j\omega)\,d\omega$$
$$= \frac{1}{2\pi}\,\text{Re}\int_{-\infty}^{\infty} \hat{h}(j\omega)|\hat{f}_T(j\omega)|^2\,d\omega$$
$$\geq \frac{\delta}{2\pi}\int_{-\infty}^{\infty} |\hat{f}_T(j\omega)|^2\,d\omega$$
$$= \delta\|\hat{f}_T\|_2^2$$

where $\hat{f}_T$ denotes the Fourier transform of $f_T(\cdot)$. Clearly (62) is the same as (60). ▬

Now we come to the following.

63 ***Theorem*** (*Generalized Popov Criterion*) Consider the system shown in Fig. 6.17. Suppose the transfer function $\hat{g}$ satisfies

64
$$g(\cdot), \dot{g}(\cdot) \in \mathcal{A}$$

Suppose the nonlinearity $\phi(\cdot)$ satisfies (49). Under these conditions, this system with inputs u_1 and z is L_2-stable if there exist constants $q \geq 0$, $\delta > 0$ such that

65
$$\text{Re}\,[(1 + j\omega q)\hat{g}(j\omega)] + \frac{1}{k} \geq \delta > 0, \qquad \forall\, \omega \in R$$

REMARKS: Theorem (63) guarantees L_2-stability, in the sense that both the "errors" e_1 and v belong to L_2 whenever the two "inputs" u_1 and z belong to L_2. However, we are interested in concluding that e_1 *and* e_2 belong to L_2, because these are the quantities of interest in the original system (shown in Fig. 6.16). This is not very difficult. If $q \geq 0$, then the function $s \mapsto 1/(1 + qs) \in \hat{\mathcal{A}}$; hence $e_2 \in L_2$ whenever $v \in L_2$. The restriction on the inputs cannot be removed; however, we can conclude that $e_1, e_2 \in L_2$ whenever $u_1, z \in L_2$. Now, if $q > 0$ and u_2 is an arbitrary element of L_2, the "input" z may or may not belong to L_2. Hence, if $q > 0$, we need to make extra assumptions about the nature of u_2. For example, we can assume that $\dot{u}_2 \in L_2$. These comments are summarized in the following.

66 **fact** Under the conditions of Theorem (63), the system of Fig. 6.16 has the property that $e_1, e_2 \in L_2$ whenever $u_1, u_2 \in L_2$, with the added assumption that $\dot{u}_2 \in L_2$ if $q > 0$.

proof of Theorem **(52)** (63) Suppose (65) holds, and suppose $u_1, z \in L_2$, e_1, $v \in L_{2e}$. Note that, because $g, \dot{g} \in \mathcal{A}$, the function $s \mapsto \hat{g}(s)(1 + qs) \in \hat{\mathcal{A}}$.

Next, given two functions $f_1, f_2 \in L_{2e}$, define their *truncated inner product* $\langle f_1, f_2 \rangle_T$ as

67 $$\langle f_1, f_2 \rangle_T = \int_0^T f_1(t) f_2(t)\, dt$$

Note that $\langle f_1, f_2 \rangle_T$ is the inner product (in the L_2-sense) of the L_2-functions f_{1T} and f_{2T}, so that by Schwarz's inequality, we have

68 $$|\langle f_1, f_2 \rangle_T| \leq \| f_{1T} \|_2 \cdot \| f_{2T} \|_2$$

For the system of Fig. 6.17, we have

69 $$e_1 = u_1 - y_2$$

70 $$v = z + w$$

Hence

71 $$\begin{aligned} \langle e_1, w \rangle_T + \langle v, y_2 \rangle_T &= \langle e_1, v - z \rangle_T + \langle v, u_1 - e_1 \rangle_T \\ &= -\langle e_1, z \rangle_T + \langle e_1, v \rangle_T - \langle v, e_1 \rangle_T + \langle v, u_1 \rangle_T \\ &= -\langle e_1, z \rangle_T + \langle v, u_1 \rangle_T \\ &\leq \| e_{1T} \|_2 \cdot \| z_T \|_2 + \| v_T \|_2 \cdot \| u_{1T} \|_2 \\ &\leq \| e_{1T} \|_2 \cdot \| z \|_2 + \| v_T \|_2 \cdot \| u_1 \|_2 \end{aligned}$$

On the other hand, by lemma (48), we have

72 $$\langle v, y_2 \rangle_T \geq \frac{1}{k} \| y_{2T} \|_2^2$$

Now, if we define

73 $$\nu = \inf_{\omega \in R} \text{Re}\, [(1 + j\omega q) \hat{g}(j\omega)]$$

74 $$m = \sup_{\omega \in R} |(1 + j\omega q) \hat{g}(j\omega)|$$

we have

75 $$\| w_T \|_2 \leq m \| e_{1T} \|_2$$

by lemma (4). Next, by lemma (58), we have

76 $$\begin{aligned} \langle e_1, w \rangle_T \geq \nu \| e_{1T} \|_2^2 &= \left(\nu - \frac{\delta}{2}\right) \| e_{1T} \|_2^2 + \frac{\delta}{2} \| e_{1T} \|_2^2 \\ &\geq \left(\nu - \frac{\delta}{2}\right) \| e_{1T} \|_2^2 + \frac{\delta}{2m^2} \| w_T \|_2^2 \\ &= \left(\nu - \frac{\delta}{2}\right) \| e_{1T} \|_2^2 + \frac{\delta}{2m^2} \| (v - z)_T \|_2^2 \end{aligned}$$

Combining (72) and (76), we have

77 $$\begin{aligned} \langle e_1, w \rangle_T + \langle v, y_2 \rangle_T \geq \frac{1}{k} \| y_{2T} \|_2^2 &+ \left(\nu - \frac{\delta}{2}\right) \| e_{1T} \|_2^2 \\ &+ \frac{\delta}{2m^2} \| (v - z)_T \|_2^2 \end{aligned}$$

Because $y_2 = u_1 - e_1$, (77) becomes

78 $$\langle e_1, w\rangle_T + \langle v, y_2\rangle_T \geq \frac{1}{k}\|(u_1 - e_1)_T\|_2^2 + \left(\nu - \frac{\delta}{2}\right)\|e_{1T}\|_2^2 + \frac{\delta}{2m^2}\|(v - z)_T\|_2^2$$

Combining (71) and (78) gives

79 $$\frac{1}{k}\|(u_1 - e_1)_T\|_2^2 + \left(\nu - \frac{\delta}{2}\right)\|e_{1T}\|_2^2 + \frac{\delta}{2m^2}\|(v - z)_T\|_2^2 \leq \|e_{1T}\|_2 \cdot \|z\|_2 + \|v_T\|_2 \cdot \|u_1\|_2$$

Expanding $\|(u_1 - e_1)_T\|_2^2$ and $\|(v - z)_T\|_2^2$ gives

80 $$\left(\frac{1}{k} + \nu - \frac{\delta}{2}\right)\|e_{1T}\|_2^2 + \frac{\delta}{2m^2}\|(v - z_T\|_2^2 \leq c_1\|e_{1T}\|_2 + c_2\|v_T\|_2 + c_3$$

where c_1, c_2, c_3 are finite constants. Now, by (65),

$$\frac{1}{k} + \nu - \frac{\delta}{2} > 0$$

and of course $\delta/2m^2 > 0$. Thus the left-hand side of (80) is a positive definite form in e_{1T} and v_T. This implies that $\|e_{1T}\|_2, \|v_T\|_2$ are bounded independently of T. Hence $e_1, v \in L_2$ whenever $u_1, z \in L_2$. ∎

REMARKS:

1. The inequality (65) can be given a graphical interpretation, which makes it quite useful in practice. Suppose we plot $\text{Re } \hat{g}(j\omega)$ vs. $\omega \text{ Im } \hat{g}(j\omega)$, with ω as a parameter ranging from 0 to ∞. [Note that it is only necessary to do the plot for $\omega \in [0, \infty)$, because both $\text{Re } \hat{g}(j\omega)$ and $\omega \text{ Im } \hat{g}(j\omega)$ are even functions of ω.] This graph is sometimes referred to as the *Popov plot* of $\hat{g}$. The inequality (65) means that, for some $\delta > 0$, one can draw a straight line with a nonnegative slope (namely $1/q$) through the point $(-1/k + \delta, 0)$ such that the plot of $\text{Re } \hat{g}(j\omega)$ vs. $\omega \text{ Im } \hat{g}(j\omega)$ lies entirely to the right of such a line. Such a line is known as a *Popov line*. This situation is depicted in Fig. 6.18.
2. If we set $q = 0$ in (65), (65) becomes the same as (15); i.e., the Popov criterion becomes the same as the circle criterion [Theorem (8), case (ii)] with $\alpha = 0$. Because of the flexibility of choosing the constant $q \geq 0$ in (54), the inequality (54) is less restrictive than (15). This can be explained by noting that the circle criterion [case (ii)] guarantees L_2-stability for *all* feedback nonlinearities in the sector $[0, \beta]$, whereas the Popov criterion guarantees L_2-stability only for all *time-invariant* nonlinearities in the sector $[0, k]$.

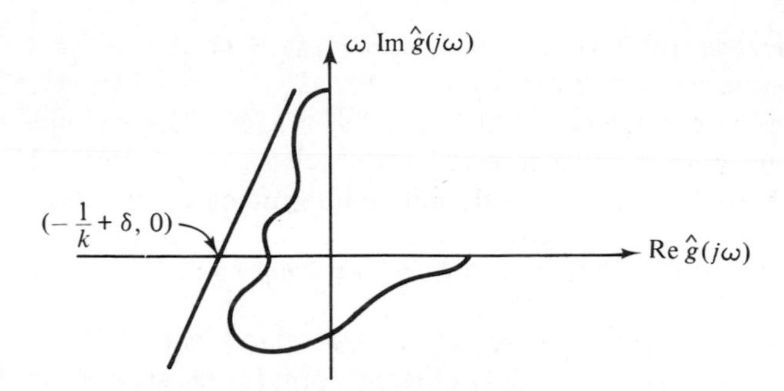

FIG. 6.18

3. With the proof given above, the Popov criterion is a special case of what is known as the *passivity theorem*.

81 Example. Let

$$\hat{g}(s) = \frac{1}{s^2 + 4s + 4}$$

The Nyquist plot and Popov plot of $\hat{g}$ are shown in Figs. 6.19 and 6.20, respectively. From Fig. 6.20, we see that no matter how small $1/k$ is (i.e., no matter how large k is) we can always construct a suitable Popov line, as indicated. Hence the system of Fig. 6.16 is L_2-stable for all *time-invariant* nonlinearities $\phi(\cdot)$ in the sector $[0, k]$, for *all finite* k. On the other hand, if we apply the circle criterion, case (ii), we see that

$$\inf_{\omega \in R} \text{Re } \hat{g}(j\omega) = -\frac{1}{32}$$

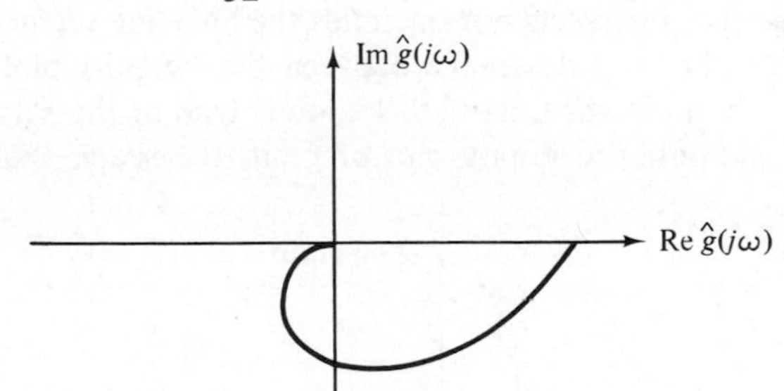

FIG. 6.19

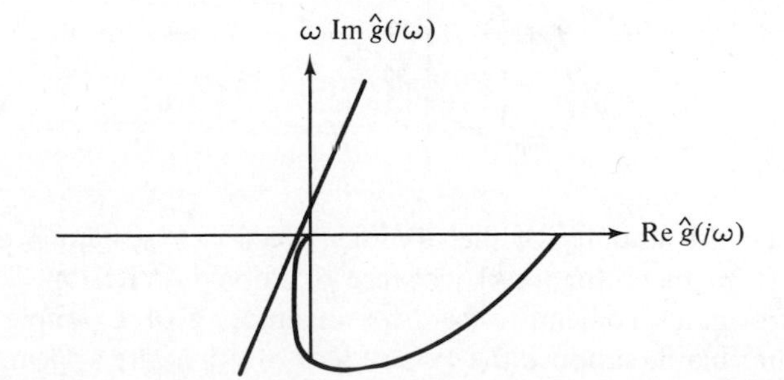

FIG. 6.20

Hence (15) is satisfied whenever $(1/\beta) > (1/32)$, i.e., $\beta < 32$. Hence the system of Fig. 6.13 is L_2-stable for *all possibly time-varying* nonlinearities $\phi(\cdot, \cdot)$ in the sector $[0, \beta]$, for all $\beta < 32$. This example shows that in general the circle criterion and the Popov criterion yield different stability bounds, which should be interpreted differently.

An Application: Aizerman's Conjecture

In Sec. 5.5, we stated Aizerman's conjecture, which we shall restate here in a slightly different form for the sake of convenience.

> *Conjecture* (*Aizerman*) Suppose that $g, \dot{g} \in \mathcal{A}$ and that $\hat{g}/(1 + h\hat{g}) \in \hat{\mathcal{A}}$ for all $h \in [0, k]$. Then the system of Fig. 6.16 is L_2-stable for all nonlinearities $\phi(\cdot)$ in the sector $[0, k]$.

Thus Aizerman's conjecture states that if the system of Fig. 6.16 is stable whenever $\phi(\cdot)$ is a constant gain of value $h \in [0, k]$, then it is stable for all nonlinearities $\phi(\cdot)$ in the sector $[0, k]$. In general, Aizerman's conjecture is false. However, Popov's criterion provides a means of identifying a large class of transfer functions $\hat{g}(\cdot)$ for which Aizerman's conjecture holds.

Suppose $g, \dot{g} \in \mathcal{A}$. If g contains any impulses, $\dot{g}$ would contain higher-order distributions and hence would not belong to $\mathcal{A}$. Thus $g, \dot{g} \in \mathcal{A}$ implies that g does not contain any impulses, i.e., that $g \in L_1$. Now, because $g \in L_1$, $\hat{g}/(1 + h\hat{g}) \in \mathcal{A}$, $\forall\, h \in [0, k]$, implies, by the graphical stability criterion of Theorem [6.6(58)], that the Nyquist plot of $\hat{g}$ neither intersects nor encircles the half-line segment $(-\infty, -1/k]$. Because the only difference between the Nyquist plot and the Popov plot is in the vertical axis, the same is true of the Popov plot as well. Now, suppose the Popov plot of $\hat{g}$ has the shape shown in Fig. 6.21.

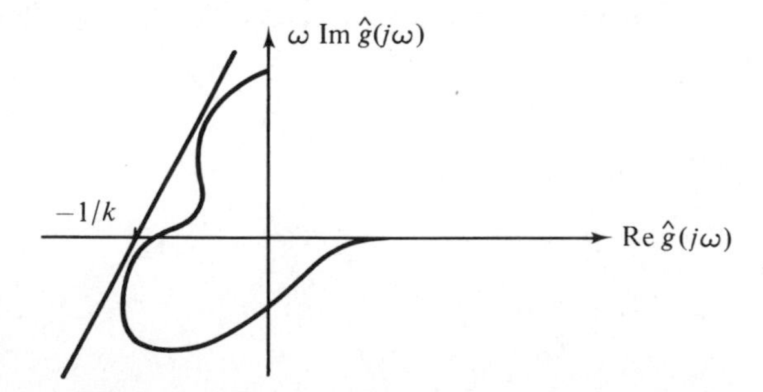

FIG. 6.21

Then the stability of the nonlinear feedback system is assured for all $\phi(\cdot)$ in the sector $[0, k]$, because of Popov's criterion. Thus $\hat{g}$ satisfies Aizerman's conjecture [see, for instance, $\hat{g}$ of Example (69)]. On the other hand, suppose the Popov plot of $\hat{g}$ has the appearance shown in

Fig. 6.22. In this case, Popov's criterion is not satisfied. However, because Popov's criterion is only a sufficient condition for stability, it still does not follow that Aizerman's conjecture is false for such a $\hat{g}$. Thus, in summary, Popov's criterion provides a readily verifiable sufficient condition for determining whether Aizerman's conjecture is valid for a particular transfer function $\hat{g}(\cdot)$.

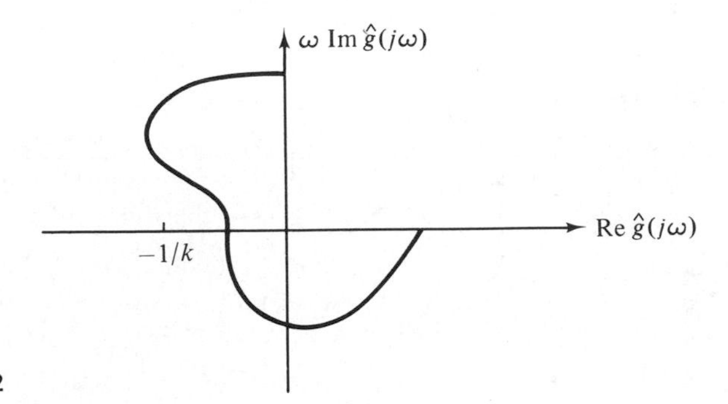

FIG. 6.22

Problem 6.14. Suppose $\hat{g}(\cdot)$ is of the form (44), and that the criteria of Theorem (8) are satisfied. Show that in this case $1 + kd \neq 0 \;\forall\; k \in [-r, r]$. Then, using the results of Problem 6.5, show that Theorem (8) contains Theorem [5.5(60)] as a special case.

Problem 6.15. Show that if the criteria of Theorem (8) are satisfied, then (23) holds.

Problem 6.16. Using the circle criterion, determine the largest range $[\alpha, \beta]$ such that the system of Fig. 6.12 is L_2-stable for all $\phi(\cdot, \cdot)$ in the sector $[\alpha, \beta]$ with

(i) $\hat{g}(s) = \dfrac{1}{\sqrt{s}\,(\sinh \sqrt{s}\,)}$

(ii) $\hat{g}(s) = e^{-3s}\dfrac{s^2 + 3s + 2}{s^2 - 4}$

(iii) $\hat{g}(s) = 2 - \dfrac{3^{-es}}{s + 5}$

Problem 6.17. Using the Popov criterion, determine the largest constant k such that the system of Fig. 6.15 is L_2-stable for all $\phi(\cdot)$ in the sector $[0, k]$, with

(i) $\hat{g}(s) = e^{-s}\dfrac{s + 4}{s^2 + 5s + 6}$

(ii) $\hat{g}(s) = e^{-2s}\dfrac{s + 2}{(s + 3)^2}$

(iii) $\hat{g}(s) = e^{-3s}\dfrac{s^2 + 9s + 20}{s^3 + 6s^2 + 11s + 6}$

Do any of these transfer functions satisfy Aizerman's conjecture?

Note:

$$\frac{d}{dt}\left[\int_{\tau=f_1(t)}^{\tau=f_2(t)} \rho(\tau)\, d\tau\right] \overset{?}{=} \rho[f_2(t)] - \rho[f_1(t)]$$

Dwight

Appendix

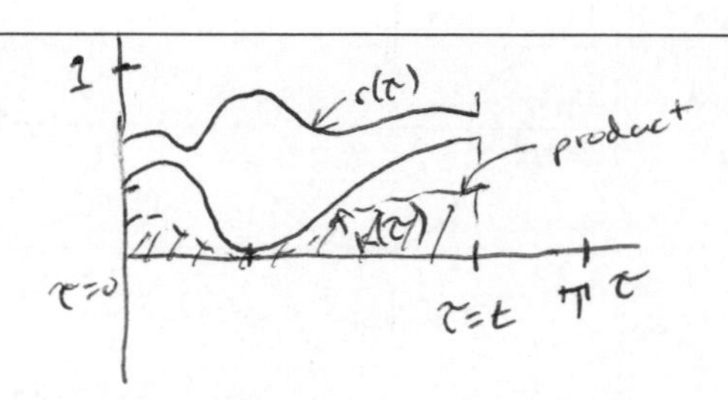

I
BELLMAN–GRONWALL INEQUALITY

1 ***Theorem*** Suppose $c \geq 0$, $r(\cdot)$ and $k(\cdot)$ are nonnegative valued continuous functions, and suppose

This implicit bound (r(t) on rt. side)

2 $$r(t) \leq c + \int_0^t k(\tau) r(\tau)\, d\tau, \qquad \forall\, t \in [0, T]$$

≡ s(t) s(0) = c

Then

implies this explicit bound →

3 $$r(t) \leq c \exp\left[\int_0^t k(\tau)\, d\tau\right], \qquad \forall\, t \in [0, T]$$

proof Let $s(t)$ denote the right-hand side of (1). Then (1) states that

4 $$r(t) \leq s(t), \qquad \forall\, t \in [0, T]$$

Further

5 $$\dot{s}(t) = k(t)\, r(t) \leq k(t)\, s(t), \qquad \forall\, t \in [0, T]$$

6 $$\dot{s}(t) - k(t)\, s(t) \leq 0, \qquad \forall\, t \in [0, T]$$

7 $$[\dot{s}(t) - k(t)\, s(t)] \exp\left[\int_0^t -k(\tau)\, d\tau\right] \leq 0, \qquad \forall\, t \in [0, T]$$

mult each side by this which is always pos.

8 $$\frac{d}{dt}\left\{s(t) \exp\left[\int_0^t -k(\tau)\, d\tau\right]\right\} \leq 0 \qquad \forall\, t \in [0, T]$$

9 $$s(t) \exp\left[\int_0^t -k(\tau)\, d\tau\right] \leq s(0) = c$$

← if $\frac{dy(x)}{dx} \leq 0$, $\int_{y(0)}^{y(x)} dy \leq \int_0^x 0\, dx$

$= \dot{s} \exp(\ ) + s(t) \exp(\ )[-k(t)]$

10 $$s(t) \leq c \exp\left[\int_0^t k(\tau)\, d\tau\right]$$

Now, (10) and (4) together imply (3).

The utility of this inequality lies in that, given the *implicit* bound (2) for $r(t)$ [notice that the right-hand side of (2) also involves $r(\cdot)$], we are able to obtain an *explicit* upper bound for $r(t)$. In particular, note that if $r(\cdot)$ is nonnegative valued and satisfies (2) with $c = 0$, then $r(t) \equiv 0 \; \forall \; t \in [0, T]$.

II SUMMARY OF MATRIX EXPONENTIALS AND STATE TRANSITION MATRICES

Given an $n \times n$ matrix $\mathbf{A}$, the polynomial $c(\lambda)$ defined by

1 $$c(\lambda) = \det(\lambda \mathbf{I} - \mathbf{A})$$

is known as the *characteristic polynomial* of $\mathbf{A}$. The zeroes of the characteristic polynomial of $\mathbf{A}$ are known as the eigenvalues of $\mathbf{A}$. The well known Cayley–Hamilton theorem states that

2 $$c(\mathbf{A}) = \mathbf{0}$$

A polynomial $m(\lambda)$ of the lowest possible degree such that $m(\mathbf{A}) = \mathbf{0}$ is called a *minimal polynomial* of $\mathbf{A}$. A minimal polynomial of $\mathbf{A}$ is a divisor of the characteristic polynomial of $\mathbf{A}$, and every eigenvalue of $\mathbf{A}$ is also a zero of a minimal polynomial of $\mathbf{A}$.

Given an $n \times n$ matrix $\mathbf{A}$, the $n \times n$ matrix-valued function

3 $$e^{\mathbf{A}t} = \sum_{i=0}^{\infty} \mathbf{A}^i t^i / i!$$

is known as the *matrix exponential* of $\mathbf{A}$. It can be shown that the infinite series in (3) converges uniformly and absolutely over every finite interval. Suppose $m(\lambda)$ is a minimal polynomial of $\mathbf{A}$, with zeroes $\lambda_1, \ldots, \lambda_k$ of multiplicities $m_1, \ldots, m_k$ respectively. Then it can be shown that $e^{\mathbf{A}t}$ is of the form

4 $$e^{\mathbf{A}t} = \sum_{i=1}^{k} \sum_{j=0}^{m_i - 1} p_{ij}(\mathbf{A}) t^j e^{\lambda_i t}$$

where the p_{ij}'s are the so-called *interpolating polynomials*. The following fact is obvious from the expansion (4).

5 **fact** $e^{\mathbf{A}t}$ is bounded as a function of t if and only if $\text{Re}\, \lambda_i \leq 0 \; \forall \; i$, and $m_i = 1$ whenever $\text{Re}\, \lambda_i = 0$ (i.e., if all eigenvalues of $\mathbf{A}$ have nonpositive real parts, and all eigenvalues of $\mathbf{A}$ having zero real parts are simple zeros of a minimal polynomial of $\mathbf{A}$); $e^{\mathbf{A}t} \longrightarrow \mathbf{0}$ as $t \longrightarrow \infty$ if and only if $\text{Re}\, \lambda_i < 0 \; \forall \; i$ (i.e., all eigenvalues of $\mathbf{A}$ have negative real parts).

Note that the infinite series

6 $$\sum_{i=0}^{\infty} \mathbf{A}^{i+1} t^i / i!$$

also converges absolutely and uniformly over every finite interval to $\mathbf{A}e^{\mathbf{A}t}$. So we see that $e^{\mathbf{A}t}$ is a solution of the matrix differential equation

7 $$\dot{\mathbf{M}}(t) = \mathbf{A}\mathbf{M}(t) \quad \forall\, t \geq 0, \mathbf{M}(0) = \mathbf{I}$$

Further, in view of the existence and uniqueness theory developed in Sec. 3.4, we see that $\mathbf{M}(t) = e^{\mathbf{A}t}$ is in fact the *unique* solution of (7). The solution of the vector differential equation

8 $$\dot{\mathbf{x}}(t) = \mathbf{A}\mathbf{x}(t) \quad \forall\, t \geq 0, \mathbf{x}(0) = \mathbf{x}_0$$

is given by

9 $$\mathbf{x}(t) = e^{\mathbf{A}t}\mathbf{x}(0) = e^{\mathbf{A}t}\mathbf{x}_0$$

In the case of the time-varying linear vector differential equation

10 $$\dot{\mathbf{x}}(t) = \mathbf{A}(t)\mathbf{x}(t) \quad \forall\, t \geq t_0, \mathbf{x}(t_0) = \mathbf{x}_0$$

where $\mathbf{A}(\cdot)$ is piecewise-continuous, the solution for $\mathbf{x}(t)$ is given by

11 $$\mathbf{x}(t) = \mathbf{\Phi}(t, t_0)\mathbf{x}(t_0) = \mathbf{\Phi}(t, t_0)\mathbf{x}_0$$

where $\mathbf{\Phi}(\cdot, \cdot)$ is the unique solution of the linear matrix differential equation

12 $$\frac{d}{dt}\mathbf{\Phi}(t, t_0) = \mathbf{A}(t)\mathbf{\Phi}(t, t_0) \ \forall\, t \geq t_0, \ \mathbf{\Phi}(t_0, t_0) = \mathbf{I}$$

The matrix $\mathbf{\Phi}(\cdot, \cdot)$ is known as the state transition matrix, and satisfies the following properties:

13 $$[\mathbf{\Phi}(t, \tau)]^{-1} = \mathbf{\Phi}(\tau, t), \qquad \forall\, t, \tau$$

14 $$\mathbf{\Phi}(t, \tau)\mathbf{\Phi}(\tau, s) = \mathbf{\Phi}(t, s), \qquad \forall\, t, \tau, s$$

For a detailed treatment of these topics, see [7] Chen.

References

[1] Arnold, V. I. *Ordinary Differential Equations*, MIT Press, Cambridge, Mass., 1973.

[2] Athans, M., and Falb, P. L. *Optimal Control*, McGraw-Hill, New York, 1966.

[3] Barman, J. *Well-Posedness of Feedback Systems and Singular Perturbations*, Ph.D. Thesis, Univ. of California, Berkeley, Ca. 1973.

[4] Bellman, R. E. *Introduction to Matrix Analysis*, 2nd ed., McGraw-Hill, New York, 1970.

[5] Bergen, A. R. and Franks, R. L. "Justification of the Describing Function Method," *SIAM J. Control*, vol. 9, pp. 568–589, 1971.

[6] Blaquière, A. *Nonlinear System Analysis*, Academic Press, New York, 1966.

[7] Chen, C. T. *Introduction to Linear System Theory*, Holt, Rinehart, & Winston, New York, 1971.

[8] Chua, L., and Lin, P. M. *Computer-Aided Analysis of Electronic Circuits: Algorithms and Computational Techniques*, Prentice-Hall, Englewood Cliffs, N.J., 1975.

[9] Desoer, C. A., and Shensa, M. J. "Networks with Very Small and Very Large Parasitics: Natural Frequencies and Stability," *Proc. IEEE*, vol. 58, pp. 1933–1938, 1970.

[10] Desoer, C. A. and Vidyasagar, M. *Feedback Systems: Input-Output Properties*, Academic Press, New York, 1975.

[11] Eggleston, H. G. *Convexity*, Cambridge University Press, Cambridge, 1966.

[12] Gelb, A. and Vander Velde, W. E. *Multiple-Input Describing Functions and Nonlinear System Design*, McGraw-Hill, New York, 1968.

[13] Hahn, W. *Stability of Motion*, Springer-Verlag, Berlin, 1967.

[14] Hille, E., and Phillips, R. S. *Functional Analysis and Semigroups*, Amer. Math. Soc., Providence, R.I., 1957.

[15] Jury, E. I. *Inners and Stability of Dynamic Systems*, John Wiley & Sons, New York, 1975.

[16] Lefschetz, S. *Stability of Nonlinear Control Systems*, Academic Press, New York, 1962.

[17] Narendra, K. S. and Taylor, J. H. *Frequency Domain Criteria for Absolute Stability*, Academic Press, New York, 1973.

[18] Nemytskii, V. V. and Stepanov, V. V. *Qualitative Theory of Differential Equations*, Princeton Univ. Press, Princeton, 1960.

[19] Ralston, A. *A First Course in Numerical Analysis*, McGraw-Hill, New York, 1965.

[20] Royden, H. L. *Real Analysis*, Macmillan, New York, 1963.

[21] Taussky, O. "A Generalization of A Theorem of Liapunov," *Siam J. Appl. Math.*, vol. 9, pp. 640–643, 1961.

[22] Wilkinson, J. H. *The Algebraic Eigenvalue Problem*, Clarendon Press, Oxford, 1965.

[23] Willems, J. C. *The Analysis of Feedback Systems*, MIT Press, Cambridge, Mass., 1970.

Index

Index

A

B

C

L

M

N

O

P